Charles Boyd Kelsey

Diseases of the Rectum and Anus

Charles Boyd Kelsey

Diseases of the Rectum and Anus

ISBN/EAN: 9783337035112

Printed in Europe, USA, Canada, Australia, Japan

Cover: Foto ©berggeist007 / pixelio.de

More available books at **www.hansebooks.com**

DISEASES

OF THE

RECTUM AND ANUS

BY

CHARLES B. KELSEY, M.D.,

SURGEON TO ST. PAUL'S INFIRMARY FOR DISEASES OF THE RECTUM; CONSULTING
SURGEON FOR DISEASES OF THE RECTUM TO THE HARLEM HOSPITAL AND
DISPENSARY FOR WOMEN AND CHILDREN, ETC., ETC.

NEW YORK
WILLIAM WOOD & COMPANY
56 & 58 LAFAYETTE PLACE
1882

PREFACE.

In preparing the following pages for publication, I have endeavored to condense into convenient form, for both student and practitioner, as great an amount as possible of practical information concerning diseases of the rectum and anus.

The advances which have been made during the past few years in this special branch of surgery have been very great. The whole pathology of malignant disease has been rewritten; and the close relationship of the so-called benign polypoid growths to epithelial cancer has been worked out after careful study with the microscope. The operation of excision of cancerous growths in the lower part of the rectum has again become a legitimate surgical procedure, and what is better, its range of applicability has been definitely determined. The advances in abdominal surgery, for which the present century will always be famous, have also had a bearing on diseases of the rectum, and the operation of excision of cancer of the upper part of the rectum and the sigmoid flexure, through an incision made in the abdominal wall, has brought within treatment a class of cases formerly beyond the reach of art.

New methods of treatment of benign stricture have been devised as a substitute for colotomy. New methods of palliation in benign and malignant stricture have been devised as substitutes for the repulsive operation of colotomy, and that operation bids fair in the immediate future to assume deservedly a place of less prominence than it has occupied in the past. New and effectual methods of curing hæmorrhoids and prolapse without cutting operations have been added to the surgeon's resources.

The advances have been due to the efforts of no one man or nation. The records of them are scattered through English, Continental, and American periodical literature, and many of them are practically beyond the reach of the busy practitioner.

I have tried, therefore, as far as possible to condense what was positively known within the following pages, and to reduce it to a form suitable for ready reference by student and practitioner, giving not only the results which have been reached by experiment and clinical experience, and which may be relied upon as the basis of practice, but in many questions marking out by foot-notes and references the way for any who may desire to go over for himself the ground which I have followed with no little difficulty.

In addition, I have endeavored, whenever possible, to illustrate the subject under consideration by the reports of cases either from my own practice or that of others, knowing that a lesson is often conveyed to the student in this way better than by any other.

My thanks are especially due to the librarians of the New York Hospital for their unvarying kindness and assistance, which has rendered my work a far from unpleasant one.

CHARLES B. KELSEY.

"THE MADISON," No. 25 MADISON AVE.,
 CORNER OF 25TH STREET,
 NEW YORK, September, 1882.

CONTENTS.

CHAPTER I.

PRACTICAL POINTS IN ANATOMY AND PHYSIOLOGY.

Rectum.—Position and Measurements.—Curves.—Divisions.—Relations.—Anus.—Parts in Detail.—Peritoneum.—Relations to Three Portions of the Rectum.—Distance of Peritoneal *Cul-de-Sac* from Anus.—Muscular Layer.—**Arrangement of** Fibres.—Submucous Layer.—Mucous Membrane.—Sustentator **Tunicæ** Mucosæ.—Columnæ Recti.—Glands of Mucous Membrane.—Muscles **of the** Rectum and Anus.—External Sphincter.—Internal Sphincter.—**Recto-Coccygeus.**—**Levator** Ani.—Transversus Perinei.—Arteries.—**Superior Hæmorrhoidal.**—Middle Hæmorrhoidal.—Inferior Hæmorrhoidal.—Veins.—Superior Hæmorrhoidal.—Middle Hæmorrhoidal.—Inferior **Hæmorrhoidal.**—**Minute Anatomy** of Veins.—General and Visceral **Venous Systems.**—**Nerves.**—Cerebro-Spinal and Sympathetic Nerve Supply.—Tonic **Contraction of Sphincter.**—Explanation of Wandering Pains in Rectal Disease.—Lymphatics.—**External and Internal** Lymphatic Vessels.—Physiology.—Anatomy **of the Third Sphincter.**—Valves **of** Mucous Membrane.—Plica Transversalis Recti **of Kohlrausch.**—Lack of Uniformity in Different Subjects.—Physiology **of Defecation.**—Explanation of Retention of Fæces after Destruction of the **Sphincter.**—**Conclusions** Resulting from Study of Third Sphincter 1

CHAPTER II.

CONGENITAL MALFORMATIONS OF THE RECTUM AND ANUS.

Separate Development of Rectum and Anus.—Narrowing of the Anus or Rectum without Complete Occlusion.—Congenital Stricture.—Closure of the Anus by a Membranous Diaphragm.—Entire Absence of the Anus, the Rectum Ending in a Blind Pouch at a Point more or Less Distant from the Perineum.—Rectum Same **as in** Last Variety **and the** Anus Normal.—Anus Absent and Rectum Opening by an Abnormal **Anus** at Some Point in the Perineal **or Sacral** Regions.—Cases.—Anus **Absent and** Rectum Ending in the Bladder, Urethra, or Vagina.—Cases.—Rectum **and** Anus Normal, but Ureters, Uterus, or Vagina Empty into Rectum.—Total Absence of Rectum.—Absence of Large Intestine.—Obliteration from Intra-**Uterine** Disease.—Treatment.—Operation Should Always be Performed and Without Delay.—Attempt Should First be Made to Establish an Anus

in the Anal Region.—Measurements of Pelvis at Birth.—Use of Trocar
not Justifiable.—Useful Anus Seldom Obtained by Means of Incision
Alone.—Objections to Cutting Operation Without Plastic Operation.—
Proctoplasty.—If Attempt to Establish New Anus in Anal Region Fail,
Colotomy at Once to be Performed.—Inguinal Preferable to Lumbar
Colotomy.—History of Colotomy.—Callisen.—Amussat.—Description of
Operation of Colotomy.—Dangers of Operation.—The Inguinal Opera-
tion.—Description.—Attempts at Establishing Anus in Anal Region after
Colotomy Generally Unsuccessful.—Cases.—Closure of Artificial Anus.—
Operation of Dupuytren.—Modifications of Dupuytren's Operation . . 30

CHAPTER III.

GENERAL RULES REGARDING EXAMINATION, DIAGNOSIS, AND OPERATION.

Necessity for Physical Examination.—Questions which may lead to Diagnosis.—How to make Examination.—Table.—Lamp.—Instrument Case.—
Position of Patient.—Necessity for Enema before Examination.—What
may be learned by simple Inspection.—Rectal Touch.—What may be
discovered by it.—Bougies; Varieties; Author's Bougies.—Rectal Specula:
Van Buren's; Fenestrated; Bivalve; Objections.—Colonoscope.—Stretching the Sphincter; Proper Method of Performing the Operation; Results.
—Difficulties of Diagnosis of Disease high up in the Rectum.—Manual
Examination.—What may be Learned by this Method.—Preparation of
Patient for Operation.—Assistants.—Primary Anæsthesia.—Thermo-Cautery.—Hæmorrhage.—Rules for Controlling Hæmorrhage.—Cold.—Styp-
tics.—Packing the Rectum.—Treatment after Operation.—Dressings.—
Necessity for Rest.—Retention of Urine.—Case of Fatal Retention . . 48

CHAPTER IV.

INFLAMMATION OF THE RECTUM.

Cases of Proctitis.—Varieties : Acute, Chronic, Primary, Secondary, Localized, General.—Symptoms and Course of each Variety.—Causes of Proctitis : Direct Propagation, Foreign Bodies, Drastic Cathartics, Gout,
Pederasty, Gonorrhœa.—Treatment 66

CHAPTER V.

ABSCESS AND FISTULA.

Abscess divided into Superficial and Deep.—Superficial Abscesses.—Simple
Furuncles; Causes; Characters; Results; Treatment.—Suppuration of Ex-
ternal Hæmorrhoid.—Suppuration of Internal Hæmorrhoid.—Diffuse In-
flammation of Subcutaneous Tissue, Causes; Symptoms; Treatment.—
Form of Incision.—Deep Abscesses.—Divided into Abscess of the Ischio-
Rectal Fossa and of the Superior Pelvi-Rectal Space.—Causes; Symp-
toms; Diagnosis.—Dangers of Deep Abscess.—Formation of Deep and
Extensive Fistulæ.—Horse-shoe Abscess.—Idiopathic Gangrenous Cel-
lulitis.—Reasons why Abscesses do not Heal Spontaneously.—Prognosis.
—Treatment.—Incisions and Subsequent Treatment of Deep Abscesses.—

Incontinence of Fæces.—Relief of Incontinence resulting from Operation.—Fistula.—Generally due to Abscess.—Divided into Superficial and Deep.—Complete Fistula.—External Fistula.—Internal Fistula.—Description of Superficial Fistulæ.—How to Detect an Internal Opening.—Location of Internal Opening.—Description of Track of Fistula.—Symptoms of Superficial Fistula.—Deep Fistula.—Fistula with Numerous External Openings.—Blind Internal Fistula.—Ulceration of Rectum Causing Internal Fistula.—Treatment.—Spontaneous Cure.—Advisability of Operation.—Fistula in Relation to Phthisis.—Contra-indications to Operation.—Treatment by Cauterization.—The Ligature.—The Elastic Ligature.—Galvano-Cautery.—How to Pass Ligature.—Incision.—Description of Operation.—Author's Knife for Fistula.—Division of Deep Tracks.—Treatment of Track running up the Bowel.—Treatment of Blind External Variety; of Horse-shoe Variety; of Fistula with Numerous External Openings.—Dressing after Incision.—Packing the Incision.—Hæmorrhage in Operation.—Treatment of Blind Internal Variety.—Incurable Fistulæ.—Treatment of Deep and Extensive Tracks.—Fistula with Stricture . 71

CHAPTER VI.

HÆMORRHOIDS.

Definition.—Division into External and Internal.—Differences between the two Varieties.—External Hæmorrhoids.—Pathology.—Inflamed Hæmorrhoids.—Treatment.—Means of Prevention.—Palliative Treatment.—Excision.—Internal Hæmorrhoids.—Division into Capillary, Arterial, and Venous.—Description of Capillary Variety, of Venous Variety, of Arterial Variety.—Symptoms of Internal Hæmorrhoids.—Strangulation.—Diagnosis.—Treatment of Internal Hæmorhoids.—Palliative Treatment.—Constitutional and Local Means of Palliation.—Treatment of Strangulation.—Curative Treatment.—Hæmorrhoids Associated with Uterine Disease.—Symptomatic Hæmorrhoids.—Radical Cure.—Caustics.—Dangers of Nitric Acid.—Vienna Paste.—Treatment by Carbolic Acid Injections; Cases and Cures.—Advantages of this Treatment.—Treatment by Ligature.—Description of Operation.—Operation with Clamp and Cautery . 91

CHAPTER VII.

PROLAPSE.

Four Varieties.—First Variety: Prolapse of the Mucous Membrane Alone.—Second Variety: Prolapse of all the Coats of the Rectum.—Third Variety: Prolapse of the Upper Part of the Rectum into the Lower, or Invagination.—Fourth Variety: Invagination in the Continuity of the Bowel.—Prolapse of the Mucous Membrane alone.—Causes.—Symptoms.—Treatment: Palliative and Curative.—Prolapse with Hæmorrhoids.—Treatment by Injections.—Cauterization.—Description of Operation.—Smith's Clamp.—Dupuytren's Operation.—Prolapse of the Second Degree.—Pathological Changes.—Presence of Peritoneum.—Strangulation.—Dangers in Forcible Reduction.—Fatal Case of Reduction.—Advisability of Reducing Inflamed or Gangrenous Prolapse.—Excision of Prolapse after the Formation of a Slough.—Dangers of Operation of Excision in Extensive

Prolapse.—Operation by Elastic Ligature.—Third and Fourth Varieties.
—Differences between Third and Fourth.—Degrees of Invagination.—
Anatomical Appearances. — Pathology. — Relative Frequency. — Symptoms.—Physical Signs.—Acute and Chronic Forms.—Diagnosis.—Differential Diagnosis from Volvulus; from Stricture; from Internal Hernia;
from Obstruction by Pressure from without the Bowel; from Foreign
Bodies; from Peritonitis with Perforation.—Treatment.—Replacement by
Manipulation; by Injections.—Treatment by Puncture.—Laparotomy.-
Description of Operation 110

CHAPTER VIII.

NON-MALIGNANT GROWTHS OF THE RECTUM AND ANUS.

Polypus.—Definition.—Hypertrophy of Villi.—Characteristics.—Villous Tumor.—Adenomatous Polypus.—Fibrous Polypus.—Structure; Characteristics.—Symptoms of Polypus.—Diagnosis.—Diagnosis from Malignant
Disease.—Treatment.—Vegetations. — Definition.—Description. — Microscopic Appearances.—Relation to Syphilis.—Symptoms of Vegetations.—
Diagnosis.—Treatment.—Condylomata.—Distinction between Condylomata and Vegetations.—Description.—Syphilitic and Non-syphilitic Condylomata.—Benign Fungus.—Gummata.—Rarity and Literature.—Anorectal Syphiloma.—Definition of Fournier. — Fibromata. — Lipomata.—
Characteristics.—Enchondromata.—Cysts.—Dermoid Growths.—Characters.—Pilo-Nidal Sinus.—Hydatids.—Fœtal Inclusions.—Spina Bifida.—
Congenital Cysts 135

CHAPTER IX.

NON-MALIGNANT ULCERATION.

Varieties.—Simple Ulcers.—Generally due to Traumatism.—Various Forms
of Injury to which Rectum is Subject.—Sodomy.—Injury of Rectum in
Labor.—Ulcers due to Surgical Interference.—Fissure or Irritable Ulcer.
—Nothing Distinctive in the Ulcerative Process.—Characteristics of Irritable Ulcer.—Theories concerning this Form of Ulcer.—Description.—
Herpes.—Tubercular Ulceration.—Distinction between True Tubercular
Ulcer and a Simple Ulcer in a Tuberculous Person.—Description of Each.
—Scrofulous Ulceration.—Esthiomène.—Rodent Ulcer.—Dysentery.—A
Cause of Stricture.—Venereal Ulceration.—Gonorrhœa.—Chancroids.—
Chancroidal Stricture.—Discussion.—True Chancre.—Secondary and Tertiary Syphilitic Ulcerations.—Diagnosis of Syphilitic Ulcers.—Ano-rectal
Syphiloma as a Cause of Ulceration.—Ulceration Secondary to Stricture.
—Gangrene.—Symptoms of Ulceration.—Gravity of the Disease.—Diagnosis.—Treatment.—General and Local Measures.—Treatment of Fissure.
—Fissure Complicated with Polypus.—Treatment by Rest, Fluid Diet and
Incision of the Sphincter.—Local Applications 158

CHAPTER X.

NON-MALIGNANT STRICTURE OF THE RECTUM.

Stricture due to Changes in the Rectal Wall and to Pressure from Without.—
Spasmodic Stricture.—General Division into Venereal and Non-Venereal

Strictures and into Fibrous and Cicatricial.—Frequence of Syphilis in Connection with Stricture.—Non-Venereal Strictures.—Congenital, Dysenteric, Traumatic, Varieties.—Stricture from Hypertrophy of Valves.—Pathological Anatomy.—Changes in Rectal Wall above and below the Stricture.—Changes in Parts around the Stricture.—Symptoms.—Value of Flattened Passages as Symptom.—Signs of Obstruction.—Obstruction with Stricture of Considerable Calibre. — Diagnosis. — Dangers to be Avoided in Examination.—Difficulty when Disease is Situated high up in the Bowel.—Use of Bougie for Diagnosis.—Treatment.—Advisability of Anti-Syphilitic Medication.—Palliative Treatment.—Medicinal Treatment of Threatened Obstruction.—Surgical Measures.—Dilatation, Gradual or Sudden —Rules for Gradual Dilatation.—Divulsion, Dangers of, and Methods of Performing.—Treatment by Free Division.—Description of Operation.—Collection of Cases.—Results of this Treatment.—Comparison with Colotomy.—Cases from Author's Practice.—Knife for Operation.—Excision of Non-Malignant Stricture.—Colotomy.—Restrictions to the Operation.—General Considerations Regarding it.—Treatment of Stricture High Up 181

CHAPTER XI.

CANCER.

General Characters of Malignant as Distinguished from Benign Growths.—Malignant, Semi-Malignant, and Benign Adenoma.—Encephaloid.—Colloid.—Melanotic Cancer.—Osteoid Cancer.—Age at which Cancer occurs.—Symptoms.—Diagnosis.—Treatment.—Excision: History and Results of Operation.—Conclusions Regarding Excision.—Modes of Performing the Operation.— Excision of Cancer of the Sigmoid Flexure.—Palliative Treatment 218

CHAPTER XII.

IMPACTED FÆCES AND FOREIGN BODIES.

Impacted Fæces.—Intestinal Concretions. - Diagnosis and Treatment of Impaction.—Foreign Bodies Swallowed. — Results which may Follow the Swallowing of a Foreign Body.—Ulceration and Abscess.—Foreign Bodies introduced per Anum. — Cases. — Prognosis.—Treatment.—Dangers of Attempts at Removal.—Laparotomy for Removal.—Cases Successful . 252

CHAPTER XIII.

PRURITUS ANI.

Pruritus Generally a Symptom of some other Disease.—Description.—Causes.—Relations of Internal Hæmorrhoids, Fistula, Worms, Parasites, and Eczema to Pruritus.—Treatment of Eczema.—Herpes and Erythema.—Constitutional Conditions causing Pruritus.—Dependence upon Constipation.—Treatment of Constipation.—General Treatment of Pruritus . 269

CHAPTER XIV.

SPASM OF THE SPHINCTER, NEURALGIA, WOUNDS, RECTAL ALIMENTATION.

PAGE

Spasm without other Disease. — Cases. — Authorities. — Symptoms. — Treatment. — Neuralgia. — Cases. — Diagnosis. — Treatment. — Wounds. — Complications. — Spontaneous Rupture. — Treatment of Wounds. — Alimentation. — Physiology of Absorption. — **Nutritive Enemata.** — Nutritive Suppositories 277

LIST OF ILLUSTRATIONS.

		PAGE
FIGURE	1. Antero-posterior curve of the **rectum**,	2
"	2. Section of normal rectal wall,	8
"	3. Section of rectal mucous membrane,	9
"	4. Rectal veins seen from without,	14
"	5. Rectal veins seen from within,	15
"	6. Nerves of the anus,	18
"	7. Third variety of congenital malformation,	32
"	8. Fourth variety of congenital malformation,	33
"	**9.** Fifth variety of congenital malformation,	34
"	10. Sixth **variety of** congenital malformation,	35
"	11. Condition **of bowel after** colotomy,	46
"	12. Idem,	46
"	13. **Enterotome of Dupuytren in position,**	**47**
"	14. Examining table, closed,	50
"	**15.** Examining table, opened,	50
"	**16.** Lamp for rectal examinations,	51
"	**17.** Case for rectal instruments,	52
"	**18.** Blunt-pointed bougie,	**55**
"	**19.** Sharp-pointed bougie,	**56**
"	**20.** Bougie à boule,	**56**
"	21. Van Buren's rectal speculum,	58
"	22. Fenestrated rectal speculum,	59
"	23. Bivalve rectal speculum,	59
"	24. Rectal depressor,	59
"	25. Endoscope,	60
"	26. Thermo-cautery,	63
"	**27.** Varieties of fistula,	78
"	28. Fistula with double track,	79
"	29. Idem,	79
"	30. Allingham's ligature holder,	84
"	31. Helmuth's ligature holder,	84
"	32. Author's fistula knife,	86
"	33. Gorget,	86
"	34. Spring scissors,	87
"	35. Forceps for hæmorrhoids,	107
"	36. Smith's clamp,	109
"	**37.** First variety of prolapse,	111
"	38. Second variety of prolapse,	111
"	39. Third variety of prolapse,	112

LIST OF ILLUSTRATIONS.

		PAGE
FIGURE 40.	Rectal supporter,	115
" 41.	Rectal polypus,	136
" 42.	Villous polypus,	137
" 43.	Microscopic section of villous polypus,	137
" 44.	Glandular polypus,	138
" 45.	Vertical section of glandular polypus,	139
" 46.	Vegetations around anus,	142
" 47.	Condylomata,	147
" 48.	Stricture of the rectum,	184
" 49.	Rectal dilator,	200
" 50.	Wales's dilator,	201
" 51.	Knife for proctotomy,	205
" 52.	Cancer of the rectum—Malignant adenoma (Stimson)	219

DISEASES OF THE RECTUM AND ANUS

CHAPTER I.

PRACTICAL POINTS IN ANATOMY AND PHYSIOLOGY.

Rectum.—Position and Measurements.—Curves.—Divisions.—Relations.—Anus. —Parts in Detail.—Peritoneum.—Relations to Three Portions of the Rectum.— Distance of Peritoneal *Cul-de-Sac* from Anus.—Muscular Layer.—Arrangement of Fibres.—Submucous Layer.—Mucous Membrane.—Sustentator Tuniceæ Mucosæ.—Columnæ Recti.—Glands of Mucous Membrane.— Muscles of the Rectum and Anus.—External Sphincter.—Internal Sphincter. —Recto-Coccygeus.—Levator Ani.—Transversus Perinei.—Arteries.—Superior Hæmorrhoidal.—Middle Hæmorrhoidal.—Inferior Hæmorrhoidal.— Veins.—Superior Hæmorrhoidal.—Middle Hæmorrhoidal.—Inferior Hæmorrhoidal.—Minute Anatomy of Veins.—General and Visceral Venous Systems.—Nerves.—Cerebro-Spinal and Sympathetic Nerve Supply.—Tonic Contraction of Sphincter.—Explanation of Wandering Pains in Rectal Disease.—Lymphatics.—External and Internal Lymphatic Vessels.—Physiology, —Anatomy of the Third Sphincter.—Valves of Mucous Membrane.—Plica Transversalis Recti of Kohlrausch.—Lack of Uniformity in Different Subjects. —Physiology of Defecation.—Explanation of Retention of Fæces after Destruction of the Sphincter.—Conclusions Resulting from Study of Third Sphincter.

THE rectum is the terminal portion of the large intestine extending from the sigmoid flexure to the anus. In its natural position its length varies in different persons from six to eight inches. When dissected out of the body and straightened, it will be found to measure about two inches more. Its position in the true pelvis is comparatively fixed; and its fixity renders it the more liable to those displacements, such as invagination and prolapse, which are due to straining at stool; and accounts also for the fact that, when denuded by the destruction of the surrounding cellular tissue, it remains separated from the walls of the pelvis, and cannot come in contact with the adjacent soft parts and thus undergo healing.

The upper limit of the rectum is difficult to determine with accuracy, except from the fact that it is separated from the sigmoid flexure by a slight constriction which becomes more apparent when attempts are made at dilatation. From this upper point it gradually expands into a pouch, the ampulla, and then again suddenly contracts under the grasp of the muscles which close its lower end.

Curves.—The curves of the rectum are exceedingly important in a practical point of view. There are two, one antero-posterior, the other lateral. The former is double. From above downwards it follows the curve of the sacrum and coccyx, being concave in front, and convex behind. When it reaches a point opposite the tip of the coccyx it suddenly reverses its direction, turns sharply backwards, and ends at the anus about one inch in front of the tip of that bone.

By this backward curve of its lower end, which is represented in an exaggerated form in Fig. 1, it is separated from the vagina in the female,

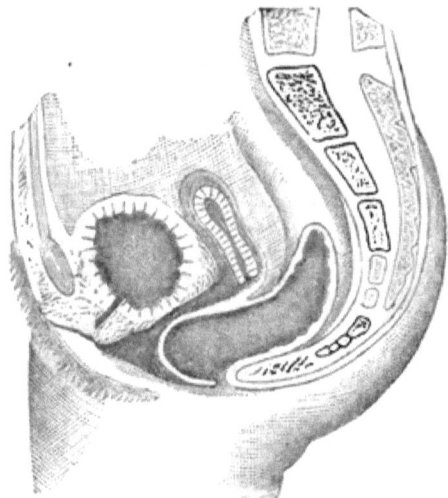

Fig. 1.—Exaggerated antero-posterior curve of rectum.

and from the urethra in the male, by a triangular space having its base at the perineum, its upper wall at the vagina or urethra, and its lower at the upper wall of the rectum. The angle of junction of these two curves is well marked, measuring from twenty to thirty degrees; and the curve is not without influence in the function of defecation, since, by it, an obstruction is formed to the downward course of the fæces.

The lateral curve is generally a single one from left to right, starting at the left sacro-iliac synchondrosis and ending at the median line at a point opposite the third sacral vertebra, from which point it generally passes straight on to the anus. This curve may, however, pass beyond the median line to the right in its lower portion, and again return to the median line at the anus. It is subject to many variations, and the upper portion may be more or less twisted on itself like the sigmoid flexure.

The sigmoid flexure may occupy an unnatural position, and the rectum, instead of commencing at the left sacro-iliac junction and curving towards the right, may commence at the right and curve toward the left. In one case, reported by Cruveilhier,[1] where the sigmoid flexure was in the natural position, the rectum passed almost transversely to the right side as far as the right sacro-iliac junction, and then returned again very obliquely in the left side.

Divisions.—For convenience the rectum is usually divided into three portions, named first, second, and third, from below upward. The first extends from the anus to the tip of the prostate; is about an inch and a half long; is firmly closed by the sphincters; and gives attachment to a portion of the levator ani muscle. On account of the direction of this portion, which is the reverse of that next above, the finger should never be passed toward the sacrum, or even directly inward in making an examination; but rather toward the pubes. Bearing this simple anatomical point in mind will often save the patient much unnecessary suffering. The second portion is often described as reaching from the apex of the prostate to the recto-vesical fold of peritoneum; but, as the point of duplicature of the peritoneum is not only variable in different individuals, but at different times in the same individual, it is better to adopt a fixed bony point, as the third piece of the sacrum; in which case the middle portion will measure about three inches in length. This portion, it will be remembered, is convex backward, following the curve of the sacrum. The third portion extends from the third sacral vertebra to the left sacro-iliac synchondrosis; its lower part is partially, and its upper, completely, surrounded by peritoneum; which, in the upper part, forms the meso-rectum attaching it to the sacrum.

Relations.—The most important surgical relations of the rectum are on the anterior surface. The first portion is surrounded laterally and posteriorly by a bed of connective tissue, rich in fat and blood-vessels, and may, therefore, be incised on either side, or backward, with comparative safety. In front, however, it is directly in relation with the membranous urethra in the male, and with the vagina in the female; though at the anus it is separated from them both by its backward and downward course. This intimate relationship with the urethra is often taken advantage of in catheterism, when by passing the finger into the rectum the tip of the instrument may easily be felt; and it also explains why in all operations on the urethra or vagina the rectum should first be emptied to save it from being wounded.

In the second portion also, the lateral and posterior surfaces have no

---

[1] Anat. Path., Amer. Edition, 1844, p. 377.

special surgical relations; while the anterior is in direct contact with the prostate, the base of the bladder, the seminal vesicles, and sometimes, at its upper limit, with the peritoneal fold of Douglas. This portion is closely connected with the bladder in the male, and with the vagina in the female, by connective and muscular tissue; and the two cavities may easily be made to communicate by any morbid process or by a surgical procedure. It was at this point that the trocar was plunged from the rectum into the bladder in the old operation of puncturing the bladder through the rectum; and Hyrtl[1] speaks of a man who was only able to pass his water after first introducing his finger into the rectum and raising a calculus out of the trigone of the bladder. A somewhat analogous case is reported in which a long slender calculus perforated the bladder and projected into the rectum, from which it was easily removed.[2] The prostate, when large, may project over the sides of the rectum, or the latter may receive the prostate in a kind of groove on its upper surface.

The third, or upper portion, unlike the other two, has important surgical relations on every side. Posteriorly it is in whole or part covered with peritoneum; and is separated from the sacrum by the pyriformis muscle, the sacral plexus of nerves, and the branches of the internal iliac artery. On its sides it is in contact with the adjacent convolutions of small intestine, and lower down, with the levator ani muscle and the connective tissue of the ischio-rectal fossa. In the male it is in relation, in front, with the posterior surface of the bladder, from which it is separated by coils of small intestine. In cases of retention either of urine or fæces the two may be brought into actual contact. In the female, it is in relation, anteriorly, with the broad ligament, the left ovary and Fallopian tube, the uterus and vagina. When the rectum and uterus are empty, the coils of small intestine pass down between them to the bottom of the fold of Douglas, and they may even escape through the posterior wall of the vagina in case of injury.

From these relations it is apparent that enlargements and malpositions of the uterus may act directly upon the rectum. The vessels may be so obstructed as to cause hæmorrhoidal troubles, or interfere with operations for their relief. The rectum may be entirely occluded by the pressure of a uterine tumor; and a hasty examination of the rectum may lead to the diagnosis of a cancerous tumor when in reality the normal uterus alone is felt. The advantage of a rectal **examination in all** doubtful cases of pelvic disease is also manifest.

The Anus.—The rectum terminates below in the anus which is tightly closed by the external sphincter muscle. The skin around its border is thin and pigmented, covered with fine hair in the male, and contains a great number of sebaceous follicles and muciparous glands. The skin

---

[1] Topog. Anat., ii., p. 103.
[2] Gooch: Chirurg. Works, London, 1792, vol. iii., p. 216.

passes deeply into the anal orifice, and its point of junction with the mucous membrane is in some persons indicated by an indistinct white line.[1] This white line of junction also corresponds to the division between the external and internal sphincter muscles; and also to the point at which many of the terminal filaments of the internal pudic nerve perforate the gut. Both skin and mucous membrane at the anus are remarkable for the development of erectile tissue; the arteries coming from the inferior hæmorrhoidal, and the veins being very numerous, winding, and twisted. The presence of this erectile tissue accounts for the habit of pederasty which will occasionally be referred to as a cause of rectal disease. It is a habit to which few are addicted in this country, but which is not uncommon in some other parts of the world. In America it is chiefly seen amongst the negro race and on shipboard amongst sailors who are on a long voyage. Among the latter it was a vice whose existence was well known and which was occasionally punished by the officers during the late war. The nerves are derived both from the cerebro-spinal and sympathetic systems, as will be shown later.

After these general considerations of the position and relations of the rectum as a whole, the individual parts may be taken up more in detail. The rectal wall is composed, as are the other parts of the intestine, of four layers: an external or peritoneal; a muscular, divided into longitudinal and circular; a sub-mucous connective tissue layer; and most internally, the mucous membrane. The total thickness of these coats collectively varies greatly in different subjects, the variation being chiefly in the muscular coat, the others remaining pretty constantly of the same thickness.

Peritoneum.—The upper portion of the rectum is entirely surrounded by peritoneum, and has, beside, a fold of attachment to the anterior surface of the sacrum, known as the meso-rectum. The meso-rectum is about four inches long, blends with the meso-colon above, and extends down as low as the third or fourth sacral vertebra, from which point its two layers are reflected over the sides and anterior surface of the rectum on to the posterior wall of the uterus and upper limit of the vagina in the female; and on to the bladder in the male, forming the *cul-de-sac* of Douglas. The meso-rectum may be so short as to disappear when the rectum is distended, or it may be entirely absent; in which case the peritoneum passes directly from the sides of the rectum to the sacrum. Between its two layers may be found some loose connective tissue, the hæmorrhoidal vessels and nerves, and the lymphatics.

In passing from the limit of the meso-rectum behind, to form the *cul-de-sac* in front, the peritoneum covers more or less of the lateral and anterior surfaces of the middle portion of the rectum. As before men-

---

[1] Hilton: Rest and Pain. Wood's Library of Standard Medical Authors, p. 166.

tioned, the point at which the peritoneum leaves the anterior surface of the middle portion of the rectum to be reflected upon the posterior surface of the bladder in the male, or of the vagina or uterus in the female, varies in different subjects, and at different times in the same subject; and hence the differences in its distance from the anus as given in different works on anatomy. In new-born children the bottom of the *cul-de-sac* touches the upper edge of the prostate and approaches to within about an inch of the anus. At five years it rises in the pelvis with the development of the seminal vesicles and internal organs of generation; and in old people with enlargement of the prostate, it is carried still higher. In women it generally extends to the upper border of the posterior vaginal wall; so that the latter is separated from the rectum by peritoneum for about one-third of an inch. By every expansion of the bladder or rectum as well as by tumors of the pelvis the fold is carried further away from the anus, as may easily be demonstrated on the cadaver by forcible injections of the bladder.

The average distance from the anus of the point at which the serous coat leaves the anterior wall of the rectum is, therefore, very difficult to determine; and yet it is of the greatest importance in all surgical operations on the part; since the fact of opening or not opening the peritoneal cavity may make all the difference between life and death in the result of an operation. Dupuytren gives the distance as seventy mm., and less when the organs are empty; Lisfranc gives six inches in the female, and four in the male, but does not state in what condition of the organs the measurements are taken; Sappey, Velpeau, and Legendre give five and a half cm. when the bladder is empty and eight when distended; Quain says four inches; Allingham from two to five or more. Cripps,[1] acting on the idea that the fold is not easily displaced downward by traction on the rectum, has experimented by filling the peritoneal cavity with plaster, and then thrusting a needle through the skin of the perineum till its point struck the plaster. In this way he has obtained an average measurement of two and a half inches when the bladder and rectum are both empty, and an additional inch when distended.[2]

Muscular Coat.—In the fact that the muscular coat is arranged in two layers, an external longitudinal and an internal circular, the rectum resembles the other portions of the alimentary canal; but in the further arrangement of its fibres it resembles the œsophagus more closely than the intermediate portions. The fibres are spread out into two uniform

---

[1] Cancer of the Rectum. London, 1880, p. 129.

[2] The following authors give the following measurements: Malgaigne, male, 6-8 cm.; females, 4-6 cm. Luschka, 5.5-8 cm. Hyrtl, 8 cm. Lisfranc and Sanson, 11 cm. Richet, males, 10.8 cm.; females, 16.2 cm. Blaudin, males, 8.1 cm.; females, 4.1 cm. Ferguson, males, 10.5 cm.; females, 15.4 cm. Esmarch: Die Krankheiten des Mastdarms und des Afters. Pitha u. Billroth: Chirurgie, p. 7.

layers, and are not arranged in bands crossing each other in a basket network and leaving sacculi between the meshes.

The longitudinal fibres are the direct continuation of the three longitudinal bands of the large intestine. Upon reaching the rectum, these blend into one continuous sheath which, however, is somewhat heavier on the anterior surface of the bowel than on any other. At the point of contact of the rectum with the bladder and prostate these fibres are in part reflected with the peritoneum on to the posterior wall of the latter and thus form a firm band of union between the two organs, as has been particularly described by Dr. Garson.[1] They have been named by him the recto-vesical fibres.

The ending of the longitudinal fibres is worthy of note. According to Horner,[2] when they reach the lower margin of the internal sphincter a part of them turn upwards between it and the external sphincter and ascend for an inch or two in contact with the mucous coat into which they are finally inserted; having, therefore, an obvious influence in causing protrusion of the mucous membrane. In the lower fourth of their extent these fibres become weaker and less distinct, and some of them finally blend into elastic tendinous tissue which passes between the bundles of the external sphincter, and is inserted into the subcutaneous connective tissue of the anus. Others are inserted posteriorly by means of an elastic tendon about an inch long into the anterior sacro-coccygeal ligament—an arrangement pointed out by Luschka[3] as analogous to what is found in most mammalia, in whom a considerable number of the longitudinal fibres are inserted into the base of the coccyx, giving a fixed point for the rectum in defecation.

The circular layer is reinforced at certain points; notably at the internal sphincter which is merely a collection of these fibres, and at a point higher up where they are again gathered into a bundle either partly or completely surrounding the bowel, known as the third sphincter. This muscle will be described more fully later.

Submucous Coat.—The submucous tissue forming the bed upon which the mucous membrane rests is sufficiently lax to permit of considerable sliding of the mucous membrane on the muscular coat. In it the blood-vessels ramify, and from it perpendicular processes are given off which perforate both the internal and external muscular layers and are finally lost in the sheaths of the muscular fibres, or go entirely through the muscular layer and blend with the fibrous stroma of the surrounding

---

[1] The Arrangement and Distribution of the Muscular Fibres of the Rectum. Paper read before the Brit. Med. Ass. Reported in Brit. Med. Jour., Sept. 6th, 1879.

[2] A Treatise on Special and General Anatomy. Vol. ii., p. 40, Philadelphia, 1826.

[3] Anat. des Menschen. Vol. ii., Part 2, p. 208.

fatty tissue. These processes from the submucous tissue, together with the lymph and blood-vessels, serve to bind the various layers of the rectal wall together.[1] See Fig. 2.

Mucous Membrane.—The mucous membrane of the rectum corresponds in its general characters with that of the other parts of the bowel, being modified, however, in certain particulars to suit its location and function. Its thickness is about three-quarters of a mm.; it is redder and more vascular than that of other parts of the large intestine; it glides freely on the tissue beneath; and is so ample as to be gathered into folds at various points which are of considerable surgical and anatomical interest. At its point of union with the skin of the anus it is gathered into vertical folds which diminish when the bowel is distended, but do not entirely disappear, and hence are not due solely to the contraction of the sphincter. These vertical folds have received the name of *columnæ recti*, or columns of Morgagni; and Treitz states that they contain bands of mus-

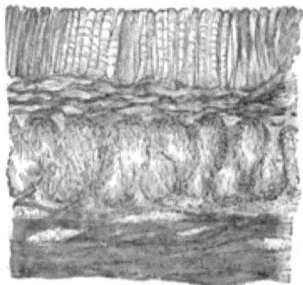

FIG. 2.—Section of normal rectal wall (Cripps).

cular fibres running longitudinally and terminating above and below in elastic tissue. Kohlrausch[2] also describes a thin layer of longitudinal muscular fibres under the mucous membrane at this point and has named it the *sustentator tunicæ mucosæ;* but most anatomists, with Henle, have failed to find anything more than the stratum of muscular tissue common to the whole mucous coat, and known as the *muscularis mucosæ*.

Between the lower ends of the *columnæ recti* little arches are stretched from one to the other, forming pouches of skin and mucous membrane. These are more developed in old people, and may retain small pieces of hardened fæces or foreign bodies in their cavities, and thus give rise to suppuration and abscess.

The mucous membrane may for the purpose of study be divided into three separate layers, the muscular, glandular, and epithelial. Fig. 3.

---

[1] Cripps, op. cit., p. 38.
[2] Anat. u. Physiol. der Beckenorgane, Leipzig, 1854. Boyer also says they are strengthened by muscular fibres. Traité d'Anat., T. iv. Paris, 1815.

The muscular layer (*muscularis mucosæ, sustentator tunicæ mucosæ*) is a layer of unstriped muscular tissue about 0.02 mm. thick, which is everywhere found in the deepest layer of the mucous membrane, extending from the œsophagus to the rectum, but is more strongly developed in the region of the anus where it serves to hold the membrane in place and prevent prolapse. It consists of bundles running in some parts both longitudinally and circularly, and in others in one direction only; and which send prolongations up between the glands to the villi.

The glandular layer is about 0.07 mm. in thickness. It consists of a layer of Lieberkuhn's follicles with an occasional solitary closed follicle below them, the situation of which is marked by a slight depression in the mucous membrane, and an absence of the tubular follicles at that point. The follicles are tubular depressions arranged with great regularity and set so closely together that the width of the intervening tissue

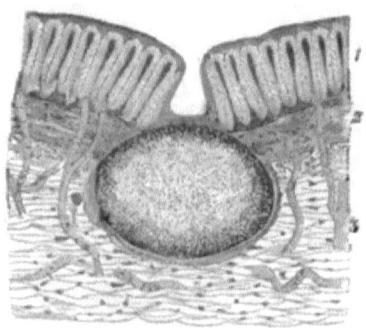

FIG. 3.—Section of the rectal mucous membrane (Esmarch). 1. Follicles of Lieberkuhn. 2. Muscular layer of mucous membrane. 3. Submucous connective tissue and vessels; with a solitary closed follicle, over which the tubular follicles are wanting.

is, on the average, about one-sixth the diameter of the follicle. The length of the tubes is four or five times their diameter, the respective measurements being: length, 0.35 mm.; diameter, 0.08 mm. These tubular depressions or follicles are lined with epithelial cells arranged with their bases resting on the connective tissue and their apices free in the cavity of the follicle; and the cells of one follicle are directly continuous with those of the next hanging freely into the lumen of the bowel as they pass over from one depression into the next. The appearance of the cells is analogous to that of a bee's honeycomb, the intervening wall being common to two cells. The intertubular tissue consists of a fine trabecular network, the meshes of which are very long in the vertical direction running parallel to the follicle (Cripps).

The follicles of Lieberkuhn are simply inverted villi and answer the same purpose of absorption. There are good reasons for the substitution of follicles for villi in this part of the canal, the former being less subject to injury from hardened fæces, and the fact of such substitution gathers great weight from the fact that in certain cases where an artificial anus has been established, the whole bowel below that point has been found in after-years covered with a growth of villi.[1]

Muscles of the Rectum and Anus.—The muscles which may properly be included in a description of the rectum and anus are the external and internal sphincters, the levator ani, ischio-coccygeus, retractor recti or recto-coccygeus, and the transversus perinei.

External Sphincter.—The external sphincter muscle is a thin layer of voluntary fibres, about half an inch broad on each side of the anus, surrounding it in the form of an ellipse, and having a narrow pointed insertion anteriorly and posteriorly. It is situated immediately beneath the skin, and extends about two centimetres up the bowel where its upper limit may sometimes be seen by the white line already mentioned. It is divided into a superficial and deep portion. The superficial is inserted both in front and behind into the subcutaneous cellular tissue. The deeper and thicker portion is inserted posteriorly by a narrow flat tendon into the posterior surface of the fourth coccygeal vertebra. Between the tendon and the bone is a bursa about the size of a pea—bursa mucosa coccygea of Luschka. Anteriorly it is inserted into the central tendon of the perineum in common with the transversus perinei and bulbo-cavernosus, and in women with the sphincter vaginæ. The action of the muscle is to close the anus and, under the control of the will, to antagonize the proper dilators of the anus, the levator ani and ischiococcygeus, as well as the peristaltic action of the bowel and the contraction of the diaphragm. The superficial band of fibres acts only in puckering the skin. The nerve-supply comes from the hæmorrhoidal branch of the internal pudic, and the hæmorrhoidal branch of the fourth sacral nerve.

Internal Sphincter.—The internal sphincter is situated immediately above and partly within the deeper portion of the external sphincter; being separated from it by a layer of fatty connective tissue. Its thickness is about two lines; its vertical measurement from half an inch to an inch; and it is a direct continuation of the involuntary circular fibres of the bowel, growing thicker and stronger as it approaches the anus. It also is supplied by the hæmorrhoidal branch of the internal pudic.

Recto-coccygeus (Retractor recti, **Trietz**;[2] Tensor Fasciæ Pelvis, Kohlrausch).—This muscle consists of two flat lateral bands of unstriped fibres, each of which is about four mm. broad, which diverge at an acute angle from the anterior coccygeal ligament at the tip of the coccyx, and passing forward and downward, embrace the lower end of the rectum on

---

[1] Specimen No. 1,288, Museum of College of Surgeons (Cripps).

[2] Vierteljahrsschrift f. praktische Heilkunde. Prag, 1863, Bd. i., S. 124. Henle. Abbildung 2, 183.

each side like a fork. It is located directly under that portion of the levator ani which forms the floor of the pelvis between the tip of the coccyx and the anus; and blends partly with the longitudinal muscular fibres of the rectum, and partly with the pelvic fascia surrounding its end. Its function is to hold the end of the rectum against the coccyx and to give it a fixed point in defecation.

Levator Ani.—The levator ani and ischio-coccygeus muscles form a true diaphragm to the pelvis by giving an uninterrupted muscular and tendinous plane from the lower border of the pyriformis, behind, to the arch of the pubes in front. That part which is named ischio-coccygeus is usually described as a separate muscle, though in no way differing in function from the larger portion, and only distinguishable from it by its more tendinous structure. It is situated just in front of the sacro-sciatic ligaments, and arises by aponeurotic fibres from the sides and tip of the spine of the ischium, from the anterior surface of the lesser sacro-sciatic ligament, and often from the posterior part of the pelvic fascia. It is inserted, also by aponeurotic fibres, into the border of the coccyx and lower part of the border of the sacrum. Owing to its tendinous origin and insertion, the greater part of the muscle is composed of aponeurotic fibres. It is in relation superiorly, by its concave surface, with the rectum; inferiorly, by its convex surface, with the sacro-sciatic ligaments and the gluteus maximus; posteriorly, its border is in contact with the lower border of the pyriformis; and anteriorly, it is directly continuous with the fibres of the levator ani. Its action is to draw the coccyx to its own side, or, when both muscles act together, to fix that bone and prevent its being thrown backward in defecation. It probably has no such action as would justify the name of levator coccygis, given it by Morgagni. Its nerve-supply is from the anterior branch of the fourth sacral nerve.

The levator ani proper, which constitutes the remaining portion of the pelvic diaphragm, is in its general shape an inverted cone, supporting the pelvic contents in its cavity and allowing the rectum and prostate to pass through its apex. Considering each lateral half of the muscle apart, we find it made up of a delicate layer of muscular fibres forming a thin, curved, and quadrilateral sheet, broader behind than in front. Its upper border is stretched across the pelvis from the pubes to the spine of the ischium, arising from both these bony points and from the tendinous line of union of the pelvic with the obturator fascia, which runs antero-posteriorly between them. Its attachment to the pubic bone is at a point on its inner surface, near the middle of the descending ramus and a little to one side of the symphysis. This attachment will be found to vary somewhat in different dissections, being sometimes a little higher or a little lower on the bone, and sometimes on the cartilage between the bones. The muscular fibres may also be traced at times upward into the pelvic fascia above its junction with the obturator.

From this extensive though delicate and in great part membranous origin, the fibres proceed downwards and inwards toward the median line. Those most anterior unite with those of the opposite side beneath the neck of the bladder, the prostate, and the adjacent portion of the urethra. These fibres are concealed by the pubo-prostatic ligament or anterior fold of the recto-vesical fascia, from which they also sometimes take origin in part. They are in relation, in front, with the posterior surface of the triangular ligament. This portion is sometimes separated from the main body of the muscle by a cellular interval, similar to those often found in other parts of this thin muscular sheet.

The fibres which arise from the tip of the spine of the ischium are inserted into the side of the tip of the coccyx; while the fibres immediately in front of these (precoccygeal) unite with those of the opposite side in the median line and form a *raphé* which extends from the point of the coccyx to the posterior border of the sphincter and thus complete the floor of the pelvis.

The fibres which arise indirectly from the upper part of the obturator foramen and from the brim of the pelvis by means of the pelvic fascia, pass downward and inward, forming a curve with its concavity upwards, and may be divided into vesical and anal. The vesical pass into the sides of the bladder. The anal fibres in part pass backward and meet behind the bowel and in part blend with those of the external sphincter at its upper border, there being no distinct line of separation between the two muscles.

The relations of the levator ani are of great surgical importance. Superiorly its surface is covered by the superior pelvic fascia which separates it from the peritoneum and pelvic organs. Its inferior surface is separated from the obturator internus muscle by the obturator fascia, and beneath this is the ischio-rectal fossa. The posterior part of the muscle is in relation with the gluteus maximus.

The actions of this muscle are various. First, it acts as a support to the pelvic organs, and antagonizes the diaphragm and abdominal muscles when they act upon the abdominal contents. Again, it prevents the rectum from being protruded, and raises the anus and opens it; being in this respect the direct antagonist of sphincter. By inclosing the neck of the bladder the muscle acts upon it also, and in the act of defecation when the muscle is contracted to open the anus, the neck of the bladder is pressed upon and the urethra closed. In this way is explained the well-known difficulty of passing urine and fæces at the same time. By inclosing the bladder, vesiculæ seminales, prostate, and anus in its grasp, the muscle produces a sympathy among these parts which will often be found very distressing in diseases of the rectum or after operations for their relief—such as impossibility of micturition, erections, and lancinating pain due to spasmodic action of the muscle. It will often happen that after a complete paralysis by free division of both sphincter muscles in an operation upon the rectum, the patient will still complain of a sharp spasmodic pain at intervals—just such a pain as is caused by spasmodic contractions of the sphincter. In such cases it is the levator ani which is at fault. The muscle also aids the longitudinal fibres of the rectum in their opposition to the dragging of the fæces; and the anal fibres also draw the rectum upwards and forwards, and compress it on the sides, and thus aid in the expulsion of its contents.

The muscle receives a filament from the fourth sacral nerve on its pelvic surface, and another from the internal pudic.

Transversus perinei.—This also has an action in defecation. Its fibres do not always blend with those of the opposite side in the median *raphé*, but the two muscles are sometimes continuous, traversing the anterior extremity of the external sphincter. In such a case the two muscles form a continuous half ring the concavity of which is directed backwards and embraces the anterior part of the rectum, assisting powerfully in defecation by pressing the anterior against the posterior wall of the bowel in conjunction with the levator-ani (Cruveilhier).

Arteries.—The rectum is supplied with blood from five arteries, one single and two pairing.

The superior hæmorrhoidal is single and is a direct branch of the superior mesenteric. It is the direct continuation of the parent trunk, passing into the pelvis behind the rectum in the fold of the meso-rectum and dividing into two branches which extend, one on each side of the bowel, to its lower end. About five inches from the anus these subdivide into smaller branches about seven in number, which pierce the muscular coat about two inches lower down. They then descend between the mucous and muscular layers at regular intervals to the end of the bowel, where they communicate in loops opposite the internal sphincter, and anastomose with the terminal filaments of the middle and inferior hæmorrhoidal arteries.

The middle hæmorrhoidal arteries—one each side—are not constant in their origin, sometimes coming from the hypogastric or the inferior vesical, and sometimes from other sources.

The inferior hæmorrhoidal arteries—also pairing—are usually given off from the internal pudic near the point where it crosses the tuber ischii. They cross through the fat of the ischio-rectal fossæ and are distributed with the middle hæmorrhoidal to the lowest part of the rectum and to the anus and adjacent skin.

Veins.—There are three sets of rectal veins, as there are three sets of arteries, the superior, middle, and inferior; and these are so arranged as to form two distinct venous systems, the one, rectal, and returning its blood to the vena portæ; the other anal, returning its blood through the internal iliac. The first, or rectal circulation, is made up of the superior hæmorrhoidal vein; the second, or anal, is made up of the middle and inferior hæmorrhoidal veins; the middle receiving its blood from the anus and the inferior from the adjacent integument. The middle hæmorrhoidal ascends obliquely into the ischio-rectal fossa; the inferior starts horizontally from the skin of the anus and empties into the internal pudic.

The middle hæmorrhoidal is formed from two venous trunks, one on the anterior, the other on the posterior aspect of the rectum, which by anastomosing with the corresponding branches from the opposite side surround the sphincter in a venous circle. From this circle spring the collateral branches which by their successive division and anastomoses

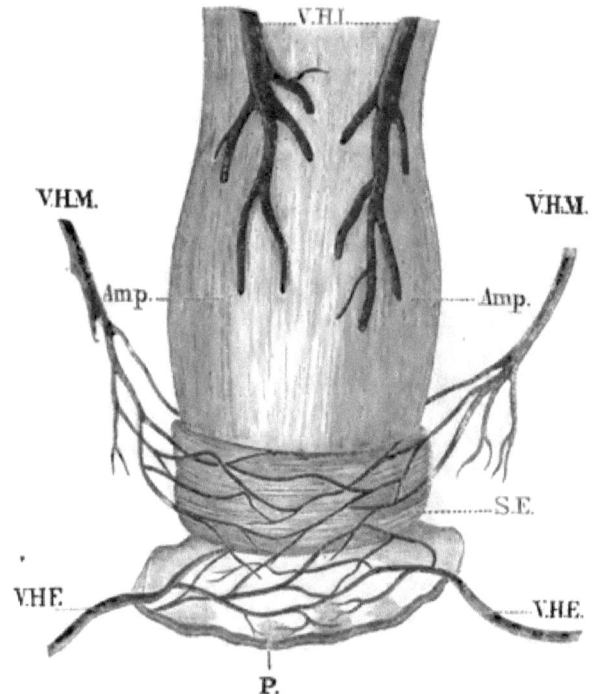

Fig. 4.—Rectal veins seen from without (Duret).[1] Amp., Rectal pouch. S. E., External sphincter. P., Skin at margin of anus dissected up and turned back. V. H. I., Internal hæmorrhoidal vein. V. H. M., Middle hæmorrhoidal vein. V. H. E., External hæmorrhoidal vein.

form a true venous plexus. The inferior hæmorrhoidal vein also has a plexiform arrangement at its origin, but its branches are situated between the skin and the inferior border of the external sphincter. The rectal pouch is not, therefore, supplied with blood from the external hæmorrhoidal veins, but only the anus and the region of the sphincters.

When, on the other hand, the venous circulation of the rectum proper

---

[1] "Recherches sur la Pathogénie des Hémorrhoïdes." Arch. Gén. de Méd., December, 1879.

is injected from the inferior mesenteric vein, three or four large venous trunks may be seen on the external surface of the rectum ascending on the sides and posteriorly, Figs. 4 and 5. These veins make their appearance suddenly by five or six branches which perforate the wall of the bowel about three inches from the margin of the anus. If the rectum be opened longitudinally and the mucous membrane dissected up to a sufficient height (about four inches), it will be seen that these five or six large veins already visible on the outside of the bowel come from within;

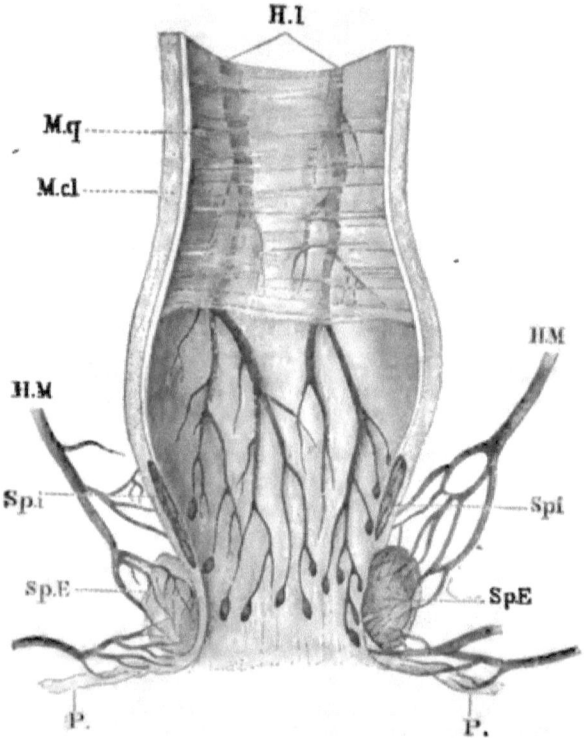

Fig. 5.—Rectal veins seen from within (Duret). M. q., Mucous membrane dissected up and cut away below. M. cl., Muscular tunic. Sp. I., Internal sphincter. Sp. E., External sphincter. P., Skin. H. I., Internal hæmorrhoidal vein. H. M., Middle hæmorrhoidal vein. H. E., External hæmorrhoidal vein.

and that they have already pursued quite a long course under the mucous membrane. They are formed by collateral branches, and especially by about a dozen primitive branches, which originate about half an inch above the anus and ascend in parallel and flexuous lines for several centimetres to unite into common trunks. Each of these little ascending branches has its origin in a minute pool of blood, the size of which varies in the normal state from that of a grain of wheat to that of a small pea.

These little sacs are arranged in a circular form around the extremity of the rectum. If carefully dissected, they may be seen to be connected with the little veins before mentioned, and also with another little vein which perforates the internal sphincter near its lower edge, and empties into one of the rudimentary branches of the external hæmorrhoidal plexus. Many of these little communicating branches between the external and internal hæmorrhoidal systems pass through the substance of the external sphincter. It results from this, that when the external sphincter is contracted, the anastomosis between the two systems is prevented.

Verneuil has laid stress upon the fact that where the internal or superior hæmorrhoidal veins perforate the rectal wall from within outwards, they pass through "muscular button-holes" surrounded by no fibrous tissue and having, therefore, the power of contracting round the vein, closing its calibre, and preventing the return of blood to the liver. In this anatomical arrangement he believes he has found the active cause of internal hæmorrhoids.

The disposition of the rectal veins into two distinct systems, the one internal and the other external, is fully in conformity with our knowledge of the development of the rectum and anus. The rectal *cul-de-sac* is at first situated at some distance from the perineum, and as it descends it carries with it its own proper vascular supply. The anal depression is of necessity provided with an independent set of veins, and when the rectum and anus are finally united into one canal the two venous systems unite.

The internal hæmorrhoidal veins also communicate freely with other branches of the internal iliac around the trigone of the bladder by means of minute branches from one-half to one mm. in diameter which pass through the prostate and vesiculæ seminales.

Nerves.—The nerves of the rectum and anus are derived from both the cerebro-spinal and sympathetic systems. The former are branches from the sacral plexus, the latter from the mesenteric and hypogastric plexuses. The spinal nerves are derived from the third and fourth sacral which supply visceral branches to all the pelvic organs, anastomosing with branches from the sympathetic. The muscular branches from the same nerves have already been spoken of in connection with the individual muscles. The fifth sacral nerve also sends a small twig to the coccygeus. The posterior branch of the superficial perineal nerve from the internal pudic, supplies the skin in front of the anus; while the anterior branch gives several small filaments to the levator ani.

The inferior hæmorrhoidal branch from the pudic supplies the lower end of the rectum, the external sphincter, and the skin of the anus. This nerve may come direct from the sacral plexus through the lesser sacro-sciatic notch. The posterior branches of the sacral nerves also supply the skin over the coccyx and around the anus.

According to a brief contribution of W. Krause,[1] the nerves end in the mucous membrane of the anus, in club-shaped bulbs, about 0.05 mm. in diameter, which lie under the bases of papillæ.

The tonic contraction of the external sphincter muscle is, in part at least, due to the influence of a nerve-centre located in the lumbar region of the spinal cord.[2] If the nerve connection of the sphincter with the spinal cord be severed, relaxation of the muscle takes place. The fact that division of the cord in the dorsal region does not affect the sphincter, except temporarily by shock or depression, proves that this centre is not located above the lumbar region. This nerve-centro is subject to various influences; and the sphincter may either be relaxed, or its tonic contraction increased, by local stimulation, or by the influence of the will or emotions.

Though the dependence of the sphincter for its tonic contraction upon the lumbar nerve-centre seems so great, still it is not absolute. In the case of a man in whom the sacral nerves were entirely paralyzed by an injury, and in whom, therefore, there was no nerve connection with the lumbar centre except perhaps through the sympathetic, Gower[3] observed the maintenance of a certain amount of tonic contraction, which could be inhibited and relaxation produced by stimulation of the mucous membrane of the rectum and anus. From this it would appear that the tonic contraction of the sphincter, as is known to be the case in the arterial system, is habitually dependent on a spinal centre, but may, nevertheless, exist without the action of that centre. The paralysis of the muscle which follows brain lesions is probably due merely to inhibition of the spinal centre, and not to the injury of any centre located in the cerebrum.[4]

The distribution of the spinal nerves serves to explain many of the reflex and so-called anomalous symptoms of pain which are encountered in diseases of the rectum and anus. Brodie[5] relates an instructive case

---

[1] Esmarch, op. cit., p. 10.

[2] Masius: Bull. de l'Acad. Royal de Belgique, xxiv. (1867), p. 312. (Foster's Physiology, p. 387.)

[3] Proc. Roy. Soc. (1877), p. 77.

[4] Foster's Physiology, Phila., 1880, p. 388.

[5] A lady consulted me, says Mr. Brodie, concerning a pain to which she had been for some time subject, beginning in the left ankle, and extending along the instep toward the little toe, and also into the sole of the foot. The pain was described as being very severe. It was unattended by swelling or redness of the skin, but the foot was tender. She labored also under internal piles, which protruded externally when she was at the water-closet, at the same time that she lost from them sometimes a larger and sometimes a smaller quantity of blood. On a more particular inquiry, I learned that she was free from pain in the foot in the morning; that the pain attacked her as soon as the first evacuation of the bowels had occasioned a protrusion of the piles; that it was especially induced by an evacuation of hard fæces; and that, if she passed a day without any evacu-

of pain in the foot over the distribution of the sciatic which was cured by curing prolapsing hæmorrhoids—the irritation being primarily at the termination of the internal pudic, and conveyed thence to the sacral plexus, to be carried to the termination of the great sciatic. In the same way a fissure of the annus may cause pain in the lumbar and iliac regions; pain, loss of sensation, and cramps in the legs; and symptoms of bladder and urethral disease, besides more general nervous phenomena. See Fig. 6.

The chief nerve supply of the rectum is at the lower portion and around the anus—the middle and upper portions possessing very little sensibility; so little in fact that the gravest diseases, such as cancer or ulceration, may exist and not manifest themselves by pain. This also explains how large masses of fæces may accumulate in the rectal pouch

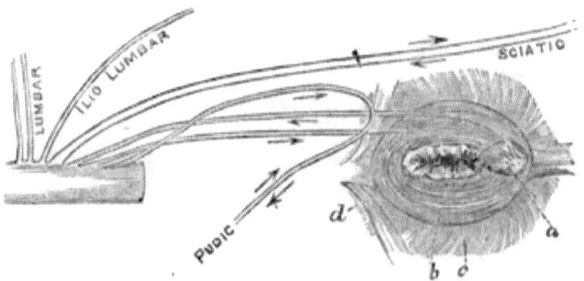

Fig. 6.—Diagrammatic view of nerves of anus. (Hilton.) *a*, Ulcer on sphincter; *b*, the filaments of two nerves are exposed on the ulcer, the one a sensory, and the other motor, both attached to the spinal marrow, thus constituting an excito-motory apparatus; *c*, levator ani; *d*, transversus perinei.

without causing suffering. Puncturing the bladder through the rectum is not a painful operation, and applications of strong acids to the mucous membrane will cause little suffering if the skin be properly protected. Exactly the opposite condition obtains at the anus, the extreme sensibility of which is well known.

The pelvic plexuses of the sympathetic are placed one on either side of the rectum and vagina. Each is composed of prolongations from the hypogastric plexus above, united with branches from the sacral ganglia.

---

ation at all, the pain in the foot never troubled her. Having taken all these facts into consideration, I prescribed for her the daily use of a lavement of cold water; that she should take the Ward's paste (confectio piperis composita) three times daily, and some lenitive electuary at bedtime. After having persevered in this plan for a space of six weeks, she called on me again. The piles had now ceased to bleed, and in other respects gave her scarcely any inconvenience. The pain in the foot had entirely left her. She observed that, in proportion as the symptoms produced by the piles had abated, the pain in the foot had abated also. Medical Gazette, vol. v.

The spinal branches to the sympathetic are mostly from the third and fourth sacral nerves. From the back part of the plexus thus formed are given off the inferior hæmorrhoidal nerves, which join with the superior hæmorrhoidal from the inferior mesenteric artery and perforate the rectal wall.

Lymphatics.—The lymphatic vessels of the rectum are arranged like those of **the intestine** generally, **in** two **layers;** one beneath the **peritoneum**; and one between the mucous **and muscular coats.** Immediately after leaving the bowel some of the **vessels** pass **through** small adjacent glands, and all finally enter the glands in the hollow **of the** sacrum, or those higher **up** in the loin.

But just **as** there is an internal **and external** system of veins, one proper to the rectum, the other to the anus, so is there another lymphatic system which comes from the integument around the anus, and passes **to** the glands in the groin; and these two sets of vessels freely communicate with each other. A knowledge of this fact is of importance in the diagnosis of the cancer of the rectum; and the glands which are deep **in** the pelvis along the **sacrum** should always be felt for, as well as those located in the groin.

Defecation.—A **study of** the anatomy of the rectum would **not be** complete without some reference **to** its physiological functions. We shall, therefore, in this place **consider the function** of defecation, postponing the question **of** absorption until **we consider** that of rectal alimentation.

In regard **to defecation the question at once arises,** how, after destruction of the lower end of the rectum, **or paralysis of the** sphincters, there still remains **a certain amount of control over the** evacuations? Such an injury **is often only** noticeable **through a constant discharge of rectal mucus, and an** occasional involuntary **escape of fluid fæces** when the **patient** is suffering from diarrhœa. This leads naturally **to a** consideration of the third or superior sphincter muscle,[1] **whose** existence has been supposed to account for such control of the evacuations as exists in this condition.

---

[1] Gosselin: "Rétrécissements Syphilitiques du Rectum." Arch. Génl. du Méd., 1854, p. 668.

Henle: "Handb. der systemat. Anat. des Menschen," 1873, Bd. ii.

Hyrtl: "Handb. der topogr. Anat.," Wien, 1857, Bd. ii., pp. 108, 109.

Sappey: "Traité d'Anat. Descriptive," Paris, 1874, t. iv.

Chadwick: "Trans. of the Am. Gynæcol. Soc.," ii., 1877.

Pétrequin: "Traité d'Anat. Topogr. Med.-Chirurg.," etc., 2me éd., **Paris, 1857,** p. **414.**

Houston: **"Dublin Hosp. Rep.,"** v., 1830.

O'Beirne: **"New Views** of the Process of Defecation," etc., **Dublin, 1833.**

Bushe: "Treatise **on the** Malformations, Injuries, and Diseases **of the Rectum and** Anus," New York, **1837.**

Kohlrausch: "Anat. u. Physiol. der Beckenorgane," Leipzig, 1854.

Rosswinkler: "Wien. med. Woch.," 1852, p. 485.

Foster: "Text-Book of Physiology," Philadelphia, **1880, p. 387.**

It is now about half a century since Nélaton first described the third sphincter muscle, and, in spite of all that has been written concerning it since that time, it is only recently that Van Buren [1] summed up the general knowledge of anatomists and surgeons in regard to it, by characterizing it as an organ to which anatomy and physiology had been equally unsuccessful in assigning either certainty of location or certainty of function. For the original description of the muscle by Nélaton we are indebted to Velpeau, who writes that he has verified the existence of a sort of sphincter of the rectum, lately discovered by Nélaton, and goes on to say that it is a muscular ring situated about four inches above the anus, just in the place where retractions of the rectum are most often found. If, after turning the rectum so that its mucous surface is external, it is moderately distended by insufflation, the muscle will be seen to be made up of fibres collected into bundles. Its breadth is from six to seven lines in front, and about an inch behind. Its thickness, on the contrary, is much greater in front, where the fibres appear to be collected in the angle which corresponds to the union of the first and second curves of the rectum, while behind they are scattered over its convexity. After thus adopting the description of Nélaton, Velpeau [2] brings out one other anatomical point—the attachment of the muscle posteriorly to the front of the sacrum. The functions ascribed to the muscle by Nélaton were those of keeping the rectum empty until a short time before the act of defecation; separating the fæcal mass and preventing its regurgitation during defecation; and of opposing the continuous and involuntary escape of fæces after the destruction of the lower sphincters.

Hyrtl refers to this description, and himself describes the muscle as being six or seven lines in breadth anteriorly and an inch posteriorly, but does not always find it present. He also in one case demonstrated the attachment to the sacrum. Sappey admits its frequent existence, and locates it at the level of the base of the prostate, in the middle portion of the rectum, six, seven, eight, or sometimes nine centimetres from the anus. It never completely surrounds the rectum, but only one-half or two-thirds of its circumference; and it appears to him to be caused by a grouping of the circular muscular fibres, some being gathered from below upward, and others from above downward, to the same point. Its breadth is one centimetre, and its thickness two or three millimetres. Situated sometimes in front, sometimes behind, and again laterally or antero-laterally; it is constant in nothing except its direction perpendicular to the axis of the bowel. In place of one, he has sometimes found two bands at opposite points and different levels, and in one specimen which he has preserved there were three. Henle adopts Sappey's description in the main. Pétrequin found the muscle irregularly oblique,

---

[1] "On Phantom Stricture," etc., Am. Jour. of the Med. Sci., October, 1879.
[2] Velpeau: "Traité d'Anat. Chirurg.," 3me éd., 1837, introduction, p. 39.

less marked in the front wall than in the back, and consisting of a collection of weak bands of fibres. Chadwick asserts that no distinct muscle exists, but describes in place of it two agglomerations of the circular muscular fibres, one on the anterior and one on the posterior wall, corresponding to two semi-circular constrictions which may be felt by digital examination, and whose effect is to give the rectum its sigmoid curve.

The third sphincter muscle and the valves of mucous membrane in the rectum are not, as might be supposed, one and the same thing, though it is true that they have become almost hopelessly confounded in surgical and anatomical literature, and are often spoken of as identical. As far as possible, we shall try to consider them separately, without doing violence to the text of the authorities. The valves of the rectum (we use the word simply as expressing the folds of mucous membrane) were first described by Houston at about the same time that Nélaton described the superior sphincter; and it is worth remembering that the two authors were writing about two entirely different things, and two things which stood in no necessary relation to each other, so far as we may judge from their descriptions. Houston's method of preparation was by filling and distending the gut with spirit before its removal from the body, and then laying it open longitudinally. He states that the folds disappear if the bowel is first removed from its natural position and then distended, but that they may be seen in the natural condition of the parts soon after death before the tonic contraction has disappeared; and that they are then found to overlap each other so effectually as to require considerable manœuvering in order to pass a bougie or the finger along the bowel. It is also remarked that this is just the arrangement necessary to prevent the fæces from urging their way toward the anus, where their presence would excite a constant sensation demanding their discharge.

According to this first and clearest of all the descriptions—for the whole article is written with a force and clearness of style which have perhaps had an undue weight in disarming criticism as to the facts—the valves exist in all persons, but vary much in different individuals as to location and number. Three is the average number, though sometimes four, and again only two are well marked. The largest and most constant is about three inches from the anus, opposite the base of the bladder; the next most constant is at the upper end of the rectum; the third is about midway between these; and the fourth, or the one most rarely present, is attached to the side of the gut about an inch above the anus. The first one generally projects from the right wall; the one next above from the left; the uppermost from the right; and the one nearest the anus, when present, from the left and posterior wall; the arrangement being such, in spite of variations, as to form a spiral tract down the gut. The folds are described as semilunar in form, with the convex border attached to the side of the bowel, and occupying from one-third to one-half of its circumference. The surfaces are sometimes horizontal, but more often oblique, with the sharp, concave, floating margin generally directed a little upward. In breadth they vary from one-half to three-quarters of an inch or more in the distended state of the gut; and they are said to be composed of a duplicature of mucous membrane inclosing some cellular tissue and a few of the circular muscular fibres.

The palpable weak points in Houston's article were very soon pointed out by O'Beirne, in a work of marked and almost amusing originality. The views were indeed "new," but they are to-day accepted in many points by those whose judgment is worthy of the most confidence in these matters. O'Beirne seems rather to regret that he is unable to accept Houston's statements as to an anatomical condition which would account so fully and so easily for the physiological emptiness of the rectum and fulness of the sigmoid flexure on which his own views depend; but nevertheless he sets himself to the task of demolishing them with great vigor and considerable success. Although he believes the rectum to be normally empty, except just at the time of defecation, he believes that condition to depend upon the anatomical arrangement of the sigmoid flexure, joined with the narrowing of the upper end of the rectum, which is entirely independent of any folds of mucous membrane. He not only denies the existence of any such folds, but states flatly that Houston is altogether incorrect in his statement that Cloquet or any other anatomist before his time makes even the slightest allusion to them.[1] He believes the folds to have been produced by the method of making the preparations—distending and hardening all the parts with spirit before making the incision—and asserts that this method is anything but natural, and nothing more or less than an attempt to exhibit natural appearances by placing the parts in an unnatural situation—such a situation, indeed, as is not known to be necessary for the exhibition of the valvulæ conniventes or any other valves of the body. He meets the statement, that by the ordinary procedure of distending the rectum after removal from the body the valves are made to disappear, by the question, why, if such valves really exist, and if muscular fibres enter into their

---

[1] Regarding this question of fact, it may be well to quote Cloquet's description from Bushe, op. cit., p. 60: "The inner surface of the rectum is commonly smooth in its upper half, but in the lower there are observed some parallel longitudinal wrinkles, which are thicker near the anus, and are variable in length. These wrinkles, whose number varies from four to ten or twelve, and which are called the columns of the rectum, are formed by the mucous membrane and the layer of the subjacent cellular tissue. Between these columns there are almost always to be found membranous semilunar folds, more or less numerous, oblique or transverse, of which the floating edge is directed from below upward toward the cavity of the intestine. These folds form a kind of lacunæ, of which the bottom is narrow and directed downward." It seems evident that the sinuses of Morgagni are here referred to.

structure, they should not be discoverable at any time after death, or in any state of the intestine—a question very difficult of solution.

Four years later, the voice of a New York surgeon is raised against these folds, and in almost the same language as O'Beirne's, though from an entirely independent stand-point. Bushe declares that he has never, in the living body, been able to detect any valve of such firmness, and capable of exerting any such influence upon the descent of the fæces as Houston describes, though he has frequently met with accidental folds produced by the partial contraction of the bowel; and the proof that they are accidental is that, in the same subject, he has on different days found them to occupy different situations, but always they were unresisting and easily displaced by the extremity of the finger. He points out that, by the method of hardening the rectum after distending it with spirit, these accidental folds are rendered permanent by the induration resulting from the action of the alcohol; and that, by the method of inflation and drying, the projections resembling valves are produced by the angles formed by the setting of the intestine during the process of desiccation.

Kohlrausch describes and figures one important fold, the plica transversalis recti, which he locates at the same point as Houston's most constant one, projecting well into the lumen of the bowel from the right side. It forms rather more than a semicircle, and runs further on the anterior than on the posterior wall. Here also we meet the direct statement that this fold is now known as the sphincter ani tertius, though Kohlrausch does not consider such a title justified by the anatomical condition, inasmuch as the circular muscular fibres do not enter into its texture, and are not more developed here than elsewhere. For, though both these things may happen, as a rule neither is the case.

Sappey says he has found in the empty state various folds of the mucous membrane, but that these have no determinate direction, and are generally only slightly marked. Three times only, in thirty recta which he examined, has he met with anything which at all answered to Kohlrausch's plica transversalis, or to Houston's chief valve. There is nothing to prove that they persist when the rectum is full; on the contrary, it is probable that they are effaced by the simple fact of distention of the latter, at least in great part. The name of valve is not, therefore, applicable to them, and, admitting even that it might be used by one of those abuses of language so frequent in anatomy, Houston would still incur the discredit of having presented as normal a fact which is only observed very exceptionally.

Henle divides the valves into two varieties, the temporary and the permanent. Of the former, he describes several, which may be present or absent in the same individual at different times or in different states of the bowel. Of the permanent variety, there is only one—the plica tranversalis—and this one is only present in a minority of subjects.

Hyrtl describes two folds, both constant: one on the right wall lower down, and one on the opposite side. Rosswinkler also describes two folds, but locates them on opposite sides to those of Hyrtl.

There would be little profit in following these descriptions of different writers, each of them an authority on the subject treated, any further;[1] and so far as we have gone, we have carefully endeavored to avoid any violence to the meaning of the text in thus separating the thickening of the muscular fibres, which can alone constitute a sphincter, from the projections and redundancies of the mucous membrane which Houston first described under the name of valves. It will readily be seen that Van Buren was correct in speaking of the third sphincter as an organ to which anatomy and physiology had been equally unsuccessful in assigning certainty of location, for we have seen it described, on equally good authority, as both mucous membrane and muscle; as on all sides of the rectum, and at almost all distances between two and four inches from the anus; as single, double, and triple; as composed of mucous membrane and cellular tissue without muscular fibre, and of well-marked muscular bands located at the base of the mucous folds, and extending into their substance. From these very differences, perhaps, the true anatomy of the part may best be deduced. It is the old question of the gold and silver shield. There are bands of the circular muscular fibres of the rectum located at various points in its upper portion. These bands are more or less developed in different subjects, and are also found in no constant location; being sometimes lower or higher, and sometimes more marked on the anterior or again on the posterior wall. There are also found various folds and duplicatures of the mucous membrane, which stand in no constant relation to the thickened portions of the muscular fibre, and have no definite or constant situation, but may alter their shape with the varying condition of the bowel, and are found at different points in different subjects. These folds vary also in their structure in different people, being larger and firmer in some than in others, and occasionally containing a few fibres of the circular muscle of the bowel.

This is also the conclusion reached by Gosselin, who says: "I do not find the line of demarcation (between the upper and middle portions of the rectum) established by a special sphincter analogous to that which some authors have indicated by the name of sphincter superior. I am convinced, indeed, by the examination of a large number of specimens that the sphincter does not exist as an isolated muscle, and that, when we are led to admit its existence, we have to do with subjects in whom the bands of the circular layer are more developed than in others. I

---

[1] Morgagni ("De Sedibus et Causis Morborum") says he found valves in two subjects, situated about an inch above the anus, in one of a circular, in the other of a crucial form. The references of Portal ("Anat. Méd."), Glisson, and Boyer "Traité d'Anat.," Paris, 1815, t. iv., p. 377) probably all refer to the sinuses of Morgagni.

have often met this isolated development of some of the circular fibres, but it is by no means always present, and for this reason the superior sphincter has not always been found by those who have searched for it. When it exists, it is at a variable height, sometimes between the middle and upper portions, sometimes at some part of the circumference of the latter, or at its very upper portion; and I explain in this way why O'Beirne has placed his superior sphincter at the junction of the rectum with the sigmoid flexure, while Nélaton has placed his lower down, without assigning it a determinate position."

It will be remembered that Hyrtl argued backward from what he considered the physiology of the rectum to the existence of a third sphincter; and that Houston, in describing the valves of membrane, asserts that such an arrangement as he discovered was just the one which was *a posteriori* probable, and which best accounted for the accepted theories of the physiology of defecation. Nélaton, too, though he described the muscle before he gave it an action, assigns to it the same function as Houston does to his folds, and as Hyrtl believed it must of necessity possess. It is plain that each was led by a certain chain of reasoning to believe in the existence of an obstruction to the passage of fæces from the sigmoid flexure above to the rectum below; and that two of them found it in the muscular structure, and the third in the mucous membrane of the bowel. The facts upon which the necessity for a superior sphincter are supposed to rest are briefly these: the normally empty state of the rectum, and the ability to retain both wind and motion after destruction of the anus and its muscles. The force of this line of argument cannot be disputed, but were some other reasonable explanation found for these two facts than the existence of a third muscle, that muscle would soon be dropped from the descriptions of the anatomy of this part. The whole tendency of the physiology of the day is to furnish such an explanation.

The "new views" of O'Beirne with regard to the process of defecation were simply as follows: The repeated descent of fæcal masses causes the sigmoid flexure to become distended, and to ascend from its position in the cavity of the true pelvis into the left iliac fossa. When this occurs, the flexure, in proportion to the rapidity and degree of its distention, begins to turn upon the contracted rectum as upon a fixed point, until at length, like the stomach, it directs its greater arch forward and upward, and its lesser backward and downward. By this movement, the contents are brought somewhat perpendicular to, and so as to press directly upon the upper extremity of the contracted rectum. But as the mere weight is insufficient to force a passage downward, and as this end cannot be accomplished either by such gentle pressure as that exerted by the alternate contraction of the diaphragm and the abdominal muscles in ordinary respiration, or by the efforts of the flexure itself, in consequence of its muscular power being so inferior to that of the rectum; the fæces are compelled to remain stationary until such time as the increased accumulation and distention produce a sense of uneasiness sufficient to call into action those great expulsive agents, the diaphragm and abdominal muscles. These muscles, instead of acting alternately, now act simultaneously, compress the abdomen and its contents on all sides, urge the free and floating mass of small intestine downward and even into the cavity of the pelvis, so as to press forcibly not only upon the sigmoid flexure, but also upon the cæcum and urinary bladder. By these means, the contents of the distended flexure are acted upon in every direction, and so as to be impelled against the upper annulus of the contracted rectum, with a force sufficient to compel its parietes to separate and afford a passage. The nisus now ceases, but as soon as the rectum becomes filled, it is aroused to make an expulsive effort by which its contents are driven or impacted into its pouch. Here they produce a great sense of weight and uneasiness in the perinæum, an urgent desire to go to stool, and a still stronger nisus, by which the sphincters are forced open and dilated, and the final expulsion of the fæces is effected. This reasoning, it will be seen, is entirely based upon the normal empty and contracted state of the rectum, which O'Beirne not only states to be a clinical fact capable of easy demonstration, but gives many reasons for, the chief being the great relative thickness of its muscular wall. He clearly pointed out also (what has been frequently verified since, and especially by those who have passed the hand into the sigmoid flexure of the living subject) that the upper extremity of the rectum was absolutely the smallest part of this portion of the bowel; but that nothing of the nature of a sphincter muscle, located at this point or near it, entered into his calculation any more than did the folds of mucous membrane.

Compare, now, these teachings of O'Beirne's, in 1833, which we have already said are to-day accepted by those who have the best right to judge of these matters, with those of Foster, in 1880. He says the fæces, in their passage through the colon, are lodged in the sacculi during the pauses between the peristaltic waves. Arrived at the sigmoid flexure, they are supported by the bladder and the sacrum, so that they do not press on the sphincter ani. Defecation is a composite act, being superficially the result of an effort of the will, and yet carried out by means of an involuntary mechanism. The voluntary effort is composed of two factors—a pressure effect produced by the contraction of the abdominal muscles, and a relaxation of the sphincter ani muscle. By the pressure of the abdominal muscles the contents of the descending colon are driven onward into the rectum, but the sigmoid flexure itself is shielded by its situation from the direct force of this pressure, and a body introduced *per anum* into the empty rectum is not affected by even forcible contraction of the abdominal muscles. The sphincter muscle guarding the anus is habitually in a state of tonic contraction, capable of being increased or diminished by a stimulus applied either internally or externally to the anus. This tonic contraction is due, in part at least, to the action of a nervous centre situated in the lumbar portion of the spinal cord. By the action of the will, by emotions, or by other nervous events, the lumbar sphincter centre may be inhibited, and thus the sphincter itself relaxed; or stimulated, and thus the sphincter tightened. This relaxation **is the** second of the voluntary elements in the act of defecation. By these **two alone the** contents of the descending colon might be pressed onwards into the rectum **and** out at the anus; **but, since** the sigmoid flexure itself is subject to neither of these influences, **such a** mode of defecation would always end in leaving it full; **and therefore** there is superadded to these two voluntary elements an **entirely** involuntary increase in the peristaltic action of the **sigmoid flexure itself**. The order of events is the reverse of what we have **stated**. The sigmoid flexure and large intestine become more and more full, while stronger and stronger peristalsis is excited in their walls. By this means the fæces are **driven** against the sphincter. Through a voluntary act, or sometimes at least by a simple reflex action, the lumbar centre is inhibited and the sphincter relaxed. At the same moment the contraction of the abdominal **muscles** causes firm pressure **on the** descending colon, and the contents of the rectum are ejected.

It should be **mentioned that the one** fact on which these physiological views rest, viz., **the normal empty state of** the rectum, is not universally admitted. Indeed, **as Hyrtl says, the rectum** will be found by any one who practises **frequent digital examination, in very** different states in this regard at different times in the **same individual.** This may or may not be entirely due to changes produced **by** constipation in those examined; but even he admits that it is **more often** found **empty than any** other part of the canal; and the difficulty which an opposite **view leads to** will be seen at once by the attempt of Bushe to explain the **act of defecation**, starting from the point that the fæces accumulate slowly in the rectum, and gradually lose their thinner parts by absorption while there. He goes on to say that they give rise to no uneasiness until a considerable quantity is amassed, when a sensation is created which demands their expulsion. This sensation is, he believes, not due to the mere contact of fæcal matter, for the latter generally accumulates in large quantities before the sensation is felt. Nor is it due to any peculiar acrimony which they obtain by their stay in the rectum, **for** when the fæces are fluid, this sensation is produced as **soon** as they reach the rectum. Again, when **once** the sensation is felt and not attended to, **it** passes away, and does **not return** till the **next** accustomed period; **and the** longer it is unattended to, the less likely is it to return at all. In truth, he says, we are ignorant **of the** cause of this feeling, **and must,** in the present state **of** our knowledge, admit that it is organic, and consequently dependent upon some spontaneous change in the intestine, **of** which we know nothing. Rather **a** lame conclusion! **Nor is** the cause of this periodically recurring desire to evacuate the bowel touched upon in the exposition given by O'Beirne; and this is the weak point in his argument, and the one which renders Foster's explanation complete.

We need cite authorities no further to show that physiology no longer teaches the existence of an ever-present mass of fæces in the lower bowel, ready to escape at any moment when the active watchfulness of the sphincter muscle is relaxed, or to prove that into our present understanding of the cause of the emptiness of the rectum a third sphincter muscle does not enter as a necessary element, but that the true explanation of the condition lies in the anatomy of the sigmoid flexure, which, by its large size, great capability of expansion, loose mesenteric attachment, and position, is peculiarly fitted to act the part of a reservoir.

Nor does the phenomenon of retention of fæces after the destruction of the anus and its muscles necessitate the belief in a superior sphincter. So far as our reading goes, no one has as yet attempted to prove the existence of a fourth sphincter in the ascending colon; and yet the same control over the passages which has been noticed after extirpation of the anus, and has been supposed to indicate a third sphincter, has been observed to follow an artificial anus in the transverse colon.[1]

There are several ways of accounting for the slight control over the evacuations which many patients are found to have after extirpation of the anus, apart from the existence of a third sphincter or of the valves of the rectum. Indeed, the physiology of the act of defecation itself, which we have just described, goes far to explain why there should be a certain warning of an approaching evacuation, and this is what is generally meant when the patients are reported to have a certain amount of control over the movements. The control will be found in most cases to mean rather a consciousness of an approaching movement, a warning given in sufficient time to allow the patient to make necessary arrangements, than an ability to absolutely prevent the evacuation which is about to take place. Of actual control there is little, because the sphincter muscle, whose duty it is, under the power of the will, to prevent an evacuation, is absent. To the performance of this duty a healthy sphincter is abundantly equal, as every one has the chance to prove on his own person; and it is this ability to delay and postpone an evacuation of the bowels, rather than a constant action in preventing the escape of fæces which are ever ready to escape, which best expresses the true function of the muscle. After extirpation of the anus, this one element of natural defecation is destroyed, but several others are left. The fæces tend to remain by their own consistence unless actively urged forward by the

---

[1] The case was that of Fine, of Geneva, in 1797. "He formed an artificial anus, by which the fæcal matters escaped not continually, but once or twice a day only, and with a sensation of impending necessity which gave the patient time to make the slight preparations necessary to avoid soiling herself."—Manuel de Méd. Pratique" de Le Louis Odier, de Genève. 2me éd., 1811.

peristalsis of the bowel; and this peristalsis is not constant, but recurs periodically. The relative increase in the muscular elements in the rectum tends to keep it closed and empty until fæces are forced into it from above. Again, the pressure of the fæces, owing to the S-shaped form of the rectum, is not in the direction of the axis of the tube, but constantly against the wall, and at the points of greatest curvature the resistance is greatly increased. To these let us add the contraction of the cicatrix after extirpation, and the natural redundancy of the mucous membrane which may block up the new anus by an actual prolapse, and we have the factors which account for the clinical fact so often seen. On the other hand, the constant escape of fæces, which *at first* almost always follows these severe surgical operations upon the rectum, is best explained by the irritation of the wound and the constant reflex action which it excites.

That the folds of mucous membrane, such as have been described, are of the nature to form an obstruction to the passage of the fæces, would seem to admit of no reasonable doubt. But this obstruction is passive, and not active, and is by no means sphincteric in character. When it is sufficiently great to form a real obstruction to the descent of fæces, the condition is an abnormal one, but such a condition is sometimes seen, and is one which is not to be disregarded in the pathology of stricture of the rectum.

From a study of the literature of this question, and from the results of dissections and experiments which we have personally been able to make, we are led to the following conclusions:

1. What has been so often and so differently described as a third or superior sphincter ani muscle is in reality nothing more than a band of the circular muscular fibres of the rectum.

2. This band is not constant in its situation or size, and may be found anywhere over an area of three inches in the upper part of the rectum.

3. The folds of mucous membrane (Houston's valves) which have been associated with these bands of muscular tissue, stand in no necessary relation with them, being also inconstant, and varying much in size and position in different persons.

4. There is nothing in the physiology of the act of defecation, as at present understood, or in the fact of a certain amount of continence of fæces after extirpation of the anus, which necessitates the idea of the existence of a superior sphincter.

5. When a fold of mucous membrane is found which contains muscular tissue, and is firm enough to act as a barrier to the descent of the fæces, the arrangement may fairly be considered an abnormality, and is very apt to produce the usual signs of stricture.

CHAPTER II.

CONGENITAL MALFORMATIONS OF THE RECTUM AND ANUS.

Separate Development of Rectum and Anus.—**Narrowing** of the Anus or Rectum without Complete Occlusion.—Congenital Stricture.—Closure of the Anus by a **Membranous** Diaphragm.—Entire Absence of **the** Anus, the Rectum Ending in a Blind Pouch at a Point more or Less Distant from the Perineum.—Rectum **Same as in Last Variety** and the Anus Normal.—Anus Absent and Rectum **Opening** by an Abnormal Anus at Some Point in the Perineal or Sacral Regions.—Cases.—Anus Absent and Rectum Ending in the Bladder, Urethra, or Vagina.—Cases.—Rectum and Anus Normal, but Ureters, Uterus, or Vagina Empty into Rectum.—Total Absence of Rectum.—Absence of Large Intestine.—Obliteration from Intra-Uterine Disease.—Treatment.—Operation Should Always be Performed and Without Delay.—Attempt Should First be Made to Establish an Anus in the Anal Region.—Measurements of Pelvis at Birth.—Use **of Trocar not** Justifiable.—Useful Anus Seldom Obtained by **Means of** Incision Alone.—Objections to Cutting Operation Without Plastic **Operation.**—Proctoplasty.—If Attempt to Establish New Anus in Anal Region Fail, Colotomy at Once to be Performed.—Inguinal Preferable to Lumbar Colotomy.—History of Colotomy.—Callisen.—Amussat.—Description of Operation of Colotomy.—Dangers of Operation.—The Inguinal Operation.—Description.—Attempts at Establishing **Anus in** Anal Region after **Colotomy** Generally Unsuccessful.—Cases.—Closure of Artificial **Anus.**—**Operation** of Dupuytren.—Modifications of **Dupuytren's Operation.**

THE study of embryology has revealed the fact that the anus and the rectum are developed separately. The anus is at first represented by a simple depression in the skin of the perineum which gradually extends in depth and advances to join the rectum. The rectum is developed in connection with the other abdominal viscera, gradually separates itself from them, and ending in a blind pouch, advances to meet the anal depression. At the proper time the two coalesce and the intestinal canal is complete. This process of development of either the rectum or anus may be arrested at almost any stage and the result will be one of the various malformations which are now to be described.

These congenital malformations have been classified by different writers into various groups. We shall adopt in the following pages that of Papendorf[1] which is the one followed by Bodenhamer,[2] Mollière,[3] and Esmarch.[4]

1. *Narrowing of the Anus or Rectum without Complete Occlusion.*— A congenital stricture of the anus, or of the rectum at a point more or less removed from the anus, has been occasionally reported. Serremone[5] particularly insists upon congenital narrowness of the anus as a cause of fissure, and has himself observed such cases; and the same condition in the rectum is generally included among the causes of benign stricture.

The narrowing in these cases may be very slight, or may reach such a degree as hardly to admit of the passage of meconium. It is generally annular in form, resembling the contraction which would be caused by tying a tape tightly around the tube. There may be no symptoms caused by such a contraction, and the child may grow to adult life suffering only from obstinate constipation; nor do such contractions lead to the ordinary changes in the mucous membrane above and below the spot which are usually seen in cases of stricture of the rectum. On the other hand, when the stricture is tight it will give rise to all the usual signs of such a condition in the child—absence of free passage of meconium, distention of the abdomen, and vomiting. The diagnosis is easily made by a digital examination should the symptoms be sufficiently marked to lead the attention of the surgeon to the rectum; for the stricture is generally near the anus and may be felt as a ring with sharp edges. The treatment consists either in dilatation or in nicking.[6]

2. *Closure of the Anus by a Membranous Diaphragm.*—The membrane in these cases may be of greater or less firmness and thickness, and may be composed of skin or of mucous membrane. It is sometimes so thin as to bulge out with meconium when the child strains or coughs, and has been known to rupture spontaneously.

This is the simplest of all the forms of congenital malformation of the anus, and, unfortunately, one of the rarest. It is easily diagnosticated by simple inspection of the parts; and the treatment consists in making a crucial incision through the membrane. The remains of the

---

[1] "Dissertatio sistens observationes de ano infantum imperforato." Lugd. Batav., 1781, 4to. (Bodenhamer).

[2] "A Practical Treatise on the Etiology, Pathology, and Treatment of the Congenital Malformations of the Rectum and Anus," by Wm. Bodenhamer, New York. Wm. Wood & Co., 1860.

[3] "Traité des Maladies du rectum et de l'anus," par Daniel Mollière. Paris, 1877.

[4] Op. cit.

[5] Inaugural Thesis. Strasbourg, 1861, No. 555.

[6] See also Gosselin, "Clinique Chirurg.," 3d ed. Paris, 1870, T. iii., p. 706. Bérard et Maslieurat-Lagemar, Gaz. Méd. de Paris, 1839, p. 146. Demarquay, Journal de l'expérience, t. ix., 1842, p. 273. Ashton, "Diseases of the Rectum," London, 1854, p. 27. Devilliers, Rev. Méd. de Paris, 1835.

membrane, like those of the hymen which it strongly resembles, will shrink up so as not to cause trouble or deformity.

3. *Entire Absence of the Anus, the Rectum ending in a Blind Pouch at a Point more or less Distant from the Perineum.*

In these cases there may be a slight depression at the point where the anus should be found; or there may be no trace of the anal orifice; the *raphé* of the perineum extending over the spot and back to the coccyx. The external sphincter muscle is also sometimes present and at others entirely wanting. The pouch of the rectum in these cases may hang loose in the pelvis or abdominal cavity, or be attached to some adjacent part; and the space between it and the perineum may be filled up with

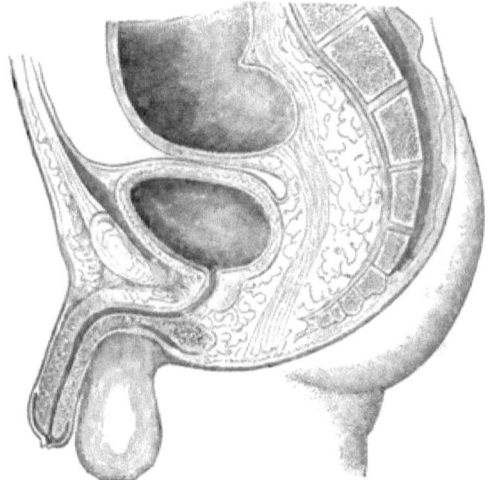

FIG. 7.—(Mollière).

cellular tissue, or in other cases a distinct fibrous cord may be traced from the rectal pouch to the skin, as is shown in the plate.

If the pouch of the rectum be not at too great a distance from the skin, a sense of fluctuation may be felt by firm pressure with one finger over the anus and the other hand on the abdomen. In females, valuable aid in diagnosis may be obtained by the introduction of a finger into the vagina. The use of a stethoscope over the anus, and of percussion on the abdomen, has been recommended to detect the rectal pouch filled with gas (Bodenhamer, Mollière); and also the irritation of the skin over the anus to provoke efforts at defecation.[1] An effort should always be made, where there is complete absence of the anus, to discover whether the rectum

---

[1] A. Copeland Hutchinson: "Practical Observations in Surgery," London, 1826.

may not have some outlet through the bladder or vagina, which shall place the case in one of the classes soon to be described.

4. *The rectum may be the same as in the last variety, and the anus be normal.*

The septum which separates the rectal and anal pouches in this case is generally within easy reach of the anus, and may be so thin as to permit a sense of fluctuation. In most cases, however, the septum is thicker, and is composed of cellular or fibrous tissue, lined both above and below by mucous membrane. It may be perforated, like the hymen,

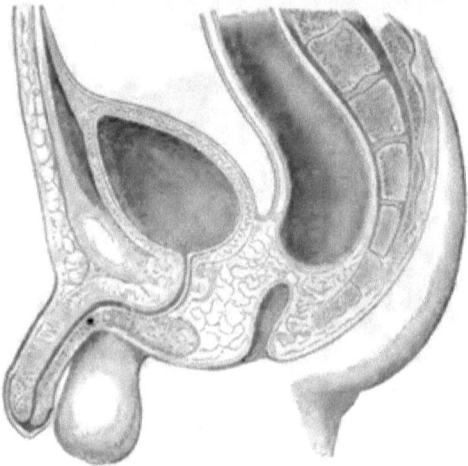

FIG. b.—(Mollière).

at some point, and allow of the slow dribbling of meconium. There may also be more than one septum. Voillemier[1] reports one case in which the rectum was divided in this way into four distinct compartments, the upper one containing meconium, and the others mucus. There is generally little difficulty in the diagnosis of these cases, provided only a digital examination be made when the infant begins to show the effects of the obstruction; but the danger lies in the fact of the normal anus, which is apt to allay suspicion as to the true nature of the difficulty.[2]

---

[1] Gaz. des Hôp., 1846.

[2] "Dr. H. G. Jameson, of Baltimore (Medical Recorder, vol. v., 1822, p. 290), divided two membranous septa, one above the other, with a button-headed bistoury, which he passed 'into the opening or ring of the septum,' and cut freely down toward the sacrum. This was done in September, 1821. The patient got well. Roser (Arch. für Physiol. Heilkunde, 1859, p. 125) mentions a circular valvular stricture an inch from the anus in a little girl of four, which he treated by division." Van Buren, "Lectures upon Diseases of the Rectum and the Surgery of the Lower Bowel." New York: D. Appleton & Co., 1881, p. 263, note.

5. *The anus may be absent, and the rectum may open by an abnormal anus at any point in the perineal or sacral regions.*

When the rectum terminates in the glans penis, the labia, or at some abnormal point in the perineum, the lower portion of it is usually of a fistulous character, as shown in the plate, but lined by true mucous membrane; and the anus, whether in the perineum or at the base of the sacrum, or tip of the coccyx, is always narrow and insufficient for its purpose. A modification of this class of abnormalities is found in those cases where the rectum terminates in two openings at a greater or less distance from each other.

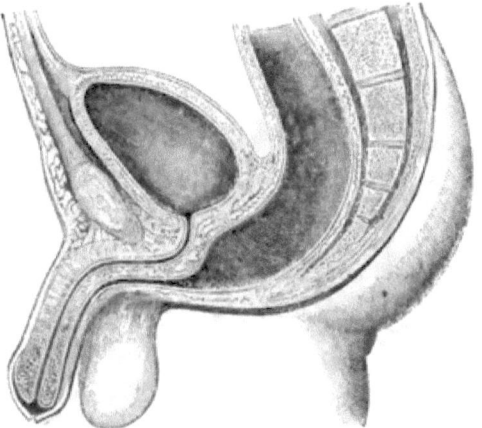

Fig. 9.—(Mollière).

Cruveilhier[1] reports a case of this nature, in which the fistulous prolongation of the rectum ran subcutaneously in the scrotal *raphé*, and terminated at the glans penis.

Mr. Morgan[2] has recently reported two modifiations of this species of deformity which are rarely met with, and are easily relieved. In the first, the anus was of the usual size and in the proper location; but there was found to be a band of tissue passing from a point corresponding to the apex of the coccyx to the median *raphé* of the scrotum, with the posterior extremity of which it was continuous. The band was about three-quarters of an inch long, and was attached at both ends, the remainder forming a thick, free cord, which lay below the aperture of the anus, while from the centre of this band there ran a small branch of similar tissue, which was attached to the skin of the left buttock, and was about half an inch in length. The skin covering the central band

---

[1] Anat. Pathologique du Corps Humain, t. i., Liv. i., Planche vi.
[2] Three Cases of Unusual Deformity of the Anus. Lancet, October 22d, 1881.

exactly resembled that of the scrotum, shrinking and contracting upon stimulation, and it was so placed that any passage of fæces must cause it to be stretched, thus accounting for the pain attending each motion of the bowels.

The second case was similar. The child was born with an imperforate anus, but the membranous septum gave way spontaneously. The child, however, continued to suffer pain on defecation, and on examination, there was seen a small, thick band passing from the median *raphe* of the perineum in front to the depression between the buttocks posteriorly, and broadest behind. At a spot corresponding to the anus, on either side of the band, was a depression; that on the right was patent, and allowed a probe to pass into the anus; that on the left, though similar in appearance, proved to be only a *cul-de-sac*.

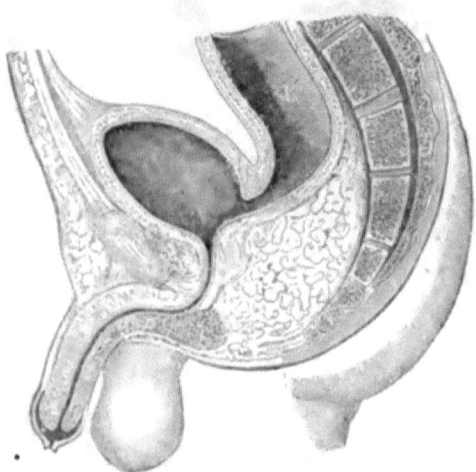

Fig. 10.—(Mollière).

In a third case, there was a depression at the usual site of the anus, and the parts around were so far natural that the skin was pigmented and puckered, but there was no communication with the rectum. The spot at which the fæces passed was in the median line half-way between this depression and the posterior commissure, but nearer the latter than the former. The opening was very small, and a probe passed up into it, showed an abundance of tissue between the passage and the vagina. The cure consisted in enlarging this abnormal opening posteriorly into the depression representing the natural one. Delans' reports an analogous case in a well-nourished child aged four and a half years. There were two openings, one on each side of a median bridle, which was con-

---

[1] Soc. de Chirurgie, March 24th, 1875.

tinuous with the *raphé* in front and behind, **and was composed only of skin and mucous membrane.** Each opening seemed to be the natural one, but the one on the left was a *cul-de-sac* fifteen millimetres deep. The septum was excised, with what result is not stated.

6. *The anus may be absent and the rectum may end in the bladder, urethra, or vagina.*

Of these varieties that in which the rectum opens into the vagina is **the most common.** In females the opening is seldom, if ever, into the bladder, but sometimes it is into the urethra. In males it is more often into the bladder than into the urethra, and in such cases the rectum may terminate either by a narrow duct running obliquely through the bladder and opening in the *bas-fond* between the orifices of the ureters, or by a free opening. The symptoms of this condition will of course vary greatly according to the location of the abnormal opening. When the communication is between the rectum and bladder the fact will be shown by the mixture of the meconium with the urine, rendering the latter thick and greenish in color. The amount of meconium present will also indicate whether the opening is large or small. This condition is generally fatal from the development of cystitis, and from intestinal obstruction unless the condition be relieved by the appropriate surgical interference.[1]

When the communication is urethral in the male, the meconium will often escape independently of the act of urination. The prognosis is not as bad in these cases as in the vesical variety; several being recorded in which life has been preserved for a number of years. Gross[2] relates one case in a man aged thirty; and Bodenhamer cites several others in which children have lived three or four years.

In the female the prognosis is more favorable than in the male, from the greater facility with which the meconium escapes.

Where the abnormal opening is between the vagina and rectum, and is of considerable size, as it generally is, the prognosis is not necessarily grave. Women have been known to live to a good old age, even to reach one hundred years in the case of Morgagni, with this malformation, and to perform all the duties of wives and mothers without even being conscious of anything abnormal (Fournier,[3] Ricord).

7. *The rectum and anus are normal, but the ureters, uterus, or vagina empty into the rectal cavity and discharge their contents through it.* This species of malformation is rare and is usually attended by other signs

---

[1] As showing what the bladder and urethra may bear, however, Rowan's case is of great interest. In it defecation took place through the penis for two months without causing any signs of irritation, though the child was several months old, and the rectum was filled with well-formed hard fæces. Australian Med. Journal, Mar., 1877.
[2] A System of Surgery. Phila., H. C. Lea, 1872, vol. ii., p. 657.
[3] Dict. des Sci. Méd., t. iv., p. 155.

of imperfect development. **It is not** incompatible with life or with conception.

8. *Total absence of the rectum.* This variety differs only from the third in the amount of the rectum which may be absent. It may or may not be attended by **an** absence of the anus, but is usually only one of the signs of arrested development. **The** blind pouch of the rectum may hang loose in the abdomen or pelvis; may be attached in the base **of the** sacrum, or to some of the adjacent **parts; or** may be continued down as a fibrous cord to the site of the anus.

9. *Absence of the large intestine.* This is **also attended** by an absence of the normal anus, the place of **which is** supplied by **an** abnormal opening in the umbilicus, or at some remote part of the body, as, for example, the side of the chest, or the face. With **this** abnormal opening the small intestine or what remains of the **colon** communicates.

Thus far only arrests or excesses of development have been mentioned. **The** rectum and anus are, however, liable to certain diseases during fœtal life which may result in narrowing or completely obliterating their calibre. Among these are enteritis and proctitis.

Treatment.—The treatment of the class of congenital contractions of the anus **and rectum, and of the class** of membranous septa, has already been referred to, **and is exceedingly** simple and generally attended by good results. The **treatment of the** remaining varieties, except the eighth and ninth which **do not admit of** surgical interference, may be guided by the following **general propositions.**

1. *An operation should always be performed and performed without delay.* There is little to be gained **even by waiting for the** rectal pouch to become distended with meconium, and there **is much to be** lost. If the obstruction **be** complete, death is a necessary result; being produced by peritonitis, by rupture of the over-distended bowel, **or** by a gradual wasting without acute symptoms. Even in cases where **a** certain amount of meconium makes its escape by a narrow orifice, and delay is not, therefore, as necessarily dangerous as in cases of complete obstruction, nothing is to be gained by delay, and an immediate operation may avoid a paralysis of the bowel from over-distention.[1]

---

[1] Cripps (Lancet, May 15th, 1880) has reported a most remarkable case bearing upon this point. The condition of imperforate rectum was diagnosticated on the third day, but operation was refused and the child taken from the hospital. *Thirty days later* she was brought back again apparently quite well; the abdomen was distended; food was taken well; but three or four times every day she vomited fæcal matter. In this case, the anus terminated in a blind pouch and a trocar was plunged upwards through it. Only a little serous fluid escaped from the peritoneal cavity, and the child died of peritonitis. At the autopsy, the rectal *cul-de-sac* was found just above the anal pouch, but the trocar had penetrated the peritoneal pouch between the two. There are two noteworthy points in the case. The first is the remarkable manner in which nature accommodated itself to the deformity; and the second is the ease with which the rectal pouch may be missed with a trocar.

2. *If there be any chance of establishing an opening at the normal site of the anus, the surgeon should* **at first direct his attention** *to this procedure.* And, since in most cases it is impossible to tell that the rectal pouch may not be within easy reach from the perineum, it is generally good surgery to make a tentative incision at this point.

Before attempting any operation on a child's pelvis, the surgeon should **remember** the exceeding smallness of **the** space in which he is obliged to **work, even in its** natural state; and **also** that the normal measurements may be decreased in any case of congenital malformation. These normal measurements, according to Bodenhamer who made them on two **newborn, well-developed, male infants, at full** term are as follows:

1. **From one tuberosity** of the ischium to the other, one inch and one line. **From the os coccygis** to the symphysis pubis, one inch and three lines. **From the os coccygis to the promontory of** the sacrum, one inch and two lines.

2. **From** one tuberosity of the ischium to the other, one inch. From the os coccygis to the symphysis pubis, one inch and **one** and a half lines. **From the os coccygis to** the promontory of the sacrum, one inch and one line.

The means **at the disposal** of the operator for reaching the rectal pouch through **the perineum and** establishing a new **outlet, consist in** puncture, incision (proctotomy), and in the formation **of a new anus** by a plastic operation (proctoplasty). **The** operation by puncture consists in plunging a trocar through the perineum in the supposed direction of the rectum, **for the purpose of establishing an** outlet. It may be **done** without a preliminary incision, or **after** a careful dissection which **has failed to reach** the desired point.

3. *The use of a trocar as an aid* **in** *finding the rectal pouch before or after incisions through the perineum, is not sanctioned by modern surgical authority.* **It is a** procedure attended with the greatest danger **to the life of the patient, and when** the rectal pouch is successfully reached, **which is rare,** the outlet thus made is of little use. The peritoneum, bladder, or uterus may each be wounded by the instrument with a fatal **result;** the opening made is not free enough to allow of easy escape of meconium; nor can such an opening be made to serve the purpose of rectum and anus by any subsequent dilatation.

4. *The results of attempts to establish* **an outlet** *for an imperfect rectum by means of incisions alone* **through** *the perineum are not favorable as regards the production of a useful anus.*

The operation consists in cutting through the perineal tissues, stroke by stroke, until the rectal pouch **is** reached and opened. The incision should **be** longitudinal, and should reach from the scrotum to the tip **of** the coccyx. Should the fibres of the external sphincter be encountered beneath the skin, they may be carefully separated as near the median line **as possible and** drawn to each side. The direction of **the** dissection, which it is needless to say should be made with the utmost care, should be backwards towards the concavity of the sacrum in the line which the rectum normally follows. Additional safety may be secured by the introduction of a sound into the male bladder or the female vagina. The finger is to be frequently used as a director in exploring for the rectal pouch, while the hand of an assistant makes pressure on the abdomen. In this way the dissection may be carried to the depth of an inch or possibly an inch and a half, but at this point, if unsuccessful, it should be abandoned for fear of wounding the peritoneum.

This operation, though it may be successful in allowing the escape of meconium, and in prolonging life, does not, in most cases, result in a useful anus for any great number of years. This is the experience of the greater number of writers upon this subject. Van Buren[1] says: "I have, in several instances, succeeded, by careful dissection, in reaching a fluctuating point of a blind rectal pouch, and in establishing a free outlet for the meconium, but in no case has it proved permanently useful. It has always been necessary to employ bougies or tents more or less constantly to keep the new canal from contracting, and the care, and pain, and trouble of fighting against the closing stricture, and the persistent tendency to obstruction and fæcal accumulation, have invariably led to early death. At present, I know of no such case treated in this way, in which a permanently satisfactory result has been attained." Amussat,[2] Sir Benjamin Brodie, Velpeau,[3] Benjamin Bell,[4] and many others, have borne testimony to the same effect. On the other hand, cases are occasionally seen where the result is more favorable, but they constitute a small minority of the whole. What the operation really accomplishes is the formation of a fæcal fistula, with all the discomforts attendant upon such a condition.

It was this difficulty, combined with the loss of two cases in which the operation had been performed from blood poisoning with jaundice, which Amussat considered to be due to the absorption of meconium and fæcal matter by the freshly-cut surface, which led him to abandon this operation, and to substitute in its place the one now to be described.

Operation of Amussat. Proctoplasty.—This operation is the same as the last, with the addition of two important features. In the first place, the rectum is drawn down and stitched to the skin; and, second, to facilitate this, when necessary, the new anus is made either just at the tip of the coccyx, or that bone is exsected, and the anus made in the place it occupied. Where much of the lower end of the rectum is deficient, it may not be possible to draw the *cul-de-sac* down to the skin

---

[1] Op. cit., p. 371.

[2] "Observation sur une Operation d'Anus artificiel," etc. Gaz. Méd. de Paris, Nov. 28th, 1835, p. 753.

[3] "Nouveau Elements de Méd. Operatoire," Paris, 1832.

[4] "A System of Surgery." Vol. ii., chapt. xix. Edinburgh, 1778.

without more traction and dissection than it is safe to employ. In such cases, the excision of the coccyx, as originally recomended and practised by Amussat,[1] and more recently by Verneuil,[2] besides adding to the chances of finding the rectal pouch, diminishes the distance over which the rectum must be stretched. Unfortunately, in the cases where the operation is most needed—those in which the rectal pouch is furthest from the skin—the operation is not always practicable; and in other cases, the adhesions of the rectum to the bladder or vagina may be an insuperable obstacle. In the latter class of cases, however, a new anus may be formed, and, if successful, the recto-vaginal fistula may be closed by subsequent operations.

5. *In case of failure to establish a new anus in the anal region, colotomy should at once be performed.*

The teachings of different authorities will vary as to the propriety of first performing the perineal operation before resorting to colotomy, according to the views of each one upon the question of the desirability of colotomy. Some follow the rule I have laid down, that it is always better to attempt the perineal operation where there is a chance of its succeeding; others limit the latter operation to cases where the rectal pouch is known to be near the skin, and in all others turn their efforts at once toward the colon. The abdominal operation is obviously the only one where the rectum ends high up in the pelvis, and it is generally to be preferred in that class of cases where it opens into the bladder or urethra.

6. *In the formation of an artificial anus, the left groin is the best site for the operation.*

The colon may be opened either in the loin or groin, and on either the left or right side. There is some uncertainty in the early history of colotomy and some ambiguity of terms, which is apt to mislead. The idea of an artificial anus was first proposed by Littre,[3] in 1710, and the incision he recommended was simply "*au ventre*" (in the abdomen); the design being to reach the sigmoid flexure. He never practised the operation which at present passes under his name—that of opening the bowel in the groin, nor did the operation he proposed involve the idea of preserving the peritoneum intact.

About the year 1770, Pillore, of Rouen, actually performed the first operation of this nature, by making an opening into the cæcum, in a case of cancer of the rectum which caused complete obstruction. The patient survived twenty-eight days, and death was not due to the operation. In 1783, Dubois operated in the same way for imperforate anus, but the operation was unsuccessful, and the child died on the tenth day.

---

[1] Troisième Mémoire sur la Possibilité d'établir une ouverture artificielle sur la colon lombaire gauche sans ouvrir la Péritoine, chez les enfans imperforés. Paris, 1842.

[2] Gaz. des Hôp. de Paris, July 29th, Aug. 5th, 1873, pp. 604, 715.

[3] Histoire de L'Acad. Roy. des Sci. de Paris, 1710, p. 36.

In 1793, Duret, of Brest, opened the sigmoid flexure of a child two days old, and this child lived to adult age. In 1794, Desault practised the same operation without success, and in 1797, Fine, of Geneva, made an artificial anus in the arch of the colon for cancer of the upper part of the rectum, which was also successful, the woman living three months and a half.[1] In 1814, the operation was successfully performed for cancer of the rectum by Martland;[2] in 1817, by Freer, of Birmingham;[3] and in 1820, by Pring.[4] In many of these cases the original operation of Littre was modified to suit the operator; but in none of them was any attention paid to wounding the peritoneum.

An undue prominence seems to attach to the name of Callisen in connection with the operation in the left loin. There was nothing original in his choice of location, nor did he bring out the idea of operating without wounding the peritoneum. He believed that the intestine could be more easily reached from this point than any other, in which he certainly was in error; and on the whole he condemned the operation in the following words:[5] "The incision of the cæcum and descending colon, *which has been proposed*, in this state of things (imperforate rectum) by means of an incision in the left lumbar region at the border of the quadratus lumborum, to establish an artificial anus, presents a very uncertain chance, and the life of the little patient can scarcely be saved; nevertheless, the intestine may be reached more easily in this place than above in the iliac region."

It is in reality to Amussat that the extra-peritoneal operation in the loin is due, and the operation which he described[6] is the one now in favor and the one usually spoken of as that of Callisen.

The guide to the descending colon is the outer border of the quadratus lumborum muscle; and the guide to the outer border of the muscle is a perpendicular from a point one-half inch posterior to the middle of the crest of the ilium; or to a point half an inch posterior to the middle of a line drawn from the anterior superior to the posterior superior spinous process. This point should first of all be accurately determined and marked with ink or iodine, for the edge of the muscle cannot easily be felt in many subjects. The descending colon is here in great part uncovered by peritoneum, being behind that membrane and in immediate contact with the transversalis fascia. The patient should be placed

---

[1] "Manuel de méd. prat. de Louis Adler de Genève." 2d Edit., 1811. Quoted by Carcopino, Thèse, No. 197, 1879. Parallel entre l'extirpation du rectum et l'établissement de l'anus artificiel.

[2] Edinburgh Med. and Surg. Jour., Oct. 1825, p. 271.

[3] Carcopino. Thèse.

[4] London Med. and Physical Journal, 1821.

[5] "Systema Chirurgiæ hodiernæ." t. i., Haffiniæ, 1813.

[6] "Quelques réflexions pratiques sur les rétrécissements du rectum." Gaz. Méd. de Paris, 1839, No. 1.

upon a hard pillow so that the loin may be brought into prominence, and the operator should stand at the back of the patient.

The incision should cross the edge of the quadratus obliquely from above downwards and from behind forwards, beginning at the left of the spine below the last rib, and extending four or five inches. In this way the middle of the outer border of the muscle will correspond to the middle of the incision, and the large branches of the spinal nerves will not be severed. The incision is then carried carefully down, layer by layer, through the latissimus dorsi, external and internal oblique, and transversalis muscles, till the outer border of the quadratus is recognized; care being taken that as the incision grows deeper it does not also grow shorter, till when the bowel is reached the operator finds himself working in the small end of the funnel. If possible the outer border of the quadratus should be distinctly recognized before the transversalis fascia is divided, under which lies the colon more or less enveloped in fat. When distended either artificially by air, or by the fæces, it is recognized either by the feel of the fæces or by its longitudinal muscular bands. When, on the other hand, it is collapsed (and Mollière[1] has called attention to the fact that it may be collapsed even in cases of prolonged retention, the accumulation being either above or below the point of operation), the patient may be turned on the back to allow it to fall into the wound, or pressure may be made on the abdomen by an assistant. Bryant recommends rolling the bowel partially forward after it has been seized, to bring its posterior surface into the wound, as an additional safeguard against wounding the peritoneum.

When the bowel has been drawn well out to the surface of the wound, it must be secured in position before it is opened, in order that its contents may not escape into the abdominal cavity. This is best done by passing a couple of ligatures through it and the lips of the wound in the following manner. The needle is entered on one side of the incision and carried through the integument alone, and not through the whole thickness of the adominal wall, for the edge of the bowel is to be attached to the skin; it is then made to transfix the bowel and brought out at the opposite edge of the abdominal incision at a corresponding point. After two such sutures have been passed and intrusted to an assistant, the bowel may be opened by a longitudinal incision about three quarters of an inch in length, over the sutures which pass across its calibre. The middle of each suture is then drawn out of the bowel and divided. In this way four sutures will be in place; and after they have been secured, one may be inserted at each end of the wound in the bowel, and as many more along the sides as may be necessary for perfect coaptation. The sutures should be of strong silk.

The operation may be modified with advantage by stitching the parietal

---

[1] Op. cit., p. 596.

and visceral layers of the peritoneum together with sutures passing down to the sub-mucous layer of the bowel, but not into its calibre. The wound may then be covered, and the opening into the bowel delayed for six or eight hours for **adhesions to occur.**

The immediate **danger** in the operation of lumbar colotomy **is that** the peritoneum may **be opened and** death result from peritonitis, **due** not so much perhaps to **the incision in** the serous **sac as** to the escape **of fluids into its cavity.** It has also happened **to good** operators to open a coil of small intestine instead of **the colon; or, by** missing the latter **at first on account** of some change in **its** position, **to become** confused in the subsequent search and fail utterly in finding **the desired** part. Both of **these** most common accidents are **best avoided by a close** adherence to the rules which have been given.

The list of mishaps in connection **with this operation is a long and curious** one. The wound is deep and it is more than probable that **in many** cases the accident which the operation is especially intended **to avoid, and** the avoidance of which is the one point in favor of the **lumbar over** the inguinal incision—a wound of the peritoneum—**is not avoided.** The portion of **the** descending colon not covered by peritoneum varies greatly **in extent in** different cases; and during the operation there is no **way of** determining whether the serous coat is or is not **under** the knife. **The kidney has more than** once been wounded at the bottom of the incision,[1] **and as good an operator as** Allingham[2] confesses to having opened **the duodenum where it embraces the** head of the pancreas, in an attempt **to find the colon on the** right **side. In** children the peritoneal investment **is** more complete than in **adults, and the operation** is contra-indicated both on this account, and **because of the greater movability of the intestine.** In one hundred and **thirty-four autopsies on** children of less than two weeks of age, Giraldis found the sigmoid flexure on the left side in 114; Curling in 100 found it **so** located in 85; and Bourcart in 117 out of 150.[3]

Inguinal colotomy is especially indicated in treating imperforate anus in children, in whom the mesocolon is so lax that the sigmoid flexure may wander even across the aorta into the opposite flank. He who attempts the extra-peritoneal operation in a child may consider himself fortunate if he finds the desired point at all; and when found it is so completely surrounded by peritoneum as to render a wound of the sac almost a certainty. The operation in the groin too is easier of performance, and when successful the resulting anus is more easily cared for by the patient. **These** facts, together with the decreasing fear of incising the peritoneum, **have led some** surgeons to advocate this operation not

---

[1] Bryant, Amussat.
[2] Op. cit., p. 230.
[3] Guyon: Dict. Encyc. **des Sci. Méd., Paris, 1863.**

only in cases of adults where disease has encroached upon the sigmoid flexure, where it is particularly indicated; but in all cases for which the lumbar incision is generally chosen. The inguinal operation is in great favor among the French, the lumbar among the English.[1]

An incision about two inches and a half long is made in the left groin parallel with Poupart's ligament, about half an inch above it, and well towards the lateral wall of the abdomen—so far that the epigastric artery should not be seen in the operation. This incision is carried down to the peritoneum, each successive layer being divided on a director as is usual in operations on this part. Before the peritoneum is opened, all hæmorrhage from the wound should be stopped and the cut rendered as dry and clean as possible. The peritoneum is then pinched up with forceps and nicked, a director is introduced, and the opening enlarged to the extent of an inch and a half. The descending colon should be in view immediately below the wound, and is recognized by the usual sign. When such is the case, the subsequent steps of the operation are comparatively simple; the incision into its wall and its union to the abdominal wound being accomplished in the same manner as already described in the lumbar operation. But when such is not the case, the bowel must be searched for, and it may be necessary to enlarge the original incision. The following case from Mollière[2] illustrates very well the difficulties which may attend the operation in an adult under such circumstances.

"An unfortunate woman was admitted to the hospital at night with symptoms of acute intestinal obstruction. The abdomen was greatly distended, but she asserted that it had been much increased in size for a long time previous. As death was imminent and punctures into the intestine through the abdominal wall gave no relief, inguinal colotomy was decided upon. Scarcely was the incision made into the peritoneum before a quantity of ascitic fluid escaped, and an enormous, white, shiny, aponeurotic-looking tumor made its appearance. This tumor was somewhat movable. The operator believing that he was dealing with an ovarian cyst, and despairing of reaching the colon, made an incision into the small intestine from which escaped a large quantity of fæces. The autopsy demonstrated later that this tumor was itself the colon, greatly distended above a contraction caused by cicatricial bands in the pelvis. The patient had succumbed to a general tubercular peritonitis."

7. *Attempts at establishing an anus in the anal region after the performance of colotomy are attended with great danger, and are generally unsuccesful.*

---

[1] For discussion as to the relative merits of the two operations the reader is referred to the following articles: Dupuytren, "Dict. en 30 vols.," Art. Anus Artificiel; Videl de Cassis, Thèse de Concours, 1842; Guyon, "Dict. Encyc. des Sci. Méd.," Paris, 1863; Giraldès, "Nouv. Dict. de Méd. et de Chir. prat.," t. ii., p. 633; Robert, "Bull. de l'acad. Roy. de Méd.," t. xxi., p. 931.

[2] Op. cit.

Perhaps the best authority on this point is embraced in the experience of Mr. Owen.[1] In two cases in which after an interval of three months he attempted to establish an anus in the natural position, the end was a fatal peritonitis due to the fact that the rectal pouch was completely covered with peritoneum. Dr. Byrd[2] has more recently reported a case in which the operation was successful. The bowel ended in this case in a sort of *cul-de-sac* with an appendix, and the operation is described as follows. "By passing my finger into the bowel through the wound, I found that the calibre of the bowel easily permitted its passage for about three inches, when it suddenly narrowed, and from that point downward it resembled the appendix vermiformis. Into this narrowed portion was passed a small sound used for searching for stone in infants, and the end of it worked downward in the narrowed bowel toward the anus.

To more easily meet the sound from below, an incision was made about two inches deep, up from the anus and back to the coccyx, large enough to permit the passage of the index finger. The sound was carried along until it could be felt only about one-eighth of an inch from the tip of the finger passed from below, when it would pass no further with ease. Force enough was then used to pass the sound through the intervening space, and the point was brought out at the anus. To the point of the sound a stout thread, running through a No. 10 Jacques catheter, was attached with a reef knot, and the sound was retracted, bringing the catheter with it. One end protruded from the anus, and the other from the artificial anus. To the end protruding from the artificial opening, a compress was tied, and by placing a bit of rubber dam under the compress and drawing the catheter down, extrusion of the bowel was prevented, and some control was exerted over the fæces. The child was very much prostrated by the shock of this operation, but, by the second day, he had fully recovered."

This plan of treatment was continued as follows: The author took "a piece of soft-rubber tubing about as large around as my little finger and one foot long. By tucking half an inch of one end up into the tube, it made a bulbous end somewhat larger than the rest of the tube; this end I fastened to the catheter, where it came out at the side, with a stout flax thread, and drew it down into the bowel by retracting the catheter. As I expected and desired, it caught against the shoulder of the narrowed bowel, and by traction upon the catheter, the mucous membrane was brought down in a fold in front of the bulb, and covered the space that otherwise would have been filled with cicatricial tissue. To-

---

[1] Surgery of Childhood. Brit. Med. Jour., February 21st, 28th; March 6th, 1880.

[2] Lumbo-Colotomy in the New-Born for Relief of Imperforate Rectum. Read before the Tri-State Med. Soc., St. Louis, Oct. 25th, 1881. (Reprint.)

day (about one month after the introduction of the rubber tube) I removed the tube, and find my little finger passes readily up the opening, which is covered throughout with mucous membrane."

Unfortunately the history of this case ends at this point, the author expressing the hope that the artificial anus would close "without further operative interference, except the wearing of a well-adjusted pad," and being prepared to perform a further operation for its closure should it prove to be necessary.

Kronlein[1] also reports a successful case of this operation. A child six days old had had no evacuation of the bowels since its birth. The anus was extremely narrow and ended in a pouch 2.5 centimetres long. An attempt to reach the rectum by an incision through this pouch, resulted only in opening the peritoneum, as was shown by a free discharge of peritoneal fluid. The bowel was then opened in the left groin, and the child lived and thrived. When the child had reached the age of seven months, the rectal pouch could be distinguished, and the original operation was again attempted, and the rectal pouch successfully united with lower one. At the close of the report, a stricture existed at the place of union, but the larger part of the fæces were already evacuated by the perineal opening.

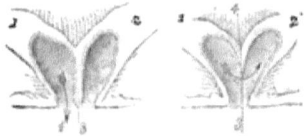

FIG. 11.   FIG. 12.
Condition of bowel after colotomy, showing septum and course of fæces (Packard).

The attempt to re-establish an anus in the anal region originated with Demarquay, and involves, if it be successful, a subsequent attempt to close the artificial opening. This is an operation of great danger and one seldom successful. The difficulties consist in re-establishing the calibre of the bowel at the point where it is partially occluded by the formation of the artificial opening, and in subsequently closing this opening by a plastic operation. The danger is of fatal peritonitis. It is well known that in cases of colotomy, the side of the bowel opposite the opening becomes sharply bent upon itself, as shown in Figs. 11 and 12. The septum thus formed is composed of two layers, each consisting of the whole thickness of the intestinal wall, and it must be destroyed before the lumen can be re-established and the opening safely closed. Dupuytren's[2] original operation consisted first in compressing this valve by an instrument invented by himself, the action of which is shown in Fig. 13.

---

[1] Berlin. Klin. Woch., 1879, No. 34-35.
[2] Leçons Orales de Clin. Chirurgicale. Paris, 1839, t. iv., p. 1.

This was applied and tightened so as at once to cause the death of the included portion. The subsequent steps in the operation consisted in closing the artificial opening. His experience extended over 41 cases, 21 of which were done by himself and 20 by others. Three cases were fatal. Of the remaining 38 the operation was unsuccessful in 8, and successful in 29 in periods varying from two to six months. It is but proper to say that considerable doubt exists as to the reliability of this very favorable showing.

Since his time, the operation of Dupuytren has been modified in various ways by different surgeons. Barker[1] has recently reported a successful operation after a plan of his own, the essential feature of which consists in introducing into the bowel through the artificial anus, after the

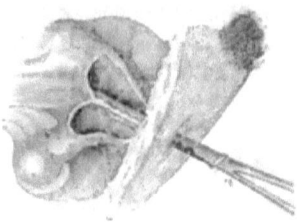

Fig. 13.—Enterotome of Dupuytren in position (Packard).

projecting spur of the bowel has been removed in the usual way, a thin and flexible strip of rubber about one and a half inches long by five-eighths of an inch broad, in such a manner as to lap up against the internal orifice; and to secure this in position by a single wire stitch at each end passed through the abdominal wall. The object is to allow the rubber to remain till the fistula is closed by paring and suturing its edges, and then by cutting the wires to allow it to pass down the bowel. In the case recorded, the rubber answered the purpose of preventing the escape of fæces very perfectly for the first few days, after which there began to be leakage, and it was removed. The fistula, however, went on to complete closure.

---

[1] "A Suggested Improvement in Dupuytren's Operation for Artificial Anus, and a Successful Case treated by it." Lancet, Dec. 18th, 1880.

CHAPTER III.

GENERAL RULES REGARDING EXAMINATION, DIAGNOSIS, AND OPERATION.

Necessity for Physical Examination.—Questions which may lead to Diagnosis.—How to make Examination.—Table.—Lamp.—Instrument Case.—Position of Patient.—Necessity for Enema before Examination.—What may be learned by simple Inspection.—Rectal Touch.—What may be discovered by it.—Bougies; Varieties; Author's Bougies.—Rectal Specula: Van Buren's; Fenestrated; Bivalve; Objections.—Colonoscope. - Stretching the Sphincter; Proper Method of Performing the Operation; Results.—Difficulties of Diagnosis of Disease high up in the Rectum.—Manual Examination.—What may be Learned by this Method.—Preparation of Patient for Operation.—Assistants.—Primary Anæsthesia.—Thermo-Cautery.—Hæmorrhage.—Rules for Controlling Hæmorrhage.—Cold.—Styptics.—Packing the Rectum.—Treatment after Operation.—Dressings.—Necessity for Rest.—Retention of Urine.—Case of Fatal Retention.

To one who has been trained in the habit of making a diagnosis before undertaking treatment it seems superfluous to insist upon the necessity of a physical examination in cases of rectal disease. The majority of patients who seek advice for this class of troubles come to the surgeon with the diagnosis of piles or fistula ready at hand, and, I am sorry to say, many of them come with the authority of some physician for that diagnosis, in whom, nevertheless, the merest inspection is sufficient to prove the existence of much more serious, and often of incurable, disease. This is not due to ignorance, but to carelessness, to too great faith in the statements of the sufferers, and often to a false modesty on the part of the practitioner which leads him to accept such statements in lieu of a thorough examination.

The following case illustrates many points in rectal diagnosis and may be as useful to others as it was to myself.

CASE I.—A young man appearing in perfect health was sent to me by Dr. N. M. Shaffer, of New York, for rectal trouble. He gave me a history of constant discharge from the bowel and of some pain after defecation, but the discharge was his chief trouble. On examination I discovered a fistula, but such an insignificant subcutaneous affair that I divided it on the spot, recomended a day's rest, and assured him that he would be entirely well in a week without further treatment. The fistula was well in a week, but the man was not. He still complained of discharge and some pain, though less than before. I made a second and more careful examination and discovered a perfectly well-marked fissure just above the external sphincter. Once more I assured him that he could easily be cured, and I divided the base of the ulcer with a bistoury. The operation was thoroughly done, for I was a little chagrined at my former carelessness and wished to make sure of the cure. The operation was not followed by the slightest relief, and six weeks were passed in the vain hope of a cure. I then did what should have been done in the first place, and set myself deliberately to make a complete diagnosis. I etherized the patient, dilated his sphincter, and made a thorough examination with artificial light. The fissure could be plainly seen and above it there was a polypus of considerable size which by its mobility had escaped me in the former examination, and by its contact with the surface of the sore had prevented a cure. This was removed, but the man was not yet cured. The pain had all disappeared, but the discharge from the bowel still remained in diminished quantity. I was about to despair, when he mentioned in the most casual way that he had had a good deal of itching at the anus for some time back, and an examination revealed a moist eczema which furnished the discharge. The skin disease had been there from the first, but as the man had asserted that it never troubled him, I had paid little attention to it. This was easily cured and I ultimately had the satisfaction of seeing my patient well. Here then was rather an unusual combination of troubles—a fistula, a fissure, a polypus, and eczema, and each one sufficient in itself to account for all the symptoms of which the patient complained. But all should have been discovered at the first examination, and the man should have been cured by one operation instead of three.

The symptomatology alone may be of great value in the diagnosis of rectal disease; it is almost never sufficient in itself for a diagnosis. There is a train of symptoms common to almost all diseases of this part and which infallibly points to trouble of some kind, but they do not tell what that trouble is. The pain of a fissure is, perhaps, diagnostic of the fissure, but it does not tell what troubles may be associated with the fissure; and so it is in every other affection. For this reason the practitioner who attempts to treat a case of disease of the rectum without first making a direct examination uselessly risks his reputation as a diagnostician, and in my own practice I am guided by the simple rule that patients, male or female, who have not yet come to the point which makes them willing to submit to an examination, have not yet reached a point which admits of treatment. An examination, especially in women, is sometimes though not often, difficult to obtain, and the dread of it keeps many sufferers from seeking relief; but still the rule I have laid down is the only safe one, and the surgeon who allows himself to be persuaded into "recommending something for piles" will sooner or later have a mistake in diagnosis laid to his charge, nor will the fact that he was moved by consideration for the patient's sensibilities save him from blame.

I have often found that the best way to secure an examination in women who otherwise could not be brought to consent to it, was to resort to ether, with the understanding that whatever surgical procedure was thought advisable should be performed at the same time. In this way a patient's sensibilities may often be spared, while both diagnosis and treatment are included in one examination.

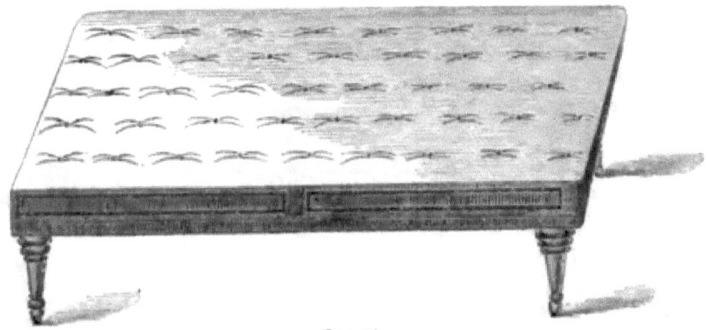

FIG. 14.

FIG. 15.

Before, however, proceeding to make the physical examination which is inevitable, certain questions and answers may give the surgeon a pretty clear idea of what he is about to find. It is generally a good plan to allow an intelligent patient to tell his or her own history, and then to supplement it with appropriate questions as to the length of time since the trouble began; the character of pain when present, whether constant or intermittent, and increased by defecation; whether it comes with the stool, immediately or some time after, and its duration. The question of

GENERAL RULES REGARDING EXAMINATION, DIAGNOSIS, ETC.   51

discharge should also be inquired into—its quantity and character, whether blood, pus, or mucus; also whether there is any protrusion of any kind, and its character. The answers to these questions and to those which relate to the presence or absence of diarrhœa, constipation, and

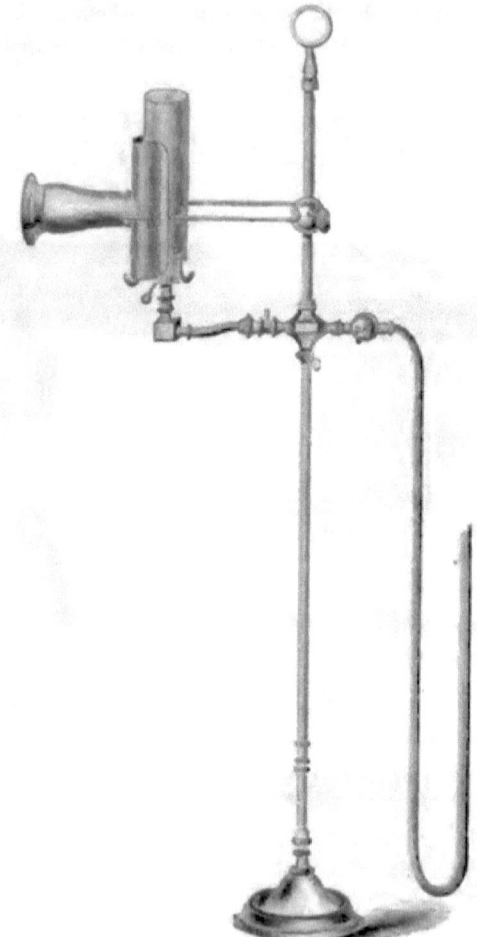

FIG. 16.—Lamp for rectal examinations.

incontinence, will generally give the surgeon a fair idea of the nature of the case before him.

How, then, to proceed to make a rectal examination which shall be at the same time thorough and as free from pain as possible? Two things are necessary above all others—a good bed or table and a good light. For a table, a strong, four-legged one, upholstered with hair and leather, answers every purpose. It should be hard, without springs, and about thirty inches in height. In place of this, any of the examining tables of the gynæcologists may be used. In my own office, I use a modification of the combined table and lounge of Dr. J. L. Little, which is represented, closed and open, in Fig. 14 and Fig. 15. Its great advantage is that, when not in use, it answers as an ordinary piece of furniture, and when raised it provides a firm, hard operating table of convenient height. Either natural or artificial light may be used, but the latter is on some accounts preferable, being always at command, and easily thrown up the bowel or concentrated upon a particular point. To do this, a forehead mirror is requisite. The lamp which I have found most convenient is a modification of Tobold's, as

FIG. 17.—Case for rectal instruments, with sliding cover A A.

represented in Fig. 16. The whole apparatus is easily moved to any part of the room, and is not cumbersome; and with the lens a very powerful illumination is always attainable.

The instruments necessary are specula of various forms, bougies, a Davidson's syringe, ointment, cotton, sponge-holders, towels, basins, etc.; and these should all be placed within easy reach of the hand. A convenient case for these things and for other surgical instruments, which is intended to stand on the floor by the side of the table or bed, is represented in Fig. 17.

The position in which the patient should be placed is a matter of some importance. For mere inspection of the anus and surrounding parts, the dorsal decubitus answers every purpose, and a digital examination of the rectum may be made either in this posture or with the patient on the side. For a speculum examination or the passage of a bougie, the patient should be placed on the side, **with the buttocks** well elevated, the thigh which is uppermost strongly flexed on the abdomen, and the breast resting on **the table.** In this way, the weight of the abdominal contents **falls upon** the front wall of the abdomen, **and not** upon the pelvis, **and the** lumen of the bowel is not so firmly closed, **nor** is the mucous **membrane** so firmly forced into the end of the speculum.

Before commencing **an** examination, **the** bowel should be emptied, either by the natural effort of the patient or by an enema, and for this reason a water-closet **in connection with the** examining **room is** indispensable to the practitioner in rectal disease. **In this way, the** patient may come directly **from** the closet to **the table with the parts** in the best condition **for** inspection; and great **additional confidence is** acquired, especially by women, that the examiner's **frequent reiteration to** "**bear down**" will not be followed by untoward consequences. The point may **seem** trivial, but the fear of an accident will frequently, in women, **result in** a firmly closed sphincter, which no word of the surgeon can overcome, and a thorough examination cannot be made while the rectapouch is filled with **fæces.** This is not merely a thing to be observed for the cleanliness of the examiner, **for the** act of defecation will bring internal hæmorrhoids **and** prolapse to the light, and may greatly assist in the diagnosis of other **maladies.** In examination with a speculum, it is indispensable to cleanliness.

A simple inspection **of the** anus and adjacent skin and mucous membrane is often sufficient **for** a diagnosis, **though it** should **never be** trusted to alone. External hæmorrhoids **and internal ones** when brought down by the use of the closet or enema, external **fistulæ, ulceration, skin diseases, many** venereal affections, pin worms, abscess, **and fissure, may** all be recognized **in this way.** A glance at the anus, too, may indicate to the **practised eye the existence of serious** disease within the rectum proper, for a discharge may flow from **it** which marks ulceration above, and it may be relaxed and patulous from over-distention **or** partial destruction of the sphincter. A sunken condition of **the** ischio-rectal fossæ, and a retracted anus surrounded by a profusion of soft, fine hair, may also properly excite a suspicion either of grave rectal disease or of some constitutional affection which is causing emaciation.

By using gentle force in pulling the anus open with the fingers, the mucous membrane may be everted to a considerable degree, especially if the patient can **be** brought to assist by an effort at bearing down. **In** this way a fissure **may** almost always be brought into view without the use of a speculum **of** any sort, and the internal opening **of** the great majority of fistulæ **may be reached, with** a good view of the **radiated** folds and lacunæ.

Dr. Storer,[1] **of Boston, has** described a method of examining the

---

[1] Lancet, May 31st, 1873.

mucous membrane just within the anus, which is applicable only in women who have a lax sphincter. It consists in everting the mucous membrane by pressing it out of the anus by the index finger in the vagina. In a case in which the manœuvre can be practised successfully and without too much pain, a small portion of the anterior wall of the rectum may be brought into view. The pessary of Gariel has also been used for the same purpose. It consists of a rubber ball, which is introduced empty into the rectal pouch, then inflated by means of a tube attached to it, and withdrawn with some force, the mucous membrane being prolapsed in front of it. But neither of these two procedures is of any great value.

After having examined the anus in this way, the surgeon next proceeds to the more difficult task of examining the rectum, an operation which may be done skilfully and almost painlessly, or awkwardly and with great suffering. The rectum may be explored either by the touch alone, or by vision alone, or by both combined. The former is the simpler and more painless method, and with practice may be made to afford all the information which can be gained by the two combined.

To practise the rectal touch, the nail of the index finger should be well trimmed, and the finger lubricated with some tenacious oil. Olive oil is much better than vaseline, the latter being too easily rubbed off by the sphincter. The condition of the spincter muscle is first to be noted. Its resistance should be overcome by a slow and steady pressure with the ball of the finger, and not by a sudden exertion of force, for such an attack is always met by increased contraction. The force of the muscle will be found to vary greatly in different people. In the aged or debilitated it is lax; in the strong and healthy it is the opposite, and the finger can scarcely be passed through it without great pain and sometimes a slight laceration of the tender mucous membrane. When inclined to spasmodic contraction, as it sometimes is in persons of nervous tendency, a satisfactory examination may be impossible without the use of ether, on account of the pain.

Unless an obstruction is encountered, the finger may be carried up the bowel its full length, and pressed as far as possible beyond this point. Additional distance may be gained by passing the three remaining fingers backward along the inter-gluteal groove, instead of closing them in the palm, as is generally done, and pressing the knuckles against the soft parts; for the knuckles prevent the full passage of the index finger.

In this way three or three and a half inches of the rectum may be carefully explored, together with the prostate, the neck of the bladder, the uterus, and the anterior surface of the coccyx and lower part of the sacrum. With an exceptionally long finger it may even be possible to feel the vesiculæ seminales and vasa deferentia. In other words, all that part of the bowel which is most subject to disease is brought within reach.

But after this is done the examiner may be no wiser than before, for to appreciate fully the condition of the rectum by the sense of touch alone requires a facility in this method of exploration which most practitioners never attain. In the majority of cases a digital examination will be made to discover whether or not the patient is suffering from internal hæmorrhoids, and in the majority of cases also the examiner will be no wiser on this point after than before, for a soft internal hæmorrhoid is a difficult thing to detect by the finger alone, being readily mistaken for the natural mucous membrane of the part, especially when the latter is abundant and gathered into folds, as it is apt to be.

Ulceration is another condition which it is sometimes difficult to detect, especially when superficial and not attended by much induration; and so is the opening of a blind internal fistula; and yet, so well educated may the finger become that other methods of examination may be almost completely discarded. To carry diagnosis to this point it is first necessary by oft repeated examinations, to become perfectly familiar with the feel of the normal bowel. After this knowledge has been gained, a gentle sweeping of the ball of the finger over the whole inner surface of the

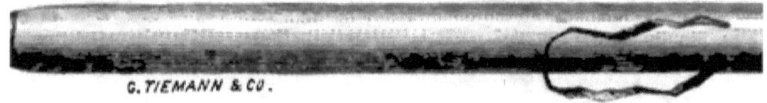

Fig. 13.

lower three inches of the rectum will detect any change in it, however slight. I wish it were possible to describe plainly the different sensations which are conveyed by the different pathological conditions, but this is a thing each practitioner must learn for himself by practice.

A stricture of small calibre cannot easily be mistaken, though one which admits the finger without constricting it may easily be overlooked. A stricture small enough to engage the end of the index finger firmly, marks the limit of safe digital examination, and the finger should not be forced through it for the sake of feeling what is above, for an attempt to do this has been followed by a fatal rupture of the bowel. In case of a tumor of any kind, advantage may be taken of conjoined manipulation through the vagina in the female, but these are the troubles most rarely met with, and most easily diagnosticated when encountered. The cervix or fundus of the uterus, when pressing upon the bowel, may be distinctly felt with the finger in the rectum, and may deceive the unwary into a diagnosis of a new growth. The prostate may do the same. The different varieties of ulceration have each their peculiar and often diagnostic feel.

For examination by the sense of touch above the reach of the finger, recourse may be had to bougies. These are of all forms, sizes, and materials, and, in general words, the softer the instrument the better it is for examination. I much prefer the black rubber instrument, with the blunt point (Fig. 18), which may readily be bent into a circle in the hand, to all others in the market, and the same instrument comes with a sharp point (Fig. 19) which sometimes answers a good purpose. These

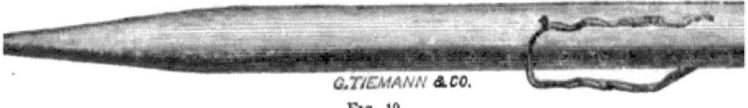

Fig. 19.

instruments are made in twelve different sizes, and for the purpose of diagnosis the medium-sized is the best. The old-fashioned red, hard-rubber bougie is unnecessarily stiff and dangerous, and should be discarded, having no advantages over the softer ones either for the purpose of diagnosis or for that of treatment. The *bougie à boule*, made of hard-rubber with a flexible whalebone handle, is a favorite instrument with many. (Fig. 20.)

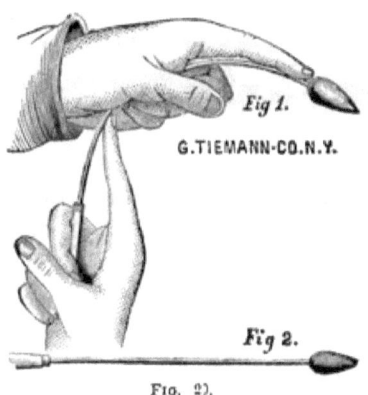

Fig. 20.

For my own use I have had a kind of bougie made by Messrs. Stohlmann, Pfarre & Co., which I prefer to all others, for the simple reason that it is softer and more flexible than any in the market. It is made of the same material as the red soft-rubber catheters, and differs from them only in size and in the thickness of its walls. With such an instrument one is pretty certain not to perforate the bowel, and for diagnosis it answers every purpose as well as the harder instruments. The better fitted a bougie is for pushing its way through a stricture the worse it is for rectal exploration.

These instruments are all used for the same purpose—that of feeling for a stricture located above the reach of the finger; and with any of them the unpractised hand will generally detect an obstruction in the perfectly healthy bowel at about four inches from the anus. I have had patients in whom I have never been able to pass any sort of a bougie without first injecting the rectum, no matter what manœuvering I resorted to; and I have seldom told a student to pass a rectal bougie that he did not at once discover a stricture. To pass a bougie into the rectum is rather a more difficult operation than to pass one into the urethra, the triangular ligament in the latter being replaced by the curves, the folds of mucous membrane, and the promontory of the sacrum in the former. Independent of Houston's valves of mucous membrane, it is not improbable that a slight degree of invagination of the upper into the lower part of the rectum may often exist; and into the sulcus formed by this condition the point of the bougie may easily pass. For the sake of overcoming these folds of membrane the most minute directions have been given as to how the bougie should be introduced and gently urged along each successive inch of the bowel by changing its direction and manipulating the handle. But such rules are of little value, for the simple reason that the obstruction is seldom of the same kind or in the same place in two different persons. Esmarch[2] gives the general rule that the patient should lie on the left side, as the chief and most constant fold of membrane, the plica transversalis recti of Kohlrausch, projects from the right wall. The instrument should be passed gently, for force is never allowable here more than in the similar operation on the urethra; and when an obstruction is met with the handle should be gently rotated, withdrawn, and again passed onward till by frequent repetitions of this manœuvre it is made to pass. If this does not suffice, a Davidson's syringe may be attached to the lower end of the bougie and a stream of warm water thrown into the bowel until it is moderately distended when the bougie will generally pass with ease.

For measuring the extent of a stricture, an ingenious instrument has been devised by Laugier, which consists in attaching a thin rubber glove-finger to the end of a perforated bougie. This is passed up the bowel empty, and then inflated and withdrawn till it reaches the upper limit of the obstruction. It is safer than the *bougie à boule*, for it may be allowed to collapse before being withdrawn, and all straining of the diseased tissues may thus be avoided.

In case disease actually exists high up in the bowel, the attempt to pass an instrument is full of danger. A patient may easily recover from a false passage made in the urethra, but such will seldom be the case with the rectum, for here when the instrument leaves the bowel it enters the peritoneum. To understand this danger it is only necessary to remember that the bowel is generally ulcerated both above and below the seat

---

[1] Houston: "Dublin Hosp. Reports," vol. v., 1830.
[2] Die Krankheiten des Mastdarmes und des Afters, Pitha und Billroth's Chirurgie.

of the contraction, and is sometimes weakened to such an extent that it will allow a bougie to pass through it without the use of any appreciable force on the part of the surgeon. The bowel may also be lacerated without being directly perforated by the bougie, for the stricture may be pushed upward or dragged downward on the point of the instrument till the bowel gives way.

Supposing, now, that a rectal bougie cannot be passed eight or ten inches up the bowel, is it safe on this account alone to make a diagnosis of stricture high up? I should hesitate long before doing so, and should make many careful attempts to pass the instrument at different times, resorting to injection if necessary, carefully exploring through the abdominal wall for induration, and watching for the usual signs of obstruction. There are one or two points worthy of remembrance in this connection. The first is that the obstruction due to a stricture will always be at the same point in the canal; and another is, that when a bougie has once become engaged in a stricture it is firmly grasped, and the resistance to its withdrawal is equal to that encountered in introducing it farther. The feel-

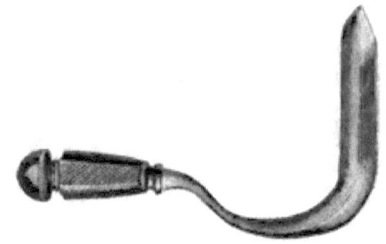

FIG. 21.—(Van Buren).

ing conveyed to the hand under these circumstances is diagnostic, and is like that which is felt when the effort is made to withdraw a sound from the grasp of a stricture in the urethra.

Should it still be necessary for diagnosis, the speculum may be used and the inside of the rectum illuminated. I have postponed any reference to this means of examination till the present, because it will generally be found useful only after the others have been tried. The thorough use of the speculum involves, almost of necessity, the administration of ether and the stretching of the sphincter muscles; to try to use it without these adjuncts is almost to inflict useless pain upon the patient. I shall not attempt any description of the infinite number of instruments which have been invented for this purpose, or any judgment upon their relative advantages, but will merely say that the best vaginal speculum is still the best for the rectum—that of Sims, with a groove where the blade joins the handle for the sphincter to rest in as suggested by Van Buren, Fig. 21. The fenestrated instrument, Fig. 22, is sometimes useful for inspecting the parts just within the anus; and a long vaginal cylindrical speculum, with the end cut at such an angle as will best expose the mucous membrane, may sometimes be of service in bringing into view a small portion of the inner surface of the bowel high up. But, after all have been tried, none will be found better for any purpose than a small-

Fig. 22.

bladed Sims's, and without ether all will be found eminently unsatisfactory.

Almost the only other speculum besides Sims's which I have found of any practical value is the bivalve shown in Fig. 23, but the same objec-

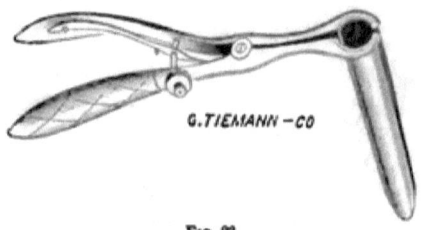

Fig. 23.

tion applies to this as to all the others, that the redundant mucous membrane prolapses between the blades to such an extent as to render it almost useless, and that when the attempt is made to dilate the blades sufficiently to overcome this, the sphincter is immediately stretched to a painful extent. With any speculum the wooden depressor, shown in Fig. 24, may be found a useful addition.

Fig. 24.—Rectal Depressor (Van Buren).

The idea of the endoscope has been applied to the rectum in the use of the instrument shown in Fig. 25. It is of little, if any, practical value, however; its introduction beyond the point which can be reached by a long vaginal speculum being exceedingly difficult, and, in case of the diseases which it is supposed to enable the surgeon to see, not devoid of danger; and the mirror quite useless.

It is almost useless to attempt to see within the rectum with any kind of a speculum without first overcoming the sphincter muscle, and the only effectual way of doing this is by stretching it. It is, therefore, my own practice to resort to this procedure in every case of doubtful character, nor was I led to this practice without many trials of the various speculæ in the market, all ending in disappointment. The stretching of the sphincter is in itself an entirely harmless proceeding, but one which necessitates the previous administration of ether. It should not, however, be done, as was at one time the usual method, and as it is often done at present, by introducing the thumbs back to back, and forcibly and suddenly separating them till they touched the tuberosities on each side. In this way, the mucous membrane is often lacerated at one or more points,

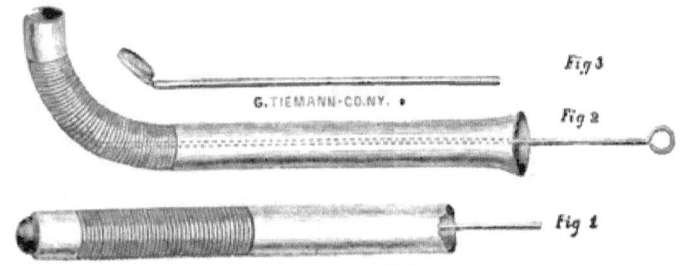

FIG. 25.—Colonoscope of Bodenhamer.

and the paralysis is not as effectual as when the stretching is done more gradually. A better way is to introduce first one finger, then two, and finally four, in the form of a funnel and gradually bore into the anus; or to introduce two fingers, and make pressure on all sides of the opening till it becomes patulous. Instead of one or two seconds, this procedure should occupy five minutes, and should be done so gently as not to lacerate the mucous membrane. The dilatation should also be made to include the internal as well as the external muscle. If this dilatation be carried to a sufficient extent, the firm, cord-like feel of the external sphincter may be made to completely disappear. The paralysis induced in this way is always temporary, and I have never known it to be followed even by a temporary incontinence of fæces. After coming out of the ether, the patients are usually conscious of only a sense of soreness in the part, but are never incapacitated for their usual duties. This stretching of the sphincters is a necessary preliminary in almost all operations within the rectum.

With the patient in the proper position on the side, under the influence of ether, with the sphincter thoroughly dilated, and with a good reflected light, the lower four or five inches of the rectum may be thoroughly illuminated and examined. A couple of inches more may be seen by the use of the cylindrical speculum, with the patient standing and bending forward over the table, and assisting the examiner by straining down; and in this way a stricture may sometimes be brought into view which could not be seen with the Sims's speculum alone.

As a rule, however, a speculum will be found of very little use in the examination of stricture, but is chiefly available for obtaining a good view of other morbid processes affecting the rectal pouch and for making applications to them or performing operations for their cure. By its aid the different varieties of ulceration may be inspected and thus differentiated, the internal openings of fistulæ may be located, and the whole rectal pouch may be brought into view.

From what has been said it may readily be seen that the diagnosis of stricture above the reach of touch or vision is a difficult matter. So difficult is it in some cases that no less an authority than Syme has written that there is good reason to suspect the honesty of a man who pretends to detect such a condition. Such is, indeed, the case, for "strictures high up" are favorites among a certain class of quacks, and the passage of a bougie two or three times a week for an indefinite period is profitable business. In reality stricture above the rectal pouch is rare; when they exist they are usually malignant, for this part of the bowel is not subject to the influences which, by exciting ulcerative action, result in the cicatricial contractions which so often affect the lower three inches of the rectum; and malignant disease of the sigmoid flexure or descending colon will manifest itself by a well-marked train of constitutional and local symptoms, and can generally be felt better through the abdominal wall than *per rectum*.

After the use of the bougie, which is at best an uncertain means of diagnosis for this condition, and after a study of the symptomatology, and a careful examination through the abdominal wall, there is still one other means of exploration open to the surgeon if he have a sufficiently small hand—the passage of the whole hand into the rectum.[1] A hand which measures seven and a half inches in circumference can generally be passed easily; one measuring more than nine is unfit for the purpose. With a small hand there is no danger of permanent incontinence of fæces, but the sphincter should be dilated gently and gradually, rather than forcibly torn open.

When the anus has been sufficiently dilated to allow the hand to enter

---

[1] G. Simon, Ueber the künstliche Erweiterung des Anus und Rectum. Arch. f. klin. Chir., xv., 1, 1872; Dtsch. Klin. f. Chir., Nov., 1882; W. J. Walsham, Some Remarks on the Introduction of the Whole Hand into the Rectum, St. Bartholomew's Hosp. Rep., vol. xii., 1876, p. 223.

the rectum, if the bladder is empty, the arch of the pubes may be felt above the prostate, if full it will be easily distinguished at the same point. The uterus and ovaries are easily made out anteriorly, and the whole curve of the sacrum may be followed posteriorly. The next point to feel for is the spine of the ischium on either side, and with this as a guide, the greater and lesser sciatic notches may be outlined. The whole brim of the pelvis may be traced, and the external and internal iliac arteries followed with the fingers. All this may be done while the hand is in the rectal pouch, and it may be done upon almost any patient, male or female, though more easily upon the female, with a small hand, without causing any unpleasant after-results. But in many persons this is all that can be gained by this method, for the anatomical reason that to pass the hand above into the sigmoid flexure is often attended with great danger from the narrowing of the bowel at this point. When the hand is met by a sense of constriction at about the level of the third sacral vertebra, where the lateral fold of Douglas is reflected from the bowel, the limit of examination has been reached, and no force should be used to overcome the constriction, which can only be accomplished by a rupture of the peritoneal coat. In many cases, however, by carefully following the natural windings of the canal, and by a semi-rotatory movement of the hand, combined with alternate flexing and extending of the fingers, this point of danger may be surmounted, and the hand be passed fairly into the sigmoid flexure, and sometimes into the descending colon. Here the comon iliacs, the bifurcation of the aorta, the left kidney, and, in fact, nearly all of the abdominal contents may be touched.

By this method of examination, a stricture situated in the sigmoid flexure, or even in the descending colon, may sometimes be discovered after all other methods of examination have failed ; but, as we have shown, the method is not always applicable, and the diagnosis of stricture high up still remains one of the most difficult things in surgery. In the great majority of cases in general practice, in which such a diagnosis has been made, it may be proved false by the introduction of a full-sized bougie after a few trials, and in the remainder the diagnosis will be confirmed sooner or later by the well-marked symptoms of intestinal obstruction.

Before attempting any surgical operation upon the rectum, the bowels should be thoroughly emptied by a cathartic. It is well to begin with three compound cathartic pills, or with five grains of mass. hydrarg. on the second evening before the operation where the patient's general condition admits of these remedies; to follow them with a slight saline or a dose of castor oil on the night immediately preceding; and finally to clear out the rectum with a simple enema on the morning of the day of the operation. After this the bowels may easily be confined for a week if desirable without inconvenience to the patient, and the passage of hard masses of fæces over a wounded surface is avoided.

In all operations in which ether is used, three assistants will be necessary and four are preferable. Each assistant should have his place assigned to him—one for the anæsthetic, one to keep each leg of the patient in position and to hold the speculum, and one to assist the operator in whatever way may be necessary. A state of profound anæsthesia will generally be necessary, though with intelligent patients I have often taken advantage of the primary anæsthetic state which ether produces for opening abscesses, dividing fistulæ, and cutting off external hæmorrhoids.

Accidents are not common in operations about the rectum, but there is one for which the surgeon should always be prepared—hæmorrhage. For this reason a bottle of dry persulphate of iron, and a Paquelin's thermo-cautery should always be at hand. The thermo-cautery as now made, Fig. 26, is not at all cumbersome, and is exceedingly useful in many operations about the rectum. The bulb containing the sponge for the benzine should never be filled with an excess of fluid which may run down into the point and interfere with the working of the instrument;

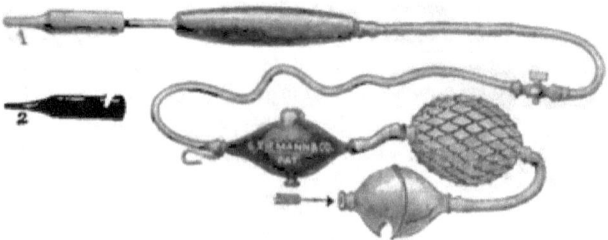

FIG. 26.—Paquelin's thermo-cautery.

and the platinum point should be *thoroughly* heated before the assistant begins to use the bulb to drive the air over the sponge. If proper regard be paid to these points the instrument is a most reliable one, and in every case where hæmorrhage is to be apprehended it should be ready for use, and an alcohol lamp or gas jet should be ready to heat the point—which is sometimes forgotten.

A hæmorrhage seldom occurs from the rectum after a surgical operation—so seldom as to be almost unknown—which cannot be controlled by the cautery or by packing the rectum. The rectum may be packed with either sponges or lint, and these may be used either with or without the persulphate of iron. Most cases of bleeding may, however, be controlled by the use of simple ice-water and a moderate amount of pressure properly applied to the bleeding surface without the necessity for a systematic packing of the whole rectal cavity. It is not long since I was called in the middle of the night to stop the bleeding from an incision which I had made into an abscess of the ischio-rectal fossa about eight hours before. I found, as is too often the case, that the patient was thoroughly immersed in a mixture of blood and persulphate of iron which covered him from the pubes to the middle of the back and had thoroughly permeated the bed. On entering the room I was informed that the wound had been carefully stuffed with lint and persulphate of iron "several times," and that the case was undoubtedly one of the hæmorrhagic diathesis. A case like this is easily managed. The treatment consists first of all in providing a good light, next in cleaning up the general nastiness, then in finding the bleeding point and making pressure upon it. In this case the bleeding came from a small spouting cutaneous vessel and was at once controlled by filling the incision I had made with picked lint *thoroughly* pressed home into the wound. Most cases of bleeding may be controlled in the same way, but where the hæmorrhage is within the bowel it is not always easy to make pressure upon the right point without packing the entire rectal cavity. For this purpose Allingham[1] recommends the following procedure which is equally simple and effectual.

Take a medium-sized bell-shaped sponge and pass a strong double ligature through the apex from within outwards and back again so as to include a considerable part of the sponge in the bite of the ligature—enough so that when the cord is pulled upon strongly from below it will not tear out. After wetting the sponge and squeezing it out it should be powdered with the persulphate of iron and passed as far up the rectum as possible with the aid of a rectal bougie, the apex being upwards. The whole of the rectum below the sponge should then be carefully filled with pledgets of cotton-wool powdered over with the iron, each roll being carefully and firmly packed away. An exceedingly large quantity of cotton may be crowded into the rectum in this way, and when the cavity is filled the sponge should be drawn down by means of the string hanging out of the anus, so that the whole mass may be tightly compressed. If the bowel has been thoroughly emptied as recommended, such a plug may be left in for a week or more without causing any discomfort and no bleeding can occur while it is in place. If, however, it is intended to leave the packing in for such a length of time it is better to pass a large-sized, stiff rubber male catheter through the apex of the sponge and pack the cotton around it. In this way a chance is given for wind and fluid fæces to escape. By this simple means, when properly used, any hæmorrhage after an operation upon the rectum may be controlled.

After operations upon the rectum or anus, a suppository of one grain of opium may generally be placed in the rectum with advantage, and the surgeon should always be provided with them. The usual dressing consists in placing a pad of lint and a soft towel over the anus and fastening them in place with a T bandage. This form of bandage will generally be found the best in any case where a continuous dressing is needed.

---

[1] Op. cit., p. 154.

Lister's impervious dressing has been **applied to** wounds of the rectum in some of the more extensive operations, such as excision of **cancer,** by the German surgeons; but it **has not** become popular, and the **use of free drainage and plenty of carbolic** acid or some other disinfectant is generally considered all that is necessary **or desirable in** this line. Verneuil recommends **the free use of a solution of chloral** as an antiseptic for this part.

Wounds of the rectum will always heal more kindly when the patient is in the horizontal position than when standing or walking, there being less tendency to venous congestion **in the** former **case.** Almost any operation may result in a sluggish open sore if **the patient be** allowed to disregard this **rule.**

Retention **of** urine is of frequent occurrence after operations upon **these parts,** both in men and women, and it should always be in the **mind of** the surgeon. It is not generally of long duration, and it **may** often be overcome by a bath and hot applications, without having recourse to the catheter. The following case conveys a lesson in this matter **which** should never be forgotten.

CASE II.—I was requested several years ago by a gentlemen **to make** an autopsy on his **brother, who had died** very suddenly and unexpectedly after being confined **to his bed about a** week with an abscess **near the** anus. Before the abscess **appeared the man** had been in perfect health, and was apparently doing **well up to the** moment of his death, as the abscess had been opened **on the day before, with great relief** to pain, and was discharging freely. **I made the autopsy, as requested,** and found a bladder distended to the point of rupture, the **urine dammed back upon** the kidneys, **which** were gorged with **blood,** and **the cerebral vessels** greatly congested. The man had died very suddenly in a **convulsion. A** little questioning revealed the fact that from the first day of the disease there had been retention of urine with dribbling from the overflow; and that for the pain arising from this condition opium had been freely given up **to** the day of death.

Once during his sickness an old woman in the house had applied a hot flannel cloth over the bowels, and the patient had passed an immense amount of urine. The condition of the bladder seemed to have entirely escaped the notice of his medical attendant, as it probably has escaped the attention of most surgeons at some time, though, fortunately, without, as in this case, a fatal result.

CHAPTER IV.

INFLAMMATION OF THE RECTUM.

Cases of Proctitis.—Varieties: Acute, Chronic, **Primary**, Secondary, Localized, General.—Symptoms and Course of each Variety.—Causes of Proctitis: Direct Propagation, Foreign Bodies, Drastic Cathartics, Gout, Pederasty, Gonorrhœa.—Treatment.

THE two cases which follow are not only interesting from their rarity, but as being good examples of two different stages of the affection under consideration.

CASE III.—Mrs. G., age thirty-seven, mother of three children. The patient, a delicate and rather anæmic lady, had not been in good health for some time past, but had never had any trouble with the **rectum until** one month before consulting me. At that time she was surprised to find that she had passed a considerable quantity of blood while at stool, and this hæmorrhage had been repeated at intervals of about a week up to the day before my visit. There had never been any pain in the rectum or anus, or any signs of hæmorrhoids, and a careful examination failed to reveal any source of the hæmorrhage. The lady complained, however, of a good deal of discomfort in the back and pelvis; had missed her last menstrual period, and was decidedly constipated. An examination showed a uterus enlarged and retroverted, and a considerable mass of fæces in the sigmoid flexure and descending colon, and treatment was begun for these conditions. The bowels were unloaded of many scybalous masses by means of frequent enemata; and the uterine condition was so far improved by treatment that the menses soon reappeared, and the pain and discomfort passed away. The bleeding from the rectum never recurred, nor has the patient ever again had her attention called to that part up to the present time—four years later.

The diagnosis in this case was a simple congestion of the rectal mucous membrane, brought about by the retained fæces and by the uterine disorder, relieving itself by a discharge of blood from the over-distended veins. Had the conditions remained, other symptoms would in all probability have soon developed, such as heat and tension at the anus, possibly a slight mucous discharge, pruritus ani, and, finally, hæmorrhoids. There are various other causes of such a condition, besides impacted fæces or menstrual disorders, such, for example, as excess at table, prolonged horse-back exercise or carriage riding, pregnancy, drastic purgatives, and, in short, anything which tends to produce hyperæmia of the pelvic viscera.

In most cases of **bleeding from the rectum** a diagnosis of congestion alone would be **an error; for a** congestion sufficiently marked to **cause** hæmorrhage is **rare, and bleeding is** in most cases **a** symptom either **of** hæmorrhoids, polypus, **or** some more serious affection. But in this case there was no such cause, and the subsequent history of four years with no other rectal **symptoms** tends to strongly confirm the diagnosis.

NOTE.—While speaking of hæmorrhage from the rectum, **it may be well** to refer to two cases of bleeding which have recently **been** reported in **the** New York Medical Record. The first (N. Y. Med. Record, Sept. 27th, 1879) is by Dr. Manley, of Lawrence, Mass. It occurred in an apparently healthy infant three days old, and ended fatally. A *post-mortem* examination showed that the blood came from an opening in one of the rectal veins about three inches from the anus, which admitted of the introduction of a bristle.

The second case (N. Y. Med. Record, Jan. 17th, 1880) **is** reported by Dr. McGuire, of Salem, Ohio, and is very similar, the child being about the same age. Notwithstanding suitable treatment by styptic applications, this also terminated fatally; but no autopsy **was** obtained, and the precise source of the hæmorrhage is unknown.

The second case **is one in which** congestion had ended in **actual inflammation** or proctitis.

CASE IV.—Woman, **married, age twenty-three,** mother of two children: youngest **six months old. Patient has** always been constipated, and for years **has been in** the habit **of using purgatives** whenever she desired an evacuation. For the past six **months has** noticed occasional discharge of blood and slime from the **rectum which is** constantly increasing. Now suffers great pain on defecation, and the amount of blood and muco-purulent matter is increasing **so** that while **at first it only came** away when at stool, it now comes several times a day. With this she has much pain in the rectum at all times, and is in poor general condition, having lost her appetite, and being unable to sleep.

A careful examination of the rectum showed it to be congested, hot, and painful as far as the eye could see; but nothing else was apparent. The amount of discharge suggested the idea of a gonorrhœa of the rectum, but there was no inflammation of the vagina, and careful questioning of the patient left no room for such a suspicion. The cause of the trouble in this case also was not difficult to find, the patient having been in the habit of taking large doses of patent cathartic remedies **two or** three times a week for a long time; and as the trouble developed **immediately** after her last confinement, this may not have been without its influence as an exciting cause.

This case gives a very good idea of the clinical history of acute inflammation of the rectum. A proctitis may be either acute or chronic, primary or secondary, localized or general. The localized variety is generally due to the injury inflicted by a foreign body or to some irritation acting upon a small part of the rectal surface. In the acute form the inflammation does not extend deeper than the mucous membrane which is congested and hyperæmic. In the chronic, the inflammation involves the submucous and muscular layers. The acute generally ends in resolution in from eight to fourteen days where the cause can be found and removed. It may, however, in severe cases go on to actual gangrene and terminate fatally. The chronic results in infiltration and consequent thickening of the rectal wall, and may end in ulceration, either superficial and confined to the epithelial layer of the mucous membrane, or deep and involving the whole thickness of the mucous layer. What is described a follicular ulceration (ulceration affecting the mouths of the tubular follicles) may result from chronic inflammation; and these ulcers, which are very minute at first, may coalesce and gain in depth till they cause perforation of the bowel. When the perforation is above the peritoneal reflection a fatal peritonitis may result; when lower down, an abscess or fistula (see Fistula). A chronic proctitis may in this way be a cause of stricture, and may result in the hypertrophy known as chronic parenchymatous proctitis.[1]

The symptoms of this affection have been partially detailed in the two cases which have been related. They are, in the acute form, a sensation of heat and weight in the part which may amount to actual pain, and may involve the bladder, uterus, and sacral region, and radiate into the loins and down the thighs. The anus also becomes painful, red and contracted, and in children the mucous membrane may become slightly everted from the swelling and tenesmus. The evacuations soon become painful and increased in number and the fæces are streaked with mucus, blood, and pus. There is apt to be also a train of symptoms referable to the bladder, and to the generative organs, such as painful micturition, cystitis, and leucorrhœa.

With these local symptoms there may be, as in the case reported, more or less constitutional disturbance, fever, and loss of appetite. As the discharge from the inflamed surface increases in amount, the desire to empty the rectum produces more frequent evacuations, so that while at first the fæces only are stained with pus and blood, later the evacuations consist entirely of the muco-purulent matter, and the anus may become excoriated by the discharge.

In the chronic form the symptoms are all less marked. The diarrhœa may alternate with constipation, and the discharge will occur only at the time of defecation. This condition may last for years. An examination of the rectum during the acute stage of proctitis will generally cause considerable pain. The rectal mucous membrane will be found intensely congested, and the temperature, as shown by the thermometer

---

[1] Dict. Encyc. des Sci. Méd., Art. Rectum.

or even by the finger, will be increased. In the chronic stage, the solitary glands may occasionally be **recognized** as small round prominences in the substance of the **mucous membrane.**

Proctitis is **generally found associated** with stricture **of the rectum** and is secondary **to it. In these cases the** mucous membrane below the stricture will **be found congested and** covered with **pus** or bloody mucus, while above it is **eroded and** destroyed; sometimes **only** superficially, **at** others for **its entire depth.** In such **cases the** other layers will be found hypertrophied, **especially** the circular muscular layer.[1]

The causes **which** may produce proctitis are numerous. **It** may result by direct propagation and continuity of **surface from** inflamed hæmorrhoids **or** prolapsus; or from **any** erosion about the **anus** such as a mucous patch or eczema. It may be, and **often** is caused by **the** presence of foreign bodies or of hardened fæces and indigestible remains **of food** which act as foreign bodies; and by irritating suppositories, injections, or medicinal applications. As in the case given above, it may be caused by the abuse of drastic purgatives such as aloes, gamboge, or even rhubarb in excess. It has been seen to result from prolonged sitting upon a cold or wet seat, and **when found** in children it will generally be due to **the** presence of worms. **It may be a** symptom of gout (Esmarch, Bushe) alternating with **the manifestation of the** disease in its usual seat, and there may be **a true diphtheria of the rectum,** as there may be of the vagina, and the **formation of a membrane similar to** that seen in the throat. Again the disease may **result both in men and women** from the habit of passive pederasty, **and in** such **cases** may **be due either** to mechanical violence or to the inoculation with gonorrhœal **pus. A true** gonorrhœa **of** the rectum, whether caused in this **way or by direct inoculation in women by** pus which is passing over the **anus from the vagina, is very rare.** Tardieu[2] has never observed a single **case. Gosselin**[3] saw only one case at Lourcine in three years. Rollet[4] reports a **case** caused by direct inoculation from the penis to the rectum in a patient who was in the habit of using a finger in the anus to provoke a passage. A. Bonnière[5] found it very difficult to inoculate the rectal mucous membrane with gonorrhœal pus placed upon it through a tube, though the anus was easily affected. On the other hand, Requin[6] believes it almost sure to follow passive pederasty with a person **suffering** from gonorrhœa. The diagnosis of gonorrhœal proctitis will rest **upon the** amount and purulent character **of the** discharge, and **upon the** existence of gonorrhœa

---

[1] Dict. Encyc. des Sci. Méd., Art. Rectum.
[2] Etudes Médico-légales sur les **Attentats aux Mœurs,** 4th ed., 1862, p. 179.
[3] Arch. Génl. de Méd., 1854.
[4] Dict. Enc. des Sci. Méd., Art. Rectum.
[5] Récherches Nouvelles sur la Blennorrhagie, Arch. Génl. de Méd., Apr., 1874.
[6] Eléments de Path. Méd. Rectite, t. i., p. 729.

of the vagina in women; or the confession of intercourse with a diseased person, in men.

The treatment of proctitis consists first of all in an endeavor to discover and remove the cause of the congestion, be it what it may. In the acute stage, the pain and tenesmus may be overcome by warm baths, and anodyne injections of starch-water with a few drops of laudanum. The bowels should be kept open by laxatives such as castor oil or preferably the saline cathartics in small doses. The patient should also be confined to the bed, and placed upon a diet chiefly of milk. In the chronic stage, astringents are indicated; such as alum and tannin, and to these may be added suppositories of iodoform (gr. v.), and the same rules with regard to rest and diet should be observed.

CHAPTER V.

ABSCESS AND FISTULA.

Abscess divided into Superficial and Deep.—Superficial Abscesses.—Simple Furuncles; Causes; Characters; Results; Treatment.—Suppuration of External Hæmorrhoid.—Suppuration of Internal Hæmorrhoid.—Diffuse Inflammation of Subcutaneous Tissue, Causes; Symptoms; Treatment.—Form of Incision.—Deep Abscesses.—Divided into Abscess of the Ischio-Rectal Fossa and of the Superior Pelvi-Rectal Space.—Causes; Symptoms; Diagnosis.—Dangers of Deep Abscess.—Formation of Deep and Extensive Fistulæ.—Horse-shoe Abscess.—Idiopathic Gangrenous Cellulitis.—Reasons why Abscesses **do not** Heal Spontaneously.—Prognosis.—Treatment.—Incisions and Subsequent Treatment **of Deep Abscesses.**—Incontinence of Fæces.—Relief of **Incontinence resulting from** Operation.—Fistula.—Generally due to Abscess.—Divided **into Superficial and Deep.**—Complete Fistula.—External Fistula.—**Internal Fistula.**—**Description of** Superficial Fistulæ.—How to Detect **an Internal Opening.**—**Location** of Internal Opening.—Description **of Track of Fistula.**—**Symptoms of** Superficial Fistula.—Deep Fistula.—Fistula **with Numerous External** Openings.—Blind Internal Fistula.—Ulceration of Rectum Causing Internal Fistula.—Treatment.—Spontaneous Cure.—Advisability of Operation.—Fistula in Relation **to** Phthisis.—Contraindications to Operation.—Treatment by Cauterization.—The Ligature.—The **Elastic** Ligature.—Galvano-Cautery.—How to Pass Ligature.—Incision.—Description of Operation.—Author's Knife for Fistula.—Division of Deep Tracks.—Treatment of Track running up the Bowel.—Treatment of Blind External Variety; of Horse-shoe Variety; of Fistula with Numerous External Openings.—Dressing after Incision.—Packing the Incision.—Hæmorrhage **in** Operation.—Treatment of Blind Internal Variety.—Incurable Fistulæ.—Treatment of Deep and Extensive Tracks.—Fistula with Stricture.

Abscesses in the region of **the anus** and rectum are best classified according **to** their anatomical **location** into superficial and deep. Of each of these **there** are several different varieties.

Considering **first** the superficial variety, the simplest form **will be** found to be that **which** involves the skin of the margin of the anus alone, and which generally **originates in one** of the minute glands of the part. Such an abscess **or, furuncle, for it is** really only **a** furuncle, may be due to traumatism, **or to any irritation, such as** the use of improper paper after defecation, prolonged walking or horse-back riding, **a** menstrual discharge, **or** a discharge due to diarrhœa or dysentery.

This **form** of disease is always distinctly circumscribed, is generally about the size of an almond, is found by preference in robust persons, more often in men than in women, seldom in old people, and almost never in children. It generally goes on rapidly to suppuration, breaks spontaneously on the cutaneous surface, and heals without the formation of fistula, though in cachectic or phthisical patients it may pursue a contrary course, the skin over it becoming thin and violet colored, and finally rupturing, leaving a permanent subcutaneous fistula.

The treatment of such an abscess consists chiefly in the attempt to avoid the formation of a fistula, and the best means for accomplishing this end is an early incision as soon as suppuration appears inevitable. Resolution is hardly to be expected, but it may be sought for by the use of laxatives, rest in the horizontal posture, and the application of a bladder of ice. The incision should be large enough to allow of the free exit of pus, and after it has been made, the part may be poulticed for a day or two, and the abscess cavity then dressed with lint, care being taken to keep the lips of the incision separated.

Another frequent cause of superficial abscess is the acute inflammation and suppuration of an external hæmorrhoid, which generally comes on after an attack of constipation and straining at stool, or may be due to the same causes as the last. The suffering caused by such a condition, as by the one last described, is out of all proportion to its apparent importance, and is sufficient to incapacitate a person of sensitive organization from all accustomed duties. The remains of former external hæmorrhoids are always liable to this accident, and by the proper abortive treatment, the inflammation may sometimes be overcome without suppuration. If, however, suppuration appears to be inevitable, a small sharp-pointed bistoury should be quickly passed through the little tumor.

There is also a form of superficial abscess which lies nearer to the mucous membrane than the skin, and is due to the acute inflammation of an internal hæmorrhoid, either just at the verge of the anus or within the sphincter. This is in reality a circumscribed phlebitis in a venous pouch which is shut off from the general circulation. A circumscribed, tense, exquisitely painful tumor is formed, varying in size from a grape to an almond, which, after a few days of suffering, ruptures spontaneously, and allows the escape of a small quantity of pus. Such an abscess, when within the bowel, is always liable, as will be shown later, to result in the formation of a blind internal fistula if left to its own course, and should, therefore, be treated by early incision.

There is still another variety of superficial abscess, more serious in its consequences than those already described, for the reason that it affects the subcutaneous tissue and not the skin, and is diffuse and not circumscribed. The causes of this variety of abscess are the same as of those already mentioned, though traumatism plays, perhaps, a more important rôle. Falls, kicks, horse-back exercise, and violence in the use of the syringe are its most frequent antecedents. Surgical interference with the rectum, as in the removal of a hæmorrhoid, may also be followed by this form of abscess, and it may arise from the perforation of the wall of the bowel just above the sphincter, by an ulceration of any kind, generally, however, that due to a foreign body. It has also been known to follow the suppuration of an internal hæmorrhoid.

The symptoms of this form of disease vary greatly in different cases. In cachectic persons, pus may form in large quantity, and break into the bowel without the knowledge of the patient, and a blind internal fistula may result. The diagnosis is generally easy. There will be the usual pain, tenderness, and swelling; and if the pain be not too severe to admit of the attempt, fluctuation may be obtained by introducing one finger into the rectum, and making counter-pressure with the other hand outside.

There is little use in hoping for resolution in an abscess of this kind, and all active attempts to cause it will be found to do harm, rather than good. The proper treatment is an early free incision. If the incision be made early, it may in itself have an abortive action, and under such circumstances it need not be very large. If pus has already formed, or the skin has begun to grow thin over the abscess cavity, the incision should be free enough to allow of the easy escape of the contents, for in this way only can the formation of a fistula be avoided. In such a case, drainage should be resorted to after the incision, and every effort should be made to secure healing from the bottom of the cavity.

When the incision is made in the early stage of such a tumor as this, while the skin is yet hard and infiltrated, a free hæmorrhage from cutaneous vessels is not uncommon, nor on account of its antiphlogistic action is it to be deprecated. Only when it has passed the bounds of safety need any steps be taken to arrest it, and this may always be done by a careful stuffing of the incision with picked lint. A word of caution against opening such abscesses as these in the surgeon's office, and allowing the patient to walk home, may not be out of place; for a small artery may commence spurting at any moment during the active exercise.

Deep Abscess.—The deep abscesses of this region differ greatly from those already described, in their location, extent, and gravity. They may with advantage be divided into those of the ischio-rectal fossa and those of the superior pelvi-rectal space.[1]

An abscess of the ischio-rectal fossa is generally bounded by the levator ani muscle superiorly, and by the skin below, with the rectum on one side, and the adjacent portion of the pelvis on the other. An abscess of the superior pelvi-rectal space, on the other hand, originates in the lax connective tissue around the upper portion of the rectum above the levator ani muscle. It may assume vast proportions, blending

---

[1] Richet: Traité d'Anat. Méd. Chir.

laterally with the subperitoneal connective tissue of the iliac fossa, and burrowing in almost any direction in the true pelvis.

The causes of deep rectal abscesses are various. Traumatism is perhaps the most frequent, and the injury is generally internal, rather than external, and is caused by the point of a syringe or a foreign body, rather than by kicks and falls. Foreign bodies, such as fish-bones, may pass entirely through the rectal wall, and be found loose in the cavity of the abscess they have caused. Such an abscess may also be due to the injury inflicted by the fœtal head in parturition, and in such a case, the diagnosis may be difficult to make from a puerperal inflammation, due to blood poisoning and involvement of the lymphatics. They may also be secondary to diseases of the urinary organs, such as acute inflammation of the prostate, or a rupture of the urethra, and extravasation of urine; and they may result from rupture, ulceration, or perforation of the rectal wall, in connection with stricture.

This explains partly, though not completely, the frequent coexistence of stricture and numerous fistulæ; for a stricture may act as the exciting cause of a deep abscess by the impairment of vitality and nutrition which it causes, as well as by producing a perforating ulcer above, as is proven by the fact that a great many fistulæ have their internal openings below, and not above the constriction.

Again, these abscesses may be due to a submucous inflammation, and production of pus, which first breaks into the rectum, and forms an internal fistula, and subsequently extends outwards, forming a large abscess; or they may be due to an acute phlebitis, or to faulty nutrition and a generally vitiated state. Finally, they may be in their origin entirely disconnected with the rectum, and due to disease of some neighboring part, or to necrosis of some adjacent bone of the pelvis or spine.

Symptoms.—In an abscess of the superior pelvi-rectal space the symptoms are often obscure and far from characteristic. There is more or less vague pain in the pelvis and lumbar region, which is seldom intense and is generally increased in defecation. Fever may be entirely absent, is seldom continuous, and chills are only occasionally met with when pus is formed. In addition there is more or less headache and general malaise.

An abscess of the ischio-rectal fossa may at its commencement be accompanied by the same symptoms, but, later, the skin becomes hard, red, and œdematous sometimes over a large portion of the corresponding buttock, the pain is very severe, and rectal touch impossible. The general symptoms are those of any acute inflammation. In abscess of the superior pelvi-rectal space, when the disease has extended to the cellular tissue of the iliac fossa, immense collections of pus may form, and this may burrow in any direction. In men it generally follows the course of the bowel, involves secondarily the ischio-rectal fossa, and makes its way through the skin at some distance from the anus. In women it is more apt to pursue a contrary direction and may appear on the surface in the region of the crest of the ilium or in the groin. An abscess of the ischio-rectal fossa may tend to discharge its contents upwards toward the superior perineal region, being less confined by fascia and muscle in this direction. In this way the prostate and urethra may be implicated, and the signs of retention of urine may be joined with those which point more directly to the rectum.

The pus from such an abscess, in time, generally breaks on the cutaneous surface and forms one or several permanent fistulous tracks. The pus from a pelvi-rectal abscess not infrequently makes its way into the rectum and is discharged with each act of defecation; before the fæces when the opening is near the anus, after them when it is above the rectal pouch. It may, however, rupture into the vagina, bladder, uterus, or peritoneum, but these internal openings are not the rule, but the exception, for the pus generally finds its way to the cutaneous surface, and fistulæ result as with ischio-rectal abscesses. Either variety may cause fistulous tracks upwards into the true pelvis, downwards into the perineum, or outwards into the thigh. When the pus reaches the rectum it may burrow for a considerable distance in the submucous connective tissue of the bowel, and separate the mucous membrane from its attachment before perforating it. In this way two large abscess cavities may be formed communicating with each other by a narrow orifice.

What is now generally known as the horse-shoe abscess or fistula is due to the formation of an abscess in each fossa and the communication of the two behind the rectum through the substance of the sphincter muscle at its attachment to the coccyx. Such an abscess generally has one opening into the bowel and two on the cutaneous surface, though the latter may be single also. By manipulation the pus may be made to cross from one fossa to the other imparting a characteristic sense of fluctuation.

There is a form of gangrenous cellulitis which sometimes affects the ischio-rectal region. It is a rare disease, and is generally idiopathic. In it there is no pus formed, but the cellular tissue and the skin over it become necrosed and slough in large, black masses. The adjacent portion of the rectal wall may be involved and the rectum be laid open for a considerable extent. The disease is attended with fever and great prostration; the tendency to relapse and extension is marked, and the cavity left after separation of the slough closes very slowly.[1] This form of disease may be fatal.

The reasons why abscesses in this region so seldom heal spontaneously are to be found in the anatomy of the part, and the fixedness or mobility of the walls of the abscess cavity. In the ischio-rectal variety the skin is

---

[1] A Clinical Lecture on Idiopathic Gangrenous Cellulitis around the Rectum. Furneaux Jordan, Brit. Med. Jour., Jan. 18th, 1879. Also, Jackson, Brit. Med. Jour., Feb. 8th, 1879.

hard, thickened and lardaceous; and from its rigidity cannot yield its position to allow of healing. The walls of the abscess higher up in the pelvi-rectal space, on the contrary, move with the varying fulness of the abdominal or pelvic organs with the incessant action of the levator ani, and with the fulness or vacuity of the abscess cavity, which depends on the intermittent discharge of pus through its small opening.

Diagnosis.—The diagnosis of these conditions should be made with great care, for on a correct appreciation of the extent of the disease will depend the prognosis and treatment; and this class of fistulæ are not always proper cases for operation.

A fistulous track communicating with a pelvi-rectal abscess may generally be recognized by its length and by the amount of tissue between it and the bowel, which may easily be estimated with one finger in the rectum and a probe in the track. The probe does not approach the rectum, but either runs parallel with it, or recedes from it. The flow of pus from the opening is also apt to be intermittent and to occur at the time of defecation, being caused by the same muscular effort. Sometimes, when the cavity has not been recently emptied, a soft tumor may be felt by rectal touch, and pressure upon it may cause a flow of pus. With the pus bubbles of gas may also appear, but in a large abscess in the neighborhood of the bowel this is not a proof of an internal opening, but may be due merely to the proximity of the intestine.

Prognosis.—The prognosis is necessarily grave. In the beginning the patient is exposed to all the dangers of pyæmia, peritonitis, and phlebitis; and should the abscess go on to a favorable termination in an external opening, there is still the dread that it may at any time seek another opening toward the peritoneum with a fatal result. The immediate results being favorable, the ultimate ones may still be disastrous; being those which always attend upon prolonged suppuration—visceral complications, amyloid degeneration of the liver and kidneys, and tubercular deposits. In the comparatively small number of cases of pelvi-rectal abscess in which healing occurs, the patient still has to meet the results of extensive cicatricial contraction. These may be stricture on the one hand, or incontinence on the other; with the subacute inflammatory tendency which is always apt to attend upon a cicatrix at the anus and cause pain and uneasiness. In females especially, such a cicatrix may be the cause of grave trouble with the genito-urinary canal.

Treatment.—It may be considered as a rule to which there are few exceptions, that an acute inflammation in this region will go on to suppuration; and hence that antiphlogistic measures adopted with a view to securing resolution are useless. Early incision is, therefore, the only rational treatment, and, where properly performed, this may result in cure without the formation of fistula. Allingham[1] goes so far as to say

---

[1] Op. cit., p. 16.

that by this means he can almost guarantee that there shall be **no fistula**. The incision should radiate from the anus to avoid as far as possible the section of nerves; **and** should **be free** enough to secure the escape **of pus,** not only at the time, **but while the** abscess is healing. If there be burrowing in any direction, **the incision** should be prolonged to correspond; and the finger **should be passed as far as** possible **into all** parts **of the cavity** to break **down all** partitions. **The** wound should then be stuffed with lint **wet with** carbolized **oil, and a** drainage **tube** inserted. The secret of **success will be** found **to lie in** securing **a free** outlet for pus, and thus **preventing** burrowing.

These abscesses should not be **laid open into the** rectum—a point which **is** generally misunderstood in **practice, because of the** confounding of an abscess which which may ultimately **result in a fistula** with fistula itself. The treatment is that of abscess, **and not that of fistula,** and is especially directed toward the prevention of fistula.

Even should the abscess have already opened into the bowel, healing **may** still be secured by following this line of treatment, with **suitable means** for keeping the rectum empty, and a laying open of the lower end **of** the rectum **may be** avoided. After a fistula is fully formed and all attempts at closure **have failed,** the usual operation of dividing the track into the bowel may **be necessary, but it** should always be undertaken with the expectation of disastrous **consequences to** the retentive powers of the sphincters. **Incontinence to a greater or less** extent is almost sure to follow such **a free division of both sphincters and** of the bowel above them.

Incontinence depends more upon **division of the internal than** of the external sphincter, and is more apt **to follow a double division of** the fibres than a single one. For this reason the **surgeon** should **always endeavor** to leave a few fibres at least of the **internal** muscle **in** any operation, and the incision should always be directly and not obliquely across the fibres of the muscle. It is also well to remember that incontinence is always more apt to result from division of the muscles in the female **than in the** male.

Even when incontinence has resulted, **the case** may be capable of relief in this regard by an operation with **the** cautery, which will be described in speaking of prolapse. I have **seen** marked benefit in this sad condition **result** from this simple **operation** combined with the persistent use of a rectal bougie and such other measures as are calculated to increase the power **of the** sphincter, and I am much less inclined **to** despair of giving **relief in these cases** than formerly. In **one case** sent me by Dr. McCready, **of New York, in** which a considerable degree of incontinence resulted from **an** ischio-rectal abscess, this mode of treatment patiently followed for some months has almost entirely relieved the condition; so that where solid fæces at first **escaped** him there is now a good degree of contractile power, and the patient is only troubled with an occasional discharge of the rectal mucus in small quantity.

Fistula.—A fistula which is not due to a perforation of the rectal wall from within is the result of a previous abscess, and, therefore, in enumerating the causes of abscess those of fistulæ have also been given. Like the abscesses from which they arise, they may well be divided into superficial and deep; or into those of the anus, which are subcutaneous, and involve at the most only a few fibres of the external sphincter, and those of the rectum and pelvis, which open into the bowel at a higher point. Both the superficial and deep may also be divided into the complete, or those which open both on the skin and into the bowel; the external, which open only on the skin, and the internal, which have an opening only within the bowel (Fig. 27).

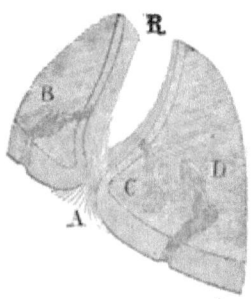

FIG. 27.—Varieties of fistula (Gosselin). A, anus; R, rectum; B, complete fistula; C, blind internal fistula; D, blind external fistula.

Superficial Fistulæ.—On account of the special laxity of the submucous connective tissue in this region, already noticed, abscesses show little tendency to spontaneous closure, and fistula is the common result when left to their own course. In the subcutaneous fistula, the external orifice may be at some distance from the anus, or in the radiating folds. It may be so small as to escape the eye in a cursory examination, unless a drop of pus chance to be squeezed out of it by the pressure of the fingers in pulling open the parts; and when discovered, it may not admit the end of an ordinary probe. The surgeon should, therefore, always be provided with a probe of small size and of pure silver, which admits of being readily bent, for using in these examinations.

The presence of more than one external orifice is rare in subcutaneous fistulæ; and an internal opening will be found in the great majority of cases, if properly searched for. The only way to settle the question of the presence or absence of an internal opening in any doubtful case is by opening the anus with a speculum and injecting milk through the external orifice. In the vast majority of cases the milk will be found in the rectum, and the internal orifice will be found just within the external sphincter.

It may sometimes be felt in this location by the educated finger as a small tubercle, and in other cases it is marked by a distinct loss of substance. In some the internal opening will be found in the radiating folds entirely below the fibres of the sphincter, and in others it may be much higher up the bowel.[1]

The internal orifice does not in all cases mark the superior limit of the fistulous track. This may run several inches up the bowel under the mucous membrane, when the internal orifice is just within the external sphincter (Figs. 28, 29).

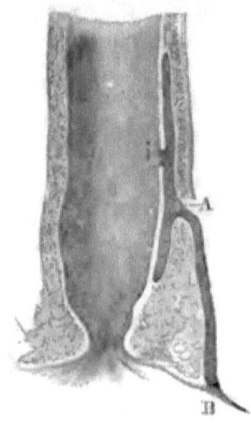
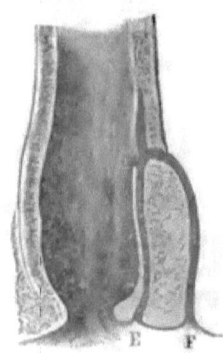

Fig. 28.   Fig. 29.

Figs. 28, 29.—Fistulæ with double tracks (Mollière).

Fig. 28.—AB, deep submuscular track resulting from an ischio-rectal abscess. AI, submucous track running up and down the bowel.
Fig. 29.—DE, Subtegumentary and submucous fistula with internal and external opening. DF, deep submuscular track, having same internal, but separate external opening.

The track of a fistula is sometimes straight, extending directly from one orifice to the other; in other cases a track, properly speaking, does not exist and both orifices open directly into the original abscess cavity. If the external orifice be very small, the cavity may at any time become distended with pus and give rise to all the symptoms of a fresh abscess, till the pus finds an exit either through the old opening or a new one. The external orifice of a true, straight fistulous track is generally large and sometimes free enough to allow of the escape of gas. The track is lined with lardaceous tissue the result of chronic inflammation, and in this may be found numerous blood-vessels of new formation. This tissue, by preventing all contact of the walls, necessarily prevents healing. On the other hand, the track is sometimes lined with healthy granulations which

---

[1] Ribes: Recherches sur la situation de l'orifice interne de la fistule de l'anus. Rév. Méd., t. i., 1820.

are capable of being formed into new tissue, and for this reason a fistula will sometimes heal spontaneously.

The history will sometimes afford valuable information as to the general character of the case. The history of a slight abscess and the escape of a small amount of pus generally means an insignificant fistula with external and internal openings near the margin of the anus; while, on the other hand, the history of a prolonged inflammation and a free discharge of pus means a large abscess cavity mounting to a considerable height, and with its internal orifice at a correspondingly high point.

The symptoms caused by this class of fistulæ vary greatly. At first they are those of the abscess in which they originate. After that the one great symptom is the incessant discharge, sometimes slight, at others abundant; sometimes purulent, at others serous; always fœtid; sometimes containing fæces and gas. It is generally the stoppage of the discharge and the consequent filling of the track or abscess cavity which induces the patient to seek the surgeon. Besides the discharge there may be no symptoms at all, or there may be more or less uneasiness in the part, and pain on defecation, with the constipation which arises from the fear of a passage, and the symptoms to which it gives rise. Such a state of affairs may exist for many years without aggravation, or without causing the patient to seek relief.

Deep Fistulæ.—Deep or submuscular fistulæ differ greatly in their extent and gravity from those last described. In them the track is large and often double or branching, and the external opening may be far away from the anus. The whole perineum and gluteal region will sometimes be found to be perforated by openings. In a case sent to me by Dr. R. W. Taylor, of New York, I easily counted between twenty and thirty of these discharging points, and the whole perineum and surrounding region were hard, brawny, and infiltrated. The man, under the pressure of his sufferings probably, had become a confirmed opium eater and was in a deplorable plight.

The track in some of these cases has been known to take a remarkably irregular course. Sir A. Cooper[1] mentions an autopsy where a fistula opened in the groin, followed the course of the spermatic cord, and ended in what seemed like an ordinary fistula in ano; and cases in which the pus has burrowed under the gluteal muscles and finally opened in the thigh or even nearly at the popliteal space, are not uncommon.

Blind Internal Fistula.—Fistulæ with internal openings alone have a somewhat special pathology. When caused by an abscess it is generally by one of the deep variety which has opened into the rectum high up and continues to discharge in this way. The abscess causing such a fistula may, however, be a small submucous one, and the symptoms will then be pain, spontaneous discharge of pus from the bowel, and subsequently

---

[1] Lec. on Prin. and Prac. of Surg., with notes by Tyrell, t. ii., p. 326.

pain after defecation resembling that of a fissure. **There** is another, and perhaps more common class of internal fistulæ in which the opening is not the result of the breaking of an abscess, but in which the opening is first formed by ulceration **and the track** is a secondary consequence. This pathological **fact was proved by the** well-known investigations of Ribes, who believed **that the internal orifice** was always the first formed, but here he was **undoubtedly in error.**

A circumscribed **ulcer which shall** perforate the mucous membrane and result in internal fistula may be due to several causes: to the inflammation of one **of the** lacunæ just above the sphincter **from** the lodgment within it of a **particle of** hard fæces; **to rupture of an inflamed** internal **hæmorrhoid; to the** application of strong **acids** to hæmorrhoids; to operations upon the rectum generally for hæmorrhoids; **and** to the peculiar ulceration met with in tubercular patients, **but not necessarily tubercular in its** nature.

Such a condition is a very painful one. The opening which **may be** large enough to show a distinct loss of substance to the touch, catches and retains particles of fæces, causing a burning pain which may last many hours after defecation. As a result of the opening an abscess forms after a time with the usual symptoms, the induration of which may be felt externally. **When the abscess is** small and the induration not extensive a speculum examination **may reveal** the ulcer; but the fistulous track and abscess may escape—a mistake **which will** render all treatment directed toward the cure of the ulcer **of no avail.** There may indeed be several ulcers, only **one** of which has **a fistula connected with it.**

Treatment.—A **fistula** may heal **spontaneously or after a very** slight excitement to reparative action, such **as the mere** passage **of a probe in** making an examination. It has been mentioned that **the track is** sometimes lined with healthy granulations, and **that these may** result in new tissue which shall close it; but this can never occur after the usual infiltrated tissue has once been formed, which is seen in all old cases. Allingham[1] relates several cases of spontaneous cure, and estimates the proportion in which it may occur as about one per cent.

Setting aside these cases, we are at once brought to the question which will often be asked by the patient, and which the surgeon may not always be able to answer to his own satisfaction, whether or not it is always best, or even safe to try and cure a fistula. In certain cases of Bright's disease, cancer, cardiac and hepatic affections, etc., all surgical interference may be plainly contra-indicated; but the question is most apt to arise in connection with pulmonary affections. There can be little doubt that phthisical patients are especially predisposed to this affection, and the reason is probably in great measure a mechanical one, depending upon a loss of fat in the ischio-rectal fossa and a resulting

---

[1] Op. cit., p. 24.

loss of support to the hæmorrhoidal veins. From this there results a venous congestion and final dilatation or rupture of the vessels, which, with the cough and concussion, leads eventually to abscess.

I believe it to be a safe rule to operate upon phthisical patients as upon others, being led by the idea that one exhausting disease—phthisis—is better than two—phthisis and fistula. I have many times followed this rule with happy results as to improved general health after the cure of the fistula. Once only has it happened to me to see the cure of a fistula followed by a marked increase of the lung trouble, and even in such a case the relation between cause and effect cannot be established. I have also yet to meet the first case which, under suitable and careful general and local treatment, refused to heal after the operation. There are several rules which should be carefully regarded in this class of cases, however. No cautious practitioner would think of operating either in a very advanced or a rapidly advancing lung trouble. Cough, when violent and frequent, is also a decided contra-indication, interfering, as it does very certainly, with the healing of the wound. The following case will perhaps illustrate the line of treatment to be followed in a general way.

CASE V.—A theological student, aged twenty-eight, applied to me from a neighboring city for relief from a large subcutaneous abscess with an internal opening within the sphincter, and an external one at some distance from the anus. The probe could easily be passed a considerable distance in every direction beneath the undermined skin. The discharge was very profuse. This condition had existed for several months; the patient was much reduced in weight, there was consolidation in the apex of one lung, with a history of phthisis and hæmorrhages.

The internal and external orifices were connected by an incision involving the external sphincter, and the abscess cavity was laid open for a distance of four inches along the perineum, and dressed with picked lint. After a fortnight's rest in his room, the patient being partially dressed most of the time, and spending his days on the lounge or easy chair rather than in bed, reparative action seemed to come to a standstill, and with careful directions as to dressing the wound, I sent him off into the mountains. He reported at my office after an interval of three months spent in the woods, during which time he had frequently been on horse-back several hours at a time. The change in his general condition was very remarkable, he having gained nearly twenty pounds in weight. The abscess cavity was nearly, but not quite closed, and again he returned to the country, with the understanding that he should report in the city every fortnight. In just six months from the operation the wound was entirely healed, there had been no exacerbation in the lung troubles, and the patient was in better general condition than for years previous.

In cases of fistula in phthisical patients, the sphincters should be interfered with as little as possible, as they are apt to be weak at the best. The internal orifice is apt to be large and ragged, and the external may be the same. The tendency to undermine the skin is always marked, and the discharge is generally thin and watery.

Cauterization.—It **is not** necessary even to enumerate the various substances which from time out of date have been advocated for this purpose. Among those for which good results have been claimed, iodine holds the first rank.[1] There is no doubt that that by its use certain fistulæ and abscesses **may** be made to heal, but the plan is uncertain **and** not very reliable.

The operation consists in closing the internal opening **with** a finger in the rectum and then injecting the fluid with a small syringe through **the** external orifice, using pressure enough on the track to bring the fluid into contact with every part. In the place of iodine, nitrate of silver either in solution or fused upon a probe; the tincture of iron; **or** carbolic acid, may be used. The galvano-cautery wire, **or a simple hot iron** may also be employed to modify the track; and a **fine sea-tangle** tent carefully introduced will sometimes set up reparative action. By any of these means failure **will be** the rule, but success may occasionally be secured after faithful trial.

The ligature.—**Under the head of** the ligature may be included also its different modifications—*écrasement* **linéaire***,* elastic ligature, **and the** galvano-cautery **wire.**

The method of **cure** by **the simple ligature** consists in passing a strong cord through the fistula from the **external opening,** through the internal, **and** out at the anus, **then in** tying **the two ends,** and tightening the loop from day to day till the tissue included **is divided.** The **operation is** generally effectual, but it is also painful, tedious, **and uncertain.** It is a substitute for **the** knife, a concession to the fear of being **cut,** and it is free from hæmorrhage; but it only accomplishes in the **end,** and sometimes after weeks of suffering, what the knife accomplishes in a moment; and except for the single fact that by its use hæmorrhage may be avoided it would bear no comparison with the latter.

If this mode of treatment is for any reason decided upon, there are certain modifications of the operation which are much to be preferred to the simple cord. The method of immediately cutting through the tissues by attaching the ends of the cord to the handle of an écraseur (*écrasement linéaire*) is a much better way of attaining the same end which is due to Chassaignac. **There are,** however, two methods of dividing the tissues which are **still** better than this—one by the galvano-cautery wire, the other by the elastic ligature. The galvano-cautery wire has the same advantage over the **knife** as the ligature **in** preventing hæmorrhage; and **it is not particularly painful** in its application. In using it, as little heat

---

[1] Boinet: "Traité d'iodothérapie."

should be used as is possible to slowly divide the tissue, or hæmorrhage may occur and all its advantages be lost. On account of the expense of the apparatus, and the skill required for its management, this method has never become very popular with the general practitioner, but it is very successful in the hands of a few.

Probably the best of all methods next to the knife is that of the elastic ligature. The cord in this case is of solid rubber which is drawn as tightly as possible—the tighter the better—and then held on the stretch by slipping a soft metal ring over the ends and squeezing its two sides together close up against the tissues. In the course of a few days the ligature will be found to have cut its way through the included tissues, the time depending on the quantity and quality of the mass to be cut.

Various devices have been recommended for facilitating the passage of the ligature. The best known is Allingham's, Fig. 30. In using it, re-

FIG. 30.

remember that it is intended to draw the cord from the rectum out of the external orifice, and not *vice versa*. Helmuth, of New York has modified the instrument and I think with advantage, Fig. 31, but the least elaborate

FIG. 31.

and most effective instrument for the purpose in my own hands is a simple silver, eyed probe which is threaded with the elastic cord and then passed from the external orifice through the track and out at the anus. I once had an awkard accident with Allingham's instrument which broke in my hand in a moderately deep and hard track.

After the ligature is in place, the patient is allowed to go about his ordinary pursuits, and this is claimed as one great advantage of this method. I have never been able to understand why cutting with a string should permit of any more liberty than cutting with a knife. The patient, it is true, will generally get well if he goes about while the string is doing its work, and so he will after the operation with the knife; but in both cases the healing will be facilitated by rest. The operation is said to be painless. I have not found it so. **Both** the passing of the cord, and its tension for the first forty-eight hours have been bitterly complained of in some of my own cases. The healing has already begun before the **ligature** comes away; but **with the** dropping out of the cord there will **sometimes** be found a **considerable** slough **in the** line **of** strangulation **which** may require some days for its separation.

The elastic ligature has undoubted advantages over the knife in cases where the latter is contra-indicated by the fear of hæmorrhage; as in a fistula running high up the bowel where hæmorrhage may be **a serious** matter; or where the patient refuses to submit to a cutting operation. Of all the methods **of** cutting with a string **it is the** best, but after all, it is only a substitute for the knife, and for my own **part I must plead guilty to** a preference for cutting with a knife when cutting **is necessary.**

Incision.—The operation for fistula by incision may **be** greatly facilitated by the observance of several minor details. In this as in **other** operations on the part, the bowels should be thoroughly emptied **on the** previous day. Care must be exercised, lest in the endeavor **to free the** alimentary canal a diarrhœa be excited, for this will prove anything **but an** agreeable complication for the operator. In all cases in which the track is of any considerable depth, or in which on account of the sensitiveness of the patient **the surgeon** has not been able to assure himself of the exact extent **of the disease and the absence** of any side tracks **or** diverticula, ether **should be given and the** anus gently and completely dilated before the operation. **It is only in the simplest cases** that the incision **may** be made without ether, **and then the best chance** of a thoroughly satisfactory exploration is **missed, and the way is opened** for **an** incomplete and therefore unsuccessful **operation.**

With regard to position the operator may choose **between** placing the patient on the affected side or on the back. In women **the** former is generally preferable. A director with probe point should **be** passed through the external orifice into the bowel and brought **out at** the anus by the index finger of the other hand, which should in any case be passed into **the bowel** before the probe is inserted into the external opening. **The** track should now be carefully and thoroughly explored and its extent discovered. This should be done deliberately and without haste, and hence the advantage of an anæsthetic. When the patient is not etherized there is always a temptation when the end of the probe is felt against the finger in the rectum, to bring it out at the anus, follow it instantly with the bistoury, and quiet the sufferer with the cheering assurance that all is finished; but a seemingly insignificant case may have a deep track connected with it which must be divided before a cure can be effected.

Having by careful **examination decided** just how **much** cutting is to done, the choice of the instrument rests with each operator. In simple cases where the track is superficial, I frequently use a knife of my own invention which (like most new inventions) I found after having it manufactured, exactly resembled those in use in the fourteenth and fifteenth century, though somewhat smaller and less formidable in appearance.¹ It is represented in Fig. 32, and consists of a flexible probe at the end of

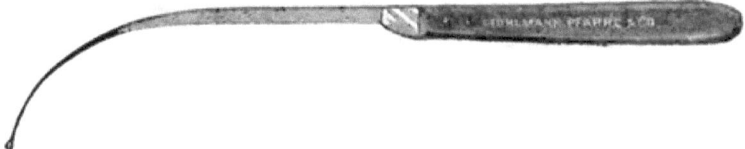

FIG. 32.—Author's Fistula Knife.

a curved bistoury. The probe point should blend as gradually as possibly with the cutting edge, as anything like a shoulder at the junction of the two interferes greatly with its use. I have thought that in suitable cases the operation was rendered more speedy and less painful by the use of this combined instrument; but it is not well adapted to those cases in which the track runs any distance up the bowel; and where the patient is etherized it has no advantages over the director on which the bistoury is generally passed. It is especially adapted for operating without ether.

In subcutaneous fistulæ the track should be divided from the external to the internal orifice. If there be at the same time any undermining of the skin with tracks leading off in different directions, these also should be laid open, so that all may be converted into an open wound. For deep fistulæ the knife or scissors should be strongly made, for it is not a very difficult matter to break an ordinary scalpel in a deep fistula. A

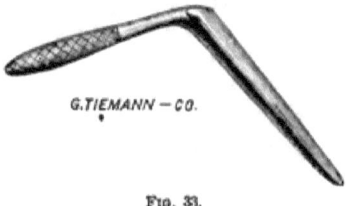

FIG. 33.

heavy steel director may also be snapped in an attempt to bring the end out of the anus preparatory to making the incision; and should the internal orifice be high up, and the external at some distance from the anus, so that the amount of tissue to be divided is large, it is often better to use the wooden gorget to guard the opposite side of the rectum and dispense with the director after the knife has been passed. (Fig. 33.) The end of the knife may be firmly fixed into the wood and both with-

---

¹ I am indebted to my friend, Dr. James L. Little, for calling my attention to the plates in Heister's Surgery, showing these instruments.

drawn simultaneously, or the incision may be made by cutting on the gorget. Allingham prefers a pair of spring scissors, one blade of which runs in a director the groove of which is more than a semicircle, for cutting deep tracks. (Fig. 34.)

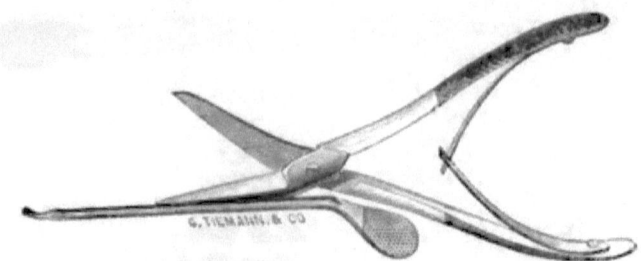

Fig. 34.—Allingham's **Spring-Scissors for Fistula.**

Some difference of opinion exists among different writers as to the proper method of treating the track that will often be found running along the bowel above the internal orifice, and directly contrary opinions are taught as to the necessity for its complete division. The operation is of course rendered more severe by the division of such a sinus in addition to the fistula, and the danger of hæmorrhage is increased; but one can never be sure that the operation will be successful when such a track is left, though no doubt many cases have turned out well. With regard to hæmorrhage in such cases, it will be found that the sinus has generally burrowed under the mucous membrane, and that the vessels have remained in the deeper layers of the bowel, so that the division of the sinus does not of necessity involve that of any large vessel, though it extend far up the bowel.

Many of these sinuses may best be divided with the scissors, and the hæmorrhage, if it be profuse, dealt with according to the rules already given. If, however, hæmorrhage be feared beforehand, the track may be divided with the écraseur, or a small canula may first be passed, through this a wire, and finally by means of the wire an elastic ligature.

When no internal orifice can be found, but the mucous membrane feels undermined, and the probe can be felt by the finger in the rectum, separated only by a thin layer of mucous membrane, it is a good plan to force an internal opening and treat the fistula as though it were complete. When there are two internal openings, both should be included in one incision. When, after the incision, the diseased integument is found to overlap the cut, and hang into it, it should be cut away, and in old tracks the healing may be hastened many days by thoroughly scraping out the lardaceous wall with the handle of the scalpel, or even scarifying it in several places, so that a healthy reparative action may be set up.

In cases of horse-shoe fistula with two **external orifices and one internal** one, it is generally best to do the usual operation on one side only, and to dilate the opening on the opposite side, so as to **allow** of free escape of pus.

Where the fistulous **tracks exist in great numbers—twenty or** thirty in some cases—two or **three operations may** be advisable at intervals, **rather** than to attempt to **do all at one sitting,** lest the patient's reparative **powers should be** unequal **to the task** thrown upon them. In such cases, **there will often** be found two or three tracks which may **be considered as primary, into which** the others run; and each of these, **with its branches, may be dealt with at** a separate operation. Many of the tracks **will be found to** run away from the bowel under the skin of the buttock **or toward** the scrotum, and these may **be** induced to heal by laying **them open,** without interfering with the **sphincters.** It will sometimes **be necessary to** divide **the** sphincter several **times,** however, before the **cure can be completed, and a certain degree** of incontinence may be **expected as a result.**

In such cases, the anal region is **generally greatly** hardened and infiltrated, and free hæmorrhage may be expected. **The best** weapon with which to meet it is the cautery of Paquelin.

In the matter of dressings after the incision, much skill may be displayed. Immediately after the operation, a dressing of dry picked lint, or if **there** be **an** abscess cavity, of lint soaked in carbolized oil, is as good **as any, and this should** be kept in place by a T-bandage. To save the **patient as** much pain **and** annoyance as possible, this should not be removed **till** suppuration has been established. Subsequent dressings **may be of the same material,** and should be changed daily. The wound **should not be tightly packed** with lint. **It will heal** from the bottom **if its surfaces are kept** apart or separated daily by the finger of the surgeon. **Care is always** necessary to prevent an immediate **union of the cutaneous edges of the incision, and I** have seen a remarkably well-**pleased patient come to** me and report **himself** as entirely cured a week after I **had divided his fistula, in consultation** with his medical attendant, **and have found on examination that the** incision had healed very **kindly by first intention** through its whole extent, and that the fistulous **track was exactly as** it was before the cut.

Healing **may be** indefinitely delayed by too frequent dressings or by stuffing the wound tightly with lint, with the intention of forcing it to heal from the **bottom.** Under such treatment, healthy granulations may entirely disappear, and the cut surface assume a mucous-membrane-like appearance, and so remain. **Standing** or walking always delays, and may sometimes entirely prevent healing.

During the treatment, the burrowing of pus and the formation of a new pocket should always be carefully watched for, and met by incision.

The **hæmorrhage** in **an ordinary operation** for **fistula** is seldom profuse enough to cause the surgeon any uneasiness, and is almost always easily controlled by packing the incision with lint, and making firm pressure with a compress held in place by a T-bandage. A free arterial hæmorrhage from a vessel well up the rectum may, however, be alarming, and if not controlled by the admission of air or the application of ice to the part, the rectum must be tamponed.

Fistulæ of the blind internal variety can only be dealt with rationally by incision. A speculum should first be introduced and a silver director bent into the form of a hook passed into the orifice and brought down to the bottom of the track; with this as a guide the fistula may be opened into the bowel.

The incision should always be continued through the sphincter and the anus, so that the wound may be properly dressed and drained; otherwise the operation will merely serve to convert a small internal opening into a larger one. An operation of this kind is always more apt to be followed by a concealed hæmorrhage into the rectum than one for a complete fistula, and this should be guarded against by a careful plugging of the wound and by the application of dry persulphate of iron if necessary.

The abscess in connection with a blind internal fistula may sometimes be detected by the induration which may be felt through the skin of the ischio-rectal fossa. In such a case, after the director has been passed into the internal orifice, a counter-opening should be made into the abscess through the skin, using the director for a guide for the incision. In this way the blind internal variety is changed into the complete, and the usual operation of division into the bowel may be performed.

After what has been said of the origin and extent of abscesses of the superior pelvi-rectal space, it is evident that there may result from them a class of fistulæ which are not to be operated upon by any of the methods we have described—fistulæ so deep and extensive as to contra-indicate all operative interference. And yet much may be done even in the worst cases of this kind, and by proper treatment some may be cured. The first attempt of the surgeon should always be toward effecting a cure without cutting the track into the bowel. External and comparatively free incisions may be made, which shall not implicate the anus, and through them drainage tubes may be passed into the abscess cavity so that it may be freely emptied. Through the drainage tube stimulating injections may be made, and the abscess treated as an abscess elsewhere would be, by rest and attention to the general health. A cure may sometimes be effected in this way in a very unpromising case.

Where the track has burrowed to great length, much may be accomplished by modified operations. In a track, for example, which has one opening near the anus and another in the middle of the thigh, a counter opening may be made between the two and the further extremity induced to heal while drainage is maintained from the middle opening by the use of injections or caustic applications. Should these means not succeed and should it appear that a free division was likely to result in a cure, the incision may be made; according to the ordinary rules of surgery. Such operations have been done, and tracks of great length extending under the gluteal muscles have been divided with the écraseur with good results. I have myself followed a track directly across the perineum and exposed the membranous urethra in the incision, dividing in the operation the sphincters four different times. Such operations may sometimes be necessary to save life, but they may be too great for the patient's powers of recuperation.

In fistula complicating stricture of the rectum, attention should always first be turned to the latter, for if this can be cured there is a prospect that the former may undergo spontaneous closure, and if the stricture be not relieved it will be of little avail to cut the fistula. Many awkward mistakes have happened to good surgeons by failing to detect this complication of diseases.

CHAPTER VI.

HÆMORRHOIDS.

Definition.—Division into External and Internal.—Differences between the two Varieties.—External Hæmorrhoids.—Pathology.—Inflamed Hæmorrhoids.—Treatment.—Means of Prevention.—Palliative Treatment.—Excision.—Internal Hæmorrhoids.—Division into Capillary, Arterial, and Venous.—Description of Capillary Variety, of Venous Variety, of Arterial Variety.—Symptoms of Internal Hæmorrhoids.—Strangulation.—Diagnosis.—Treatment of Internal Hæmorhoids.—Palliative Treatment.—Constitutional **and Local Means of** Palliation.—Treatment of Strangulation.—Curative Treatment.—Hæmorrhoids Associated with Uterine Disease.—Symptomatic Hæmorrhoids.—Radical **Cure.—Caustics.—Dangers** of Nitric Acid.—Vienna Paste.—Treatment **by Carbolic** Acid Injections; Cases and Cures.—Advantages of this Treatment.—Treatment **by** Ligature.—Description of Operation.—Operation **with Clamp and Cautery.**

HÆMORRHOIDS may **be** defined **as varicosities of the anal or** rectal vessels. They **may** present themselves under **various forms and** conditions owing to changes in their substance; but the first step in their formation is always an enlargement and dilatation of **the veins or arteries or both.**

Hæmorrhoids, for convenience, **may** be divided into **external** and internal; and these may always be distinguished from each other, though both may exist at the same time in the same patient. An external hæmorrhoid originates in the subcutaneous veins which surround the anus; it is therefore entirely below the sphincter muscle, and though it may be partially covered by mucous membrane, it does not come from the rectum proper, nor can it be forced above the external sphincter muscle. **An** internal hæmorrhoid originates, on the other hand, within the rectum, and may exist for a long time without appearing externally. When it does show itself outside of the anus, it is a result of straining, **of increase in size, or of a** lax condition of the sphincter, and after long **exposure** outside **the body it may become** changed in character and appearance, **till the mucous membrane** covering it **takes on** something of the character of integument; but it may still, with proper management, be returned within **the** bowel, though it **may not** remain there for any length of time.

The distinction between an external and an internal hæmorrhoid is not, however, a purely arbitrary one, the one being below, and the other above the external sphincter. A different set of blood-vessels is implicated in each case. An external hæmorrhoid is a varicosity of an external hæmorrhoidal vein, and is, therefore, an affection of the general venous circulation. An internal hæmorrhoid is a varicosity of the middle or internal hæmorrhoidal veins, which are parts of the visceral venous system. A glance at the venous anatomy of the rectum and anus (pages 14 and 15) will show the arrangement of these two sets of veins, and will also explain how, from the free anastomosis which exists between them, it is improbable that one should be affected without influencing the other to a greater or less extent, and how, judged by this test alone, it may be impossible to tell whether a particular hæmorrhoid belongs to one system or the other. For practical purposes, therefore, the first definition is the better one—an external hæmorrhoid is one originating outside of the external sphincter, and an internal one is within that muscle. Other secondary differences which may arise from various causes in the development and location of the tumors will be considered later.

External Hæmorrhoids.—A person of middle age who has not at some time suffered from an external hæmorrhoid is indeed a great rarity, so common is this affection. In the majority of cases, it is allowed to run its own course, and only when the pain is unusually severe, or some untoward accident has happened, does the patient consult the surgeon. It is perhaps useless to seek for the causes of a malady which is so universal beyond a few which are well recognized and manifest. Amongst these are straining at stool, pregnancy, affections of the internal organs which interfere with the return of venous blood, and constipation. Outside of these cases where a manifest cause exists, external hæmorrhoids will be found amongst all classes. Those who smoke and those who do not; the high liver and the abstemious; the laborer and the professional man; those who stand and those who sit; are all affected and about equally.

An external hæmorrhoid may appear in two different forms which bear little resemblance to each other. The first is a small, round or elongated venous tumor; the second is a tag of hypertrophied skin, sometimes improperly spoken of as a condyloma. The second is formed from the first by changes soon to be described.

The external hæmorrhoid may arise in either of two ways, by the dilatation of a vein, or the rupture of a vein and the extravasation of blood into the adjacent tissue. The dilatation may not always be of the same character. In one case it may affect the whole calibre of the vessel, in another it may be in the form of a pouch springing out from one point in the circumference. A hæmorrhoid resulting from the dilatation of a vessel is of gradual formation; but it sometimes happens, particularly after a violent straining at stool, that the patient will feel a peculiar sensation at the anus, and an examination will reveal the presence of a tense, bluish, smooth tumor, the size of a pea or a grape, situated just at its verge. In this case, a previously dilated and weakened vein has suddenly given way, and the tumor is the result of the extravasation of blood.

Such a bloody tumor as this will cause much pain and discomfort, preventing the patient from sitting down, or even from going round with any ease. It may be freely incised by transfixing its base with a small, sharp, curved bistoury and cutting outwards, the incision being in the direction of the radiating folds of the anus, and this operation is sure to give temporary relief, by allowing the escape of a small clot of blood and putting an end to the tension which is causing the suffering.

If the surgeon undertake this method of treatment, there are one or two hints which may be of value. The incision itself is extremely painful, and should therefore be done with a sharp knife of the form mentioned; and it should be done instantaneously. Whatever deliberation is required, is better exercised before entering the knife. Again, care should be exercised to empty the clot entirely out of its bed, otherwise a small wound remains which will not readily heal, because the sac is prevented from contracting, and the patient is obliged to wear a bandage perhaps for a week or longer to keep from soiling the linen with a sanious discharge. Under such circumstances also the pain is but little relieved by the operation. Again, I have in a few cases seen the incision heal by primary intention, and the sac again fill with blood, thus leaving the patient in the same condition, as regards suffering, as before operation. This is best avoided by placing a shred of lint in the cut. These, however, are untoward accidents which may attend an insignificant operation which usually gives relief to suffering, and allows the tumor to shrivel up and disappear except for a small tag of skin which may remain and form an external pile of the second variety.

When left to its own course, a bloody tumor of this variety may gradually decrease in size from the absorption of the fluid elements of the clot, the pain decreasing at the same time; and after a week or ten days of discomfort, it is changed into a cutaneous hæmorrhoid. Or the opposite course may be taken, and the tumor may show all the signs of an abscess, and finally rupture spontaneously with the discharge of a little blood and pus, and with an instantaneous ending to a week of suffering. For during this acute inflammatory process, the pain is often very severe, the discomfort constant, and there may be more or less febrile excitement; all of which will pass away the moment the tension is relieved. The treatment of such a case where the knife is not used will be described a little later.

To return to the hæmorrhoid which is due to the varicose vein, but not to the extravasation of its contents. In such a case there may be one considerable dilatation which shall cause a smooth, round, bluish tumor the size of a pea or a grape; or there may be a number of veins included in a new growth of connective tissue which shall constitute a distinct, firm, hæmorrhoidal tumor. For these dilated pouches are in themselves causes of irritation, and are subject to irritation from without; and as a result an exudation takes place in their vicinity which finally ends in the production of new tissue. It is thus easily understood why on cutting into one external hæmorrhoid a single large clot will be exposed contained in a distinct sac; while in another, several smaller clots may be seen imbedded in the surface of the section, and why there is more or less connective tissue in the tumor.

The formation of such a tumor is a gradual process due to the continuous action of the primary cause and to subsequent irritation from without. It may go on with little pain and suffering, so little that the patient will hardly care to ask for relief; and it may undergo a spontaneous cure leaving in its place only an hypertrophied tag of skin. Generally, however, during its course an attack of acute inflammation will be excited at some time, and this is very apt to bring the sufferer into the hands of the surgeon. At such a time, if the inflammation has occurred in a fleshy pile the tag will be swollen, œdematous, and exquisitely sensitive. Suppuration may occur in it and a small marginal abscess and fistula be the result. Or, if the inflammation has attacked a sanguineous tumor, it will be found hard and swollen and painful to the touch. The patient will often say that he has tried to replace the little grape-like tumor within the bowel, but has been unable, though the pressure has caused it to disappear for the moment and has given a temporary relief. This is due to emptying the vein of its blood, but the blood returns the moment the pressure is removed.

The pain is constant, often preventing sleep at night. The sufferer is unable to sit or stand and soon finds that he feels better in the recumbent posture. A motion of the bowels is feared and therefore avoided as long as possible. When after two or three days of constipation the call can no longer be delayed, the pain is greatly increased. It is astonishing how much pain and constitutional disturbance such an apparently trivial thing may cause.

Such as an attack in a sanguineous hæmorrhoid may terminate in three ways: by resolution, by induration, and by suppuration. In the former case the resolution may be complete especially when the inflammation has been of moderate intensity, and no trace of the tumor may remain, or a cutaneous tag may be left to mark its former site. When the inflammation assumes a chronic type, and the tumor becomes œdematous, and is still somewhat painful on pressure or during defecation, though not to such a degree as during the acute stage, the inflammation is said to have terminated in induration. Such a tumor is always liable on slight provocation to a fresh attack of inflammation. When suppuration occurs, the tumor discharges its pus and then shrivels up and becomes a cutaneous tag.

Treatment.—The surgeon will seldom be called upon to treat a case of external hæmorrhoids unless during an attack of acute inflammation; for at other times the annoyance caused by them is comparatively trivial. A cutaneous tag which is quiescent may as well be left undisturbed by the knife or scissors; for the removal of it will not infrequently cause an amount of suffering disproportionate to the benefit gained. The whole thought of the surgeon may then be turned first to the prevention and second to the relief of an attack of inflammation. The means of prevention are very simple and yet very effectual. They consist in the avoidance of excess in eating or drinking and in perfect regularity in defecation; for in a person affected with external hæmorrhoids a single heavy meal at an unusual hour, an evening spent in smoking and drinking, or, worst of all, the neglect to have a motion of the bowels for a single day, will give rise to a sensation of heat, pressure, and itching about the anus, which warns him that trouble has commenced. Even under such circumstances the attack may be aborted by rest in the recumbent attitude, a light diet, abstinence from wine or liquor of any kind, and a laxative, preferably one of the mineral waters, repeated every night for three or four days.

Should the attack go on and inflammation be actually excited, more active treatment will be required, and this may be either operative or medicinal. It is my own practice to try the latter first, and if it does not succeed, resort to the former. The medicinal treatment consists in keeping the sufferer on the bed or lounge, and applying a small bladder of pounded ice to the part.[1] This is generally very grateful to the patient and very effectual—much more so than warm poultices or applications of belladonna and opium; but should it not prove so, the latter may be tried. A good formula is equal parts of the extracts of belladonna and opium smeared freely over the anus. In most cases the attack will subside after forty-eight hours of this treatment, and the use of a daily laxative; but should it not, a sanguinous tumor may be incised in the manner already described, and a cutaneous tag may be seized with a sharp forceps and quickly snipped off with the scissors. Ether is not generally necessary for this operation, which, though very painful, requires but a moment; and I have generally found that attempts at local anæsthesia with the ether spray were very delusive in this part of the body. If ether be employed at all, it is much better to take advantage of the primary anæsthesia produced by the first few inhalations, the patient holding the towel or bottle in his or her own hand. This is a favorite procedure of my own in this and many other operations about the anus, and one which I cannot too strongly recommend.

---

[1] Nothing is so convenient for this purpose or causes as little pain as the rubber baudruche, which may now be procured at any druggist's.

The only caution necessary in cutting off an external hæmorrhoid is to remove neither too much nor too little tissue. If too much be removed, the wound will take a long time to heal, and if several tumors be removed, contraction to a disagreeable extent may follow; if too little, a tag of skin will still remain after cicatrization and shrinking, and, although this might be considered a matter of no importance in a male patient, I have seen ladies who did not so consider it.

Internal Hæmorrhoids.—External hæmorrhoids were described as varicosities of the external hæmorrhoidal veins; and internal hæmorrhoids may also be similarly defined as varicosities of the middle and superior hæmorrhoidal veins, but they are more than this. An internal hæmorrhoid is often an arterial tumor, as well as a venous, and the arteries may be of large size. Occasionally one will be met as large as the radial. In describing these tumors, we shall follow the division laid down by Allingham into capillary, arterial, and venous.

The capillary hæmorrhoid is in reality an erectile tumor, composed of the terminal branches of the arteries and veins and of the capillaries which join them. This form of tumor is never of large size, and never projects very far into the cavity of the rectum. To the naked eye and under the microscope they strongly resemble an arterial nævus. They may be situated high up in the rectum or low down by the sphincter; their surface is granular, and the membrane covering them is always of extreme thinness. This accounts for the chief symptom which distinguishes them clinically from the other varieties—the free arterial hæmorrhage which follows the slightest bruising of their surface even in the act of defecation. Such a tumor never appears outside of the anus unless accompanied by some other rectal affection, but it may sometimes be seen by a careful pulling open of the sphincter with the fingers, and from some part of its strawberry-like surface there is pretty sure to be a jet of arterial blood, coming *per saltem.* The disturbance caused by the gentlest examination is sufficient to start this bleeding, and it almost always occurs at defecation. This is the form of hæmorrhoid to which the name of "bleeding" most properly applies. In my own experience it is not as frequently met with as the varieties to be described later; and this probably for the reason that after existing for a longer or shorter period in this form it is changed into one of the others: and that patients do not seek relief till after such change has occurred. After a time, the mucous membrane covering such a tumor becomes thickened, and as a result of repeated irritation, there is an increase in the submucous tissue. The hæmorrhage decreases in frequency and finally ceases as the capillaries become obliterated by the increase in the connective tissue, and the capillary tumor is succeeded by the arterial or the venous one.

The one symptom of a capillary hæmorrhoid is the daily hæmorrhage; and as this hæmorrhage occurs at the time of defecation, and there is no pain at any time, the patient may be entirely ignorant of the fact that blood is daily lost. This is particularly the case with the class of patients seen in public practice who give little attention to themselves. In the higher walks of life such a loss of blood seldom occurs without the knowledge of the patient; but unfortunately it is often disregarded, especially in women who are in the habit of losing blood at every menstrual turn and who always shrink from an examination.

It is not necessary to relate in detail the train of constitutional symptoms which may follow the daily loss of a considerable quantity of arterial blood. The anæmic look, the disturbance of the heart's action, the troubles with the digestive apparatus and with the sexual organs, the cessation of menstruation, are all well known. But it is curious that, as in a recent case in my own practice, a very intelligent medical man who understood perfectly his own condition, should allow himself to be brought to a state of profound anæmia by a little hæmorrhoid of this variety rather than have anything done for himself. In his case a single application of nitric acid to the bleeding surface worked a cure which has lasted for several years.

The arterial hæmorrhoid.—In this form of tumor the capillary network has disappeared and in its place is found a mass of freely anastomosing arteries and veins bound together by connective tissue. The arteries and the veins are tortuous, often varicose and dilated into sacs and pouches, and the arteries may be of large size, especially the one which enters at the base of the tumor, the pulsations of which may often be distinctly felt by the finger. Such a tumor is often of considerable size; it is firm to the touch and smooth; it is liable to inflammation, erosion, hæmorrhage, and prolapse. The hæmorrhage which occurs is arterial in character, and apt to be abundant. When the hæmorrhoid has gained a sufficient size to become prolapsed in the act of defecation, the patient suffers the usual symptoms of the hæmorrhoidal state. If the sphincter be not tight enough to strangulate the mass after it has come out of the body, the pain will not be very severe and the patient will return the tumor by a little gentle pressure and manipulation.

The venous hæmorrhoid.—This form of hæmorrhoid may result from either of those already named or it may arise *de novo*. It consists at first of a simple dilatation of the large veins beneath the mucous membrane of the rectum; later these veins undergo certain changes due to the hypertrophy and induration of the mucous membrane and submucous connective tissue, until finally a large, bluish, hard tumor is formed which is smooth to the touch, comes out of the body on defecation, and is covered by a mucous membrane which has assumed a partially cutaneous character from exposure.

The three varieties of internal hæmorrhoids thus described may all be present in the same person, and each be distinguishable from the other. In other cases the line of distinction may not be so well marked. A venous hæmorrhoid may contain a considerable number of arteries and may bleed *per saltem*, and it is not certain that an arterial hæmorrhoid is always a later stage of the capillary variety. But the three forms are well marked and must be distinguished from each other in the matter of treatment.

Symptoms.—Usually the first symptom of internal hæmorrhoids is the loss of blood during defecation to which reference has already been made. This may be present for a long time before any other symptom is noticed by the patient except perhaps an occasional feeling of discomfort in the rectum. Pain is absent until the tumor begins to descend within the grasp of the sphincter and appears at the anus at each act of defecation. If the sphincter be firm and strong, the pain may be very severe and the tumor may become strangulated, but after the disease has existed for any great length of time, and especially in persons past middle life, there is apt to be a loss of power in the muscle which, though it facilitates prolapse, decreases the pain attendant upon it.

In ordinary cases, the patient will reduce the tumors when they come down on defecation. They may, however, become strangulated, and be entirely beyond the patient's power of manipulation. In such a case, after a period of rest, and after the relief which may follow a spontaneous escape of blood from the over-distended vessels, the hæmorrhoids may return of themselves or be put back by the patient.

If the strangulation be more intense, gangrene may set in and a part of the mass may slough; or a part may suppurate and pus be discharged. Under these circumstances there will be great pain and more or less constitutional disturbance, with fever and loss of appetite. The gangrene is very evident to the eye from the greenish or blackish color and fœtid odor of the part, and is rather a favorable termination to the trouble as it generally results in a radical cure.

Diagnosis.—It is not always an easy matter to discover an internal hæmorrhoid, even though it be far enough advanced to cause hæmorrhage and more or less uneasiness. When it has become hard, it may be detected by the accustomed finger in a simple digital examination, but when soft and not over-distended, it may escape detection. An examination should be made directly after the rectum has been emptied by an enema of warm water, when the water and the straining have brought it into prominence, and should be made with Van Buren's speculum. Under these circumstances, it may generally be brought plainly into view. An examination in a case of internal hæmorrhoids should never end at the finding of the tumor. An inch or so higher up there may be a stricture, malignant or simple, which has given no sign of its presence except the hæmorrhoids, and this is not a good thing to overlook.

Treatment.—The treatment of this most common and distressing malady may with advantage be considered under two different heads—(*a*) palliative, (*b*) radical.

(*a*) *The palliative treatment of internal hæmorrhoids.* In spite of all that the surgeon may say to his patient of the advantages of a radical cure, and the safety and facility with which it may be accomplished, he will still have many more chances in the way of palliation than will fall to him of using the knife. It is, therefore, of great advantage to know what can be done for a timid and reluctant sufferer without the knife; and, indeed, most patients may be made greatly more comfortable without any surgical interference whatever.

The first thing to be done is to secure a daily natural evacuation of bowels, and this without medicine, if possible. The diet should be plain and abundant. Highly seasoned meats, gravies, salads, old cheese, etc., all alcoholic drinks, and anything approaching excess in tobacco, should be strictly interdicted. If the bowels do not act daily with this diet, and with regularity in the time of going to the closet, a laxative must be added, and this may be either in the form of a mineral water in the morning, or of a small dose of compound licorice powder at night.

This powder may now be bought under that name at most drug stores. The formula is, however, appended for the convenience of any who may desire it:

   R. Fol. sennæ..2 parts.
   Rad. liquiritiæ......................................2 parts.
   Fruct. fœniculi pulv..............................1 part.
   Sulphuris depurati..................................1 part.
   Sacch. pulv..6 parts.

If the hæmorrhoids are in the habit of coming down when the patient has a passage, he must accustom himself for a time to the use of a bed-pan, and to having his passages while in the horizontal position. This will be considered a very objectionable remedy by most; but it is one from which great benefit will be derived.

The other treatment is local, and consists mainly in the use of astringents and of cold. After each passage, the bowel should be injected with cold water. Even ice-water will do no harm. The quantity should not exceed four ounces, and if the case is one attended with bleeding, this will be found a most valuable means of combating that symptom. The number of astringents which have been recommended for use under the circumstances we are now considering is very large. I shall content myself with naming one, the subsulphate of iron, which combines the advantages of all the others. This may be applied in the form of an ointment (3 i. — ℥ i.) to the hæmorrhoids when prolapsed, or may be given in the form of a suppository (2 gr.) and allowed to remain in the rectum over night. It will be found to act simply as an astringent, causing no pain, and destroying no tissue.

By these means, when followed with care and patience, the worst case of hæmorrhoids may be greatly improved, and when the sufferer will not submit to curative treatment, or when, from any reason, operative interference is contra-indicated, they should always be tried. Although they are given simply as palliative measures, and should be considered as such, I have had some cases where, after a few weeks of this treatment, the patients believed themselves cured, and were, at all events, so far relieved as to disappear from observation.

Treatment of strangulation.—The practitioner may at any time be called upon to treat this complication of internal hæmorrhoids, and the condition is an exceedingly painful one. He will generally find his patient in bed complaining that his piles are "down," and that he has been unable to replace them. The prolapse may have occurred at the time of defecation, or during a momentary mental excitement or physical effort. On examination, the anus will be seen to be surrounded with a mass of hæmorrhoids which are swollen, congested, livid, and more or less œdematous, and any attempt to replace them will cause exquisite pain. This is an excellent opportunity for inducing the sufferer to submit to a radical operation, and should consent be gained, ether may be given, and the usual operation, by the ligature, be at once performed. The operation, under these circumstances, does not seem to be contra-indicated, and I have never had occasion to regret performing it.

But should an operation be refused, the mass must be reduced. The patient should be turned on the face, with a hard pillow under the pelvis to raise the buttocks and allow of gravitation of the abdominal contents away from the rectum. The mass should then be well smeared with olive oil, and a gentle effort made to reduce it by the taxis. This may sometimes be done by introducing one finger into the anus and exerting pressure with the others, gradually forcing the tumors, one by one, within the bowel; at other times, the mass may be replaced by a firm and continuous pressure, with the bulbs of all the fingers directly upon it, till the blood has been crowded back, and the diminished piles slip up together. Much gentleness is required for this manœuvre, which is a very painful one under any circumstances, and one man may succeed where another would fail.

At times, however, replacement by the taxis is impossible. Under such circumstances, it is a not uncommon practice to resort to leeches; and though I have never done it, I have seen it almost immediately successful with others; and the patient himself will assure you that, if the piles would only bleed, they could be easily reduced. It is better, however, to apply cold, and to leave the patient in bed on his face, with the buttocks raised. The cold should be in the form of an ice-bag, and this will almost certainly give relief to suffering, and so reduce the œdematous swelling as to render reduction possible on a second attempt. Should this also fail, there is nothing to do but to wait for the condition to subside under the use of cold and applications of belladonna and opium in the form of a soft ointment, with rest in the position named, and the administration of laxatives. After forty-eight hours of this treatment, the patient will generally succeed by himself in reducing the mass.

(*b.*) *Curative Treatment.*—Before recommending anything in the way of a surgical operation, the surgeon must consider whether the case before him is one in which such a procedure is justifiable, and this brings us to the consideration of what have been called symptomatic hæmorrhoids, as distinguished from those which are apparently idiopathic.

Internal hæmorrhoids may be symptomatic of disease in a number of the viscera. They often indicate structural changes in the wall of the rectum itself at a higher point, such as malignant and non-malignant stricture; and under such circumstances, whatever is done in the way of relief must be done to the stricture, and not to the hæmorrhoids. Again, they are often secondary to disease of the bladder, to enlarged prostate, or to stricture of the urethra; and in these cases where it is possible to remove the cause it must always be done. If hæmorrhoids are dependent upon a calculus, or a stricture of the urethra, they will disappear when these affections are cured. I was consulted not long since by a brother practitioner in regards to a very typical external sanguineous hæmorrhoid—the size of a large pea—on the person of his four-year old child. The child had an adherent prepuce, and the pile was the result of the straining. The ordinary operation of circumcision cured the hæmorrhoid. A man with enlarged prostate is never a very desirable subject for a surgical operation, and if such a man's hæmorrhoids can be rendered endurable by the palliative treatment already described, the better way will be not to use the knife.

In women hæmorrhoids often depend upon disease of the uterus; and in every female patient this dependence should be carefully inquired into, and if found, removed before operation. The operator in rectal surgery may save himself much discredit, by postponing his operation for piles till his patient has been cured of a uterine misplacement or catarrh; for, as a rule, the co-existence of the latter diseases will prevent a favorable issue to the operation. Either the wounds will not heal readily, or the hæmorrhoids will speedily return. It will occasionally happen that a pregnant woman will suffer so severely from this complication as to demand surgical aid. Though it is better not to operate, except in a case where the hæmorrhage or the pain render it unavoidable, still pregnancy is not an absolute barrier to surgical interference in this more than in many other affections.

Hæmorrhoids may also be symptomatic of disease of the liver, kidney, heart, or lungs. There are few liver affections which need prevent operative interference in a bad case, but such interference should be preceded by general treatment pointing towards relief of the hepatic circulation. An excess of alcohol in the daily diet should be stopped, and a blue pill may be given with advantage every other day for a week before the operation. Affections of the lungs, except in a very advanced stage, need not prevent an operation. The condition which most positively stays the hand of the operator is that of albuminuria, whether dependent upon heart or kidney.

Having decided to attempt a radical cure, the surgeon finds himself embarrassed with the number of operative procedures from which he may choose. It is safe to say that no one operation is the best in all cases, and I shall make no attempt even to enumerate all of those which have, at different times, been advocated, but shall describe several which are to be relied upon, and which, together, will cover every case.

The Application of Caustics.—Chief among the caustics used for this purpose are nitric acid, pure carbolic acid, and Vienna paste. The capillary hæmorrhoid may be cured by painting it once or twice with pure nitric or carbolic acid; but large and old hæmorrhoids are not curable by this means, though the hæmorrhage from them may be stopped, and for a time they may cease to prolapse. When used upon a capillary growth, a speculum must be introduced. If used in a case of large tumors, they must first be brought outside of the body, carefully dried, and then thoroughly covered with the acid, applied with a small stick. The end of a match makes an excellent brush. The tumors should then be well oiled and replaced. The application is not painful, unless the acid is applied to the wrong surface, viz., the skin.

I have used this plan of treatment in many cases; have seen an exhausting hæmorrhage from a capillary tumor stopped forever by a single application, and have benefited old cases to an extent which convinced the patients they were radically cured in spite of my own skepticism; but it is never safe to promise anything more than temporary relief by this means. The capillary tumor is very likely to subsequently become the larger arterial one; and the old and large hæmorrhoid is more than likely to become prolapsed at some future date: so that I no longer use it in these latter cases when the patient will permit me to follow my own judgment.

There is one danger in the application of a strong acid to an old prolapsing hæmorrhoid, and that is, the occurrence of a profuse secondary hæmorrhage when the slough separates. Such an accident is not common, but it may be a fatal one, and it happens just often enough to worry the surgeon in every case in which he has employed this method in an old and debilitated subject.

The Vienna paste is a much more powerful caustic than nitric acid, and its application to the surface of a hæmorrhoid is very painful. This and the amount of tissue destroyed by it are the two great objections to its use. It has been employed to produce deep, linear, radiating cicatrices, each cicatrix running from the centre of the anus over the top of a prolapsed hæmorrhoid; and three or four such cauterizations will undoubtedly cure an ordinary case of piles; but the Paquelin cautery will do it much better, and if the patient will submit to the latter, he will submit to something better still, and that is the ligature.

Treatment by Injection.—The treatment of hæmorrhoids by injection of certain substances, chief of which is carbolic acid, may now, I believe, be accepted as a surgical procedure of a certain definite value, and one worthy of a place among the recognized means of cure at our command. Originating as it did among the quacks, it has been looked upon with suspicion, and its adoption by the profession has been followed by the accidents which generally attend a new remedy before its applicability is fully understood; but this does not diminish its real value.

The following four cases, selected from dispensary and private practice in which this plan of treatment has been adopted, will illustrate some of its advantages and disadvantages.

CASE VI.—Male. Age, thirty-nine. This was an ordinary case of prolapsing internal hæmorrhoids of about six months' duration in an otherwise healthy man. The tumors were well developed, bled freely at each motion of the bowels, and were usually reduced by the patient without much difficulty. In the course of three months four injections of carbolic acid were made into four separate tumors. Only one of them was followed by any pain or soreness, and this not very marked in character; and after three months the man was discharged cured, there being no longer any bleeding or descent of the hæmorrhoids at defecation. The man, who was a fireman, was at no time during the treatment unable to attend to the active duties of the service.

CASE VII.—Male. Age, thirty-eight. In this patient anything like a cutting operation was out of the question. He had been a hard drinker for years, and was suffering from phthisis, cirrhosis of the liver, and albuminuria. The hæmorrhoids were of long standing; the whole circle of mucous membrane prolapsed with them; and the sphincter had lost its contractile power. The man was under treatment three months, and during that time six injections of carbolic acid were made, and each one was followed by more or less pain and by sloughing of the hæmorrhoid. The pain was not, however, so great as to counterbalance the relief the patient experienced from the cessation of the bleeding and the decrease in the protrusion, and the treatment was gladly persisted in by him, till in the end he considered himself as cured and ceased to attend. I have no doubt that in this case the sloughing of the tumor, which each time left a dirty sore after the introduction of the acid, was directly due to the patient's condition; but he was sustained with generous diet and suitable tonics, and, as I say, did very well—much better than he would have done by any other plan of treatment which it was safe to try; and, but for it, I should have confined myself strictly to palliative measures.

CASE VIII.—Male. Age, fifty-two. General health excellent. Hæmorrhoids well-developed and prolapsing. Having had considerable experience with this method of treatment by this time in dispensary practice,

I ventured to try it in a private patient, and to promise an easy and painless cure. A single injection was therefore made, and for the first forty-eight hours there was little trouble; but at the end of that time I received a telegram from the gentleman that he was suffering great and constantly increasing pain—he having left me on the day following the injection to return to his home in a neighboring city. I went to him and found, to my disgust, that the injection had in his case also caused a slough, and that he was suffering intense pain at each act of defecation. Suitable treatment with laxatives and anodyne suppositories was at once instituted, but his sufferings continued for many days, and he finally went off to the mountains where he remained till the ulceration had healed. Needless to say he refused to continue this "painless" method of cure, and I lost my patient and not a little reputation, the man being rich and influential.

CASE IX.—Male. Age, fifty-three. Also a private patient, and in fair general condition, but with old and severe hæmorrhoids and partial prolapse, and weakening of the sphincter. I was first called to see him in the night when he was suffering from strangulation of the entire mass, and a week later I began the use of the acid. This was followed very cautiously and with abundant intervals of rest after each injection, and in a very short time the relief was very apparent in the diminution of the size of the protrusion. There was no pain at any time during the treatment, and only a slight nipping sensation for an hour or so after each injection. In the end he was entirely cured, all hæmorrhage and protrusion of the tumors having ceased, though the anus was still surrounded by the redundant circle of half skin and half mucous membrane which remained from the former condition of prolapse.

Here then was an old case of large prolapsing hæmorrhoids in a private patient who would submit to nothing which he considered as surgical treatment, apparently cured without any pain, without any of the usual accessories of an operation, and without a single day's detention from his ordinary pursuits—a result for which surgery has been waiting a long time. I say *apparently* cured, for the one doubt which remains in my own mind regarding this treatment is as to the *permanency* of the cure. This I have not as yet had time to test. I have seen nothing to make me doubt its being permanent; and considering what Vidal has accomplished with injections in cases of long-standing and extensive prolapse, I see no reason why it should not be permanent; but I have not as yet had a chance to examine any of my own cases after an interval of years which is the only way of positively deciding the question.

Beginning this plan of treatment, as I did, without very much confidence in it and with the fear of causing great pain and perhaps dangerous sloughing constantly before me, I can only say that the method is constantly growing in favor with me personally, and that the more I practise it the more confidence I gain in it. With solutions of the proper strength the danger of causing sloughing of the tumors is very slight; and I am not at all sure in my own mind that once more surgery is not indebted to the quacks for a valuable discovery which may do much to modify the at present accepted plans of treatment of this disease.

There are no objections to this method which do not apply equally to others. I have once seen considerable ulceration result from it in the hands of another, but I have seen an equal amount follow the application of the ligature; and I do not consider this as a danger greatly to be feared when injections of proper strength are introduced in the proper way. It is applicable to all cases; is especially adapted to bad cases; and may be used, as in the second case, where a cutting operation is inadmissible. It acts by setting up an amount of irritation within the tumor which results in an increase of connective tissue, a closure of the vascular loops, and a consequent hardening and decrease in the size of the hæmorrhoid. Except when sloughing occurs, the tumors are not, therefore, removed, but are rendered inert so that they no longer either bleed or come down outside of the body. In cases in which the sphincter has become weakened by distention, the injections will also have a decided effect in contracting the anal orifice as do injections of ergot or strychnine in cases of prolapse.

I have used this method of treatment now many times and, except in the third case reported here, have never had reason to regret using it, or to be dissatisfied with its results as far as I have been able to follow them. Although I should be very slow to advocate any one treatment of this affection to the exclusion of all others, I now often adopt this where Allingham's operation is declined by the patient, and as yet I have not known it to fail. Its advantages over all other methods, provided its results prove equally satisfactory, are manifest to all. The patient is not terrified at the outset by the prospect of a surgical operation, is not confined to his bed, and is not subjected to any suffering. The cure goes on painlessly and almost without his consciousness.

The method requires some practice and some skill in manipulation in getting a good view of the point to be injected and in making the injection properly. In the first three cases reported, the solution employed was one part of pure carbolic acid to three of glycerin and three of water; in the last, the carbolic acid was decreased one-half and this is a better solution to use. The amount injected each time was about five drops. The instrument used was an ordinary hypodermic syringe with a good sized needle through which the solution would readily pass. When the tumor to be injected is prolapsed, the needle may be thrust into it without difficulty, and after the injection is made the tumor should be gently replaced. If it be allowed to stay out of the anus for a few moments it will be seen to swell up and become black and hard with venous blood. There is seldom any hæmorrhage from the operation, but occasionally a few drops of blood will follow the puncture. If the tumor is not protruded at the time of operation it may be seized with toothed forceps and drawn out and held while the injection is made. The injection should be landed as nearly as possible in the centre of the hæmorrhoid, the needle being entered perpendicularly from the apex, and not passed upward under the mucous membrane in a longitudinal direction. If the acid be placed simply under the mucous membrane the latter will die and an ulcer result, but if placed more deeply the danger of an ulcer is much decreased. Used in this way and in the strength last indicated the acid will not be followed by any great amount of pain. Each injection should be followed by a day's rest in the horizontal position. No change need be made in the ordinary diet of the patient provided the bowels act regularly every day. Only one tumor should be injected at a time and I seldom repeat the injections oftener than once a week. It will sometimes be found necessary to inject the same tumor twice or three times when it is a large one.

It will be observed that in the cases reported the length of time during which the patient was under treatment was in each case except the second about three months. I have no doubt that this could be much shortened, were it necessary; but where the patient is at no time confined to the house, time is of little consequence, and I seldom repeat the applications oftener than once a week, preferring to see the full effect of each one before giving a second. Still, were there any reason for haste, I should not hesitate to shorten this interval, and I am led to believe that in the hands of the quacks the time is considerably shortened. I believe also that with them it is the custom to produce a sloughing of each tumor by the strength of the injection, and once or twice I have had patients come to me in this condition. But no such use of the acid is necessary to effect a cure, and this result is one which I try very carefully to avoid.

Treatment by Ligature.—This is the method of treatment which has been brought to such perfection by Allingham, and which usually passes by his name. It consists in partially cutting through the hæmorrhoid at its base, and tying the remainder. It is performed in the following manner.

As in all operations on the rectum, the bowel should be thoroughly cleared by a cathartic on the previous day and by an enema just before operating. The patient may be placed either on the side or in the lithotomy position; personally I prefer the latter. The sphincter should be carefully dilated, as already described, and this is a step of great practical importance, as the securing of complete paralysis of the muscle will do more than anything else to prevent pain and spasm after the operation. In cases where the tumors were large and prolapsed readily, I have seen this step in the operation omitted as unnecessary by good surgeons; and I have seen a week of great suffering to the patient follow the omission. So important is this step in the operation for the relief of pain, that in some cases in which the tumors were so extensive and the sphincter so dilated that they could easily be removed without it, I have first cut off the hæmorrhoids and then stretched the sphincter. It is rather a reversal of the regular order, but it illustrates the fact that stretching the muscle should not be omitted. If the muscle is forcibly and suddenly torn apart by the operator, a fissure may result, and may require a subsequent operation for its cure after recovery from the original operation. The tumors being thus brought into full view by the introduction of a speculum, one is seized and drawn down with a toothed forceps. The selection of a good forceps for this purpose is a matter of considerable importance. In my own operations, I use those figured below. The hold is firm and the handle sufficiently long for the hand of the assistant to be out of the way of the operator in the subsequent steps.

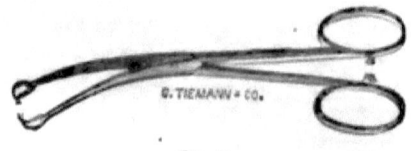

Fig. 35.

Having secured a good firm hold on the tumor, the surgeon transfers the forceps to the left hand, and with a strong and long pair of straight scissors cuts the hæmorrhoid away from its attachments for a certain distance, beginning from below and cutting upwards. In this way the mass is entirely cut off except at its upper end, where the artery or arteries which feed it enter it from above. It is to prevent hæmorrhage from these vessels that the ligature is applied instead of completely cutting off the mass; and this is done by the operator after transferring the forceps to the assistant.

The ligature should be of stout hemp, something stouter than ordinary ligature silk being necessary. The string should be tied very tightly, and after it is secured, the pile may be cut off to remove as much as possible of the dead tissue from the rectum. Each hæmorrhoid is thus treated in succession, and after all are removed, a suppository of opium is introduced, and a T-bandage tightly applied over a compress of lint and a napkin. The suppositories, which may be repeated each night for two or three days, will serve to keep the bowels confined; and when the patient begins to experience a desire to go to stool, a laxative may be administered. There may or may not be some pain when the bowels first move, and this will depend very much upon the thoroughness with which the alimentary canal has been emptied before the operation. I have seen as a result of neglecting this previous cathartic a female patient have to rid herself of a hardened mass of fæces of the size of an egg at the first motion of the bowels after the operation, and the suffering was simply atrocious. If there be a little blood with the first passage, it is a matter of no importance.

The ligatures will generally come away about the end of the first week, and the patient should be kept in bed or on the lounge for a week longer. This in an active person will sometimes be difficult to manage; but no other course should be sanctioned by the surgeon, for the reason that when the ligature comes away, an ulcerated spot is left: and under certain circumstances, the most effective of which is active exercise, these little wounds may grow larger instead of smaller. In this way a case of internal hæmorrhoids may be turned by an operation into one of ulceration of the rectum, and the change is not to the advantage of the patient. One such case I have had in my own practice in a debilitated patient in poor general health; and a long course of careful treatment was necessary to effect an ultimate cure.

Nothing has been said regarding primary or secondary hæmorrhage, for the reason that it is not a complication to be looked for. When retention of urine occurs, as it often will, it must be met in the usual way. The diet for the first few days should be chiefly fluid.

This operation, thanks to Mr. Allingham, is now so well and so favorably known, that but little need be said in addition. It is as safe as any operation in surgery, and by it the surgeon may promise his patient an absolute and permanent cure of his troubles in every case. This is saying a great deal, but not too much. It has been followed by fatal results—but so has every other minor surgical operation; and the chance of such a termination is so slight that it need not enter into the calculation of the operator. Of all the operations for the cure of internal hæmorrhoids it will be found the most satisfactory, the least liable to complications in its performance, and to unfortunate after-consequences. Once in my own practice after applying it to an old case of hæmorrhoids with slight prolapse and almost completely surrounding the whole circumference of the anus with ligatures, I have been obliged to subsequently use a bougie to prevent a threatened contraction; but this I rather expected to be obliged to do at the time, and injurious contraction need not be feared in any ordinary case. I can confirm Mr. Allingham's statement that the operation, when performed with a proper regard to minor details, is not followed by any considerable amount of suffering. I have had patients assure me that the first day following its performance was one of perfect comfort—in fact of greater ease than any they had experienced for weeks previous.

Operation with the clamp and cautery.—This is generally known as Smith's operation, because he has advocated it so forcibly and practised it with such good results. He claims no credit for introducing it, however, this being due to Mr. Cusack, of Dublin, and his own originality has been chiefly spent in improving the clamp which is shown below.

The operation consists in drawing down the tumor, embracing its base in the clamp, removing it with sharp scissors, and lastly applying the actual cautery freely to the cut surface. It is important to isolate the tumors well so as to compress them easily and completely, and in some cases where the hæmorrhoid runs, as it were, abruptly into the hypertrophied skin, Smith recommends the previous making of a slight groove with the scissors so that the compression of the neck of tumor may be the more effectual. The base should not be divided too close to the clamp lest there be not enough tissue left for the proper application of the hot iron. The latter is to be applied very thoroughly and slowly at a black heat; and the blades of the clamp may then be gradually released by the screw. Should any vessel not thoroughly cauterized bleed when the pressure is taken off, the clamp must be again screwed up and the cautery again applied. It may be necessary to do this several times.

This operation is claimed by Smith to be almost painless, provided the

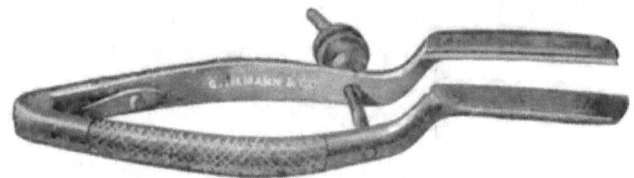

FIG. 36.

cautery does not touch the skin and the clamp is fitted with the proper ivory protectors against the transmission of heat. Next to the operation by the ligature, it is the best of all the surgical procedures, but it is much better adapted to old cases of large prolapsing tumors than to those which are less developed, for the reason that there is more for the clamp to take hold of, and more left to cut off after the clamp is in position. I can hardly imagine how the operation can be painless, especially when a previous cutting is done with the scissors, but I have not had a large experience with it, and none at all without ether.

With the means already enumerated every case of internal hæmorrhoids may be cured where a cure is desirable, or relieved when radical cure is out of the question, and I shall not, therefore, take the space necessary to describe the various others which either have been or are at present in favor; such as simple dilatation of the sphincters, the galvano-cautery wire, plunging the actual cautery into the substance of the hæmorrhoid, and cauterizing the skin of the anus in radiating lines to cause contraction.

CHAPTER VII.

PROLAPSE.

Four Varieties.—First Variety: Prolapse of the Mucous Membrane Alone.—Second Variety: Prolapse of all the Coats of the Rectum.—Third Variety: Prolapse of the Upper Part of the Rectum into the Lower, or Invagination.—Fourth Variety: Invagination in the Continuity of the Bowel.—Prolapse of the Mucous Membrane alone.—Causes.—Symptoms.—Treatment: **Palliative and Curative.**—Prolapse with Hæmorrhoids.—**Treatment by Injections.**—Cauterization.—Description of Operation.—Smith's Clamp.—Dupuytren's Operation.—Prolapse of the Second Degree.—Pathological Changes.—Presence of Peritoneum.—Strangulation.—Dangers in Forcible Reduction.—Fatal Case of Reduction.—Advisability of Reducing Inflamed or Gangrenous **Prolapse.**—Excision of Prolapse after the Formation of a Slough.—**Dangers of Operation of Excision in Extensive Prolapse.**—Operation by Elastic Ligature.—Third and Fourth Varieties.—Differences between Third and Fourth.—Degrees of Invagination.—Anatomical Appearances.—Pathology.—Relative Frequency.—Symptoms.—Physical Signs.—Acute and Chronic Forms.—Diagnosis.—Differential Diagnosis from Volvulus; from Stricture; from Internal Hernia; from Obstruction by Pressure from without the Bowel; from Foreign Bodies; from Peritonitis with Perforation.—Treatment.—Replacement by Manipulation; by Injections.—Treatment by Puncture.—Laparotomy.—Description of Operation.

Of prolapse of the rectum and invagination there are four distinct varieties.

1. *Prolapse of the mucous membrane alone.*—This, which is sometimes spoken of as "partial" prolapse because only a part of the wall of the rectum is involved in the descent, is well represented in Fig. 37.

2. *Prolapse of all the coats of the rectum including, when the disease is of sufficient extent, the peritoneum.* Fig. 38.

3. *Prolapse of the upper part of the rectum into the lower or invagination.* Fig. 39.

4. *Invagination in the Continuity of the Intestine.*—The same condition as the third variety, only occurring in a part of the bowel further away from the rectum.

The first form is a mere everting of the mucous membrane of the lowest portion of the rectum, rendered possible by the laxity of the submucous connective tissue. It is seen as an accompaniment of old cases of hæmorrhoids, and its mechanism may be studied at any time upon the horse in which it occurs naturally at the close of each act of defecation.

The second variety is an exaggeration of the first, in which, after the submucous connective tissue has yielded to its utmost, the whole thickness of the rectum begins to descend, and finally protrudes. It follows, of necessity, that after this protrusion has reached a certain length, the

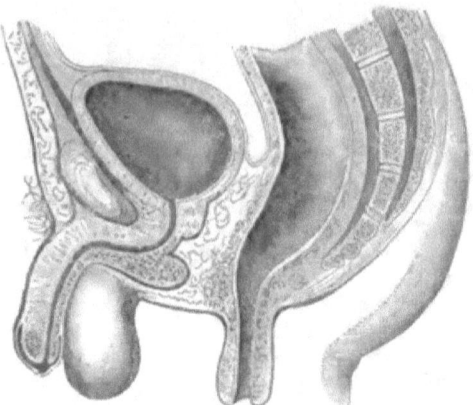

Fig. 37.—First Variety of Prolapse (Mollière).

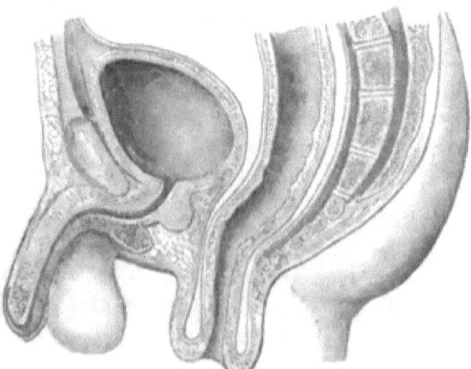

Fig. 38.—Second Variety of Prolapse (Mollière).

peritoneal coat must also descend outside of the body, and this condition is shown at a glance by reference to the plate.

In both of these forms, the protrusion begins first at the part of the rectum nearest the anus. In the third form, the part of the rectum higher up is passed through that nearer the anus, and what is known as an invagination occurs. This condition must, of necessity, cause a sulcus or groove to exist between the containing and the contained portion;

and at the bottom of this sulcus, the mucous membrane of one is directly continuous with that of the other. The depth of this sulcus must depend upon the point at which the invagination occurs; but in the variety under consideration, its bottom can generally be felt by introducing the finger by the side of the protruding portion.

In the fourth variety, this sulcus also exists, but its bottom cannot be felt, the point at which the invagination has occurred being in the continuity of the bowel, too far away from the anus. In the first three forms of the disease, there is always a protrusion of a portion of the bowel through the anus; in the fourth, there may be no such protrusion, the lower end of the invaginated bowel being still within the rectum, or, perhaps, too far up the canal to be seen or felt.

Having thus briefly defined the different varieties of prolapse and invagination, we shall consider each one in detail.

Prolapse of the Mucous Membrane Alone.—This is, perhaps, the most common of all the varieties of the disease when we take into consideration its frequent coexistence with hæmorrhoids. It is found in children

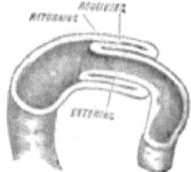

Fig. 39.—Third Form of Prolapse (Bryant).

most often between the years of two and four, and in adults it is more frequent in women than in men. Its causes are various. Among them may be enumerated the following: *a.* Those which tend mechanically to draw down the mucous membrane, such as hæmorrhoids, polypus, vegetations, and tumors. *b.* Those which tend to weaken or to destroy the action of the sphincters, such as ulcerations or incisions. *c.* Those which cause muscular spasm, such as fissures, worms, dysentery, phymosis, cystitis, calculus, stricture of the urethra, and enlarged prostate. *d.* Those which produce permanent dilatation and weakening of the sphincters, such as spinal paralysis, traumatism, chronic constipation, and sodomy. In this last connection, Mollière[1] details a very interesting case from his personal observation in a woman suffering from vesicovaginal fistula. Her husband, a brutish peasant, not daring to practise coitus in the ulcerated vagina of his wife, subjected her to unnatural intercourse daily for more than a year, with the result of producing a relaxation of the sphincter which showed itself by prolapse to an enormous extent, and by incontinence. To this lack of tonicity of the

---

[1] Op. cit., p. 202.

sphincters may be attributed the frequent occurrence of prolapse in feeble and badly nourished children. *e.* Those which produce œdema and swelling of the pelvic tissues, such as pregnancy, parturition, fæcal accumulations, and hepatic lesions. In this connection also, Mollière[1] details an instructive experiment which may easily be repeated on the cadaver. He says: "On the cadaver of a young girl, I introduced under the mucous membrane of the anus a blow-pipe, and fastened it with a ligature. By practising insufflation, the air instantly spread in the sub-mucous rectal tissue, and the mucous membrane escaped from the anus. I repeated the same manœuvre at another point of the circumference of the anus, with the same result. By dissection, I was able to assure myself that only the mucous membrane had been raised up. It was then sufficient in this case to cause tumefaction of the sub-mucous tissue, to produce prolapse; and, moreover, in this subject, the anus was still firmly closed." *f.* To these causes, it may be proper to add one anatomical one—the undeveloped sacrum in children, which, by its straightness, leaves the rectum comparatively unsupported.

Symptoms.—This first form of prolapse always comes on gradually and never suddenly. It may be partial or complete as regards the circumference of the anus, being in some cases of hæmorrhoids confined to one side of the aperture, and in others involving the whole circumference. It presents itself as a scarlet or livid mass (depending upon the state of contraction of the sphincter) projecting from the anus; covered with the natural secretion of the bowel; directly continuous with the skin on one side and with the mucous membrane on the other; and arranged in folds which radiate from the central aperture toward the circumference. It is at first spontaneously reducible, or at least easily replaced by a slight pressure, and remains reduced till the next act of defecation; but as the amount of prolapsed membrane increases, the difficulty in reduction becomes greater. At first also there is no pain, but after a time the act of defecation comes to be greatly dreaded by the patient, and the suffering continues till the tissue is replaced.

Treatment.—The first step in the treatment of prolapse of the rectum to which the surgeon will be called to attend will generally be to effect the reduction of the mass; after this has been accomplished, the treatment may be either palliative or curative. In children a prolapse may generally be reduced by laying the patient across the lap on its face and making gentle pressure on the protruded bowel with the fingers which have been well oiled; or with a soft greased rag. If this cannot be accomplished by a gentle taxis, and without bruising the parts, the child should at once be etherized and a curative procedure adopted. It is scarcely worth while in a child to stop to try the various methods of re-

---

[1] Op. cit., p. 199.

duction which have been recommended where the taxis has failed, before resorting to this step.

In an adult, however, ether and operative interference may both be declined, and the surgeon may have to tax his brain to accomplish the reduction without the aid of an anæsthetic. In such a case, after gentle taxis has been tried with the patient in the knee-elbow position, and failed, cold should be applied while the patient remains on the face in bed with a pillow under the pelvis; and this may be alternated with warm poultices and with plentiful applications of an ointment composed of equal parts of ext. of belladonna and ext. of opium. By these means the most effectual of which is position, reduction may almost always be accomplished. When by the action of the sphincter the prolapse has become gorged with blood and œdematous, the surgeon is often tempted to resort to leeches. They will generally give relief and may greatly facilitate reduction, but they are not free from the danger of a concealed hæmorrhage within the rectum after the prolapse has been replaced.

The palliative treatment is directed entirely toward diminishing the frequency and the amount of the prolapse, and in children a cure may sometimes be obtained by these means without resorting to surgical interference. The act of defecation is first to be regulated, and should be performed with the patient in the recumbent posture in bed, or while standing. One buttock may also be drawn aside so as to tighten the anal orifice with advantage; and any source of irritation which produces frequent defecation and straining in the act must be removed. After the action of the bowels, if the prolapse has occurred, the bowel should be thoroughly washed with cold water and a solution of alum (ℨ i. to ℥ viii.) before it is returned. Another favorite wash is composed of the tincture of iron, xx. to xxx. drops to four ounces of water. The patient should then be confined to the bed for some time and pressure should be applied over the anus by a pad kept in place by a T-bandage in the adult, or by a broad strip of adhesive plaster in children, applied so as to draw the buttocks into close apposition. A rectal supporter may also be worn when the patient is up and about, and perhaps the best of these is the one made by Mathieu, and represented in the figure.

After the bowel has ceased to come down with the act of defecation, an astringent injection may be given every night with advantage and allowed to remain in all night. The general health should be carefully attended to; tonics should be administered where they seem to be indicated; and if well borne, cod-liver oil may be used to fulfil the double indication of tonic and laxative. In children these measures may, as has been said, be curative, and, in fact, the disease often ceases spontaneously at about the time of puberty; but in adults they are not at all likely to be so, and more radical measures will generally be necessary. Of these there are several which are effectual, and each of them has its supporters and advocates.

In cases of prolapse attending old internal hæmorrhoids, the operation for the removal of the latter by the ligature may easily be extended so as to cure at the same time the former condition. And here a little careful discrimination may be necessary to distinguish between piles and prolapsed mucous membrane. The piles are smooth, hard, and shiny tumors; the prolapse is soft and velvety to the feel, and generally surrounds the whole margin of the anus without being divided into distinct tumors. In such a case, the proper course to pursue is to divide the prolapse into several sections with the scissors, and tie off each one exactly as though it were an internal hæmorrhoid. I have several times performed this operation with the happiest results, both as to curing the piles and the prolapse; but caution must be exercised as to the amount of tissue removed, lest too great a degree of cicatricial contraction result.

Since beginning the use of injections in the treatment of hæmorrhoids, I have also in some cases effected a cure of this form of prolapse by the use of carbolic acid in the same way as for piles. The idea of using car-

FIG. 40.—Rectal Supporter.

bolic acid for this purpose is, I believe, my own, and came naturally from my trials of the remedy in hæmorrhoids, but both strychnine and ergot have been used for the same purpose for some time.

At a meeting of the Therapeutical Society, December, 1879, reported in the *Gaz. Hebdom.*, Jan. 2d, 1880, Dr. Ferrand related the case of a lady who had suffered three years from prolapse, the tumor being nearly the size of the fist, and descending even when she walked across the room, and causing great suffering. One gramme and twenty centigrammes of a solution, composed of glycerin and water āā fifteen parts, and alkaline hydrated extract of ergot two parts, was injected into the ischio-rectal fossa beside the prolapse. Considerable benefit resulted, and three other injections were practised at intervals of twenty days, ten days, and a month, with the result of effecting a cure. The patient was seen after an interval of six months, and it was found that the prolapse was not reproduced even by such exertion as going up several flights of stairs.

Vidal[1] also has recorded three successful cases of cure with ergotine. The first was that of a man, aged thirty-nine, who had suffered for eight years. After five injections of fifteen drops of a solution of ergotine, at intervals of two days, the mucous membrane scarcely protruded at all. After the eleventh injection it only came down during defecation and returned spontaneously. The whole number of injections was twenty-two, and the man remained perfectly well four years after. The second patient, a female, aged sixty-four, was cured after twenty-four days' treatment, and remained well two years and a half after. The third patient, a female, aged forty-five, was cured in fifteen days by six injections of twenty or twenty-five drops each. The solution used consisted of fifteen grains of Bonjean's ergotine dissolved in seventy-five minims of cherry-laurel water. The injections were made at the distance of one-fifth of an inch from the anus. Acute pain always followed, and contraction of the sphincter lasting several hours. Several times an injection of twenty-five drops of the solution caused spasm of the neck of the bladder and retention of urine. In no case did the injections produce any local inflammation or abscess. Dr. Vidal has more recently expressed himself[2] as preferring Yvon's solution of ergot to Bonjean's ergotine, as causing less pain.

The danger to be avoided in this method of treatment is the use of too irritating solutions, or solutions in too great quantity which shall excite a suppurative action and produce constitutional poisonous effects.

Cauterization.—In children in whom milder measures have failed, a very effectual means of cure is the application of fuming nitric acid to the mucous membrane of the prolapsed part. The bowel should first be carefully wiped off with a towel or sponge, and the acid then applied by means of a small stick all over the mucous membrane, but not at all to the skin adjacent. After such an application the bowel should be replaced, a pad of lint firmly applied over the anus by means of broad strips of adhesive plaster, and the bowels confined by means of opium. Allingham speaks of stuffing the rectum with wool in addition, but I have always found the pad and straps sufficient when thoroughly applied, and the child kept on its bed. After three or four days the straps may be removed, and the bowels moved with castor oil. In a large proportion of cases, the cure will be found complete, though, in a few cases, I have seen a return of the disease after a few months. In any case, however, the benefit will be found to be very great, and should the disease return, a very careful search should be instituted for some existing source of irritation, such as polypus, phymosis, or calculus. In case of a recurrence, a second application will be effectual in causing a cure.

This treatment, though successful in children, is by no means so in

---

[1] Paris Médical, August 28th, 1879.
[2] Gaz. Hebdom., Jan. 2d, 1880.

adults. Allingham calls attention to the occurrence of deep sloughs in old persons with debilitated constitutions; and, as a result of **such a** slough, he has seen an almost fatal hæmorrhage. Stricture of the rectum may, without doubt, be caused by **too** free use of this remedy, but since it follows its abuse and **not its proper use** in appropriately **selected cases,** it can hardly be **considered an objection.**

Linear Cauterization.—In adults this is undoubtedly the best **means at** our command for dealing with this affection, and the best **means of applying** it is that recommended by Van Buren, with Paquelin's cautery.

The patient **is at** first etherized **and** placed **in** Sims's position. Van Buren reduces the prolapse, and applies **the** iron with the aid of a speculum. Allingham first applies the iron **and** then reduces the prolapse. In either case, from three to six vertical stripes should be made upon the mucous membrane, with the iron heated to a dull-red heat. **The cauterization** should begin about three inches up the rectum, and end at the junction of the skin and mucous membrane. They should also be deeper at the end, where there is no danger, than at the beginning, where the bowel may be perforated. Van Buren recommends that the iron be **bent** at a right angle **a** short distance from the end, so that it may be the more thoroughly **applied to the** concavity of the rectum, and that, in mild cases, a small iron **should be used,** "no thicker than an ordinary probe." Allingham, **in bad cases, burns** through the sphincter muscle at two opposite points, **after** reducing **the bowel,** and inserts a small pledget of oiled **wool.** By this **burning through the** sphincter, the patulous condition **of the anus is** overcome. **The result of** the operation is to decrease the circumference of the anal **orifice, and, in this way,** to effect a cure. **The** patient should be confined **absolutely to bed till the** wounds are entirely healed, so that a **recurrence of** the descent may be effectually avoided.

For some time after the healing, and after the patient is allowed to be up and about, in fact, until the full effect of the operation has been obtained, a bed-pan should be used. The first operation, if thoroughly performed, will probably result in permanent cure. Should it not, it may be repeated. The only danger in connection with it is the occurrence of secondary hæmorrhage when the sloughs separate, and of primary hæmorrhage from large veins at the time of the application of the iron. To avoid this, Allingham recommends the choosing of points for cauterization which are free from large venous pouches, such as may be visible on the surface of the tumor.

In old cases of extensive disease, the operation as thus described may not be effectual, and it may be necessary actually to produce a stricture at the anus to prevent recurrence of the trouble. There is, perhaps, no better means of accomplishing this than to apply the iron to the whole circumference of the anus, circularly, instead of in longitudinal stripes; but such an operation will seldom be called for.

There is one other method of dealing with this affection, which, though not as simple as the cautery-iron alone, is well worthy of trial, and that is Smith's operation with the clamp and cautery. We have already given a figure and description of the clamp and the operation in speaking of hæmorrhoids, but the operation is even better adapted to cases of prolapse than to hæmorrhoids, the mass being larger and more readily seized, cut off, and cauterized.

Having thus described the most effectual means of dealing with this troublesome affection, it is scarce worth while to describe the various cutting operations by which pieces are removed either from the mucous membrane alone, or from the sphincter muscle, with the object of accomplishing the same result that is more readily attained with the cautery iron. Dupuytren's operation consisted in removing three elliptical folds of skin and mucous membrane from the verge of the anus. The same idea has been more recently applied in Germany.[1] Robert and Dieffenbach cut out wedge-shaped pieces, and approximated the edges with deep sutures; and the latter even went so far as to cut off the whole tumor—an operation now seldom practised, except in slight cases, such as those accompanying internal hæmorrhoids.

Prolapse of the Second Degree.—As already said, the second variety of prolapse differs from the first in the fact that it is composed of the whole thickness of the bowel, and, therefore, when of sufficient length, of peritoneum also. It is probable that every prolapse of more than two inches in length may contain peritoneum; and it follows from the anatomy of the parts that the peritoneum will extend lower on the front than behind. In the peritoneal pouch thus formed in front there may be located coils of intestine, an ovary, or a part of the bladder. In this form of prolapse there is no groove or sulcus, as is shown by the figure, and the absence of such a groove is, therefore, no proof of the non-existence of a fold of peritoneum in the tumor.

It is a mistake to suppose that this second variety is not met with in children, for it is only an exaggerated form of the first, being the next step in the descent after the submucous connective tissue has yielded its utmost; and exaggerated cases of prolapse are often seen in children. It is distinguished from the first variety—first of all, by its size. The first is never very large; while the second, from the nature of the case, must be of considerable dimensions. Again, a prolapse of the first variety is seldom of long standing; while one of the second is generally so. The second generally follows the first, but a prolapse may be of this variety from the beginning; resulting, in such a case, generally from violent straining, and coming on suddenly. The first variety is not firm and thick to the feel; the folds of mucous membrane radiate from the orifice

---

[1] "Eine neue Methode der operativen Behandlung des Mastdarmvorfalls." Deutsche Med. Woch., No. 33, 1880.

to the circumference, and the opening is circular and patulous. In the second, the orifice is slit-like **and is** drawn backwards by the attachment of the meso-rectum, or in females forward by the closer attachment to the vagina. The **form of the tumor is** conical, its walls are thick and firm, and when pressed **between the** fingers, the gurgling **of** gas in a contained loop of **intestine may** sometimes be detected, **and a resonance may** be obtained **on percussion.**

If such **a tumor** be carefully **dissected, the coats of** the protruded bowel will be found enlarged; the **mucous** membrane **will** be seen to be thickened **and** dense in structure, especially at the free **extremity; and** it will also sometimes be found eroded and granular. **The** submucous areolar tissue will be seen to be infiltrated with albuminous deposit, and the **muscular layers** will be hypertrophied. Owing to these changes, **the** bowel is actually increased in size, and becomes too large **to** be retained **in** its proper place; which explains the difficulty often experienced in reducing it and in keeping it reduced, in spite of the constant straining and desire for defecation which it produces. These changes in the mucous membrane may in **rare** cases result in the production of **a** foul, hard, bleeding, eroded **mass, which** may at the first glance strongly suggest malignant growth. **The** bleeding from a prolapsed rectum is commonly in the form of **a general oozing, and** applications of astringents may be necessary for its control.

Strangulation is **rare in** infants and **in feeble** old people, but in a strong person the sphincter may be **sufficiently** powerful **to** produce such a result. A strangulation may be **only temporary when met** by the proper means, or it may continue long enough **to cause** ulceration and partial gangrene; the latter, however, is **rare. When it occurs, it is** possible for it to end fatally from the contiguity of the peritoneum; **but it more** often results in a spontaneous cure of the prolapse, **and** in a cicatricial stricture, the location of which will depend upon the length of the prolapsed portion and the point at which the sphacelus occurs.

The causes of the second variety are the same as of the first, and need **not** again be enumerated. The symptoms also are the same, with the addition of more or less incontinence of fæces in old cases; but the treatment is not the same in all respects; for certain measures which may be safe when a prolapse contains no peritoneum may be fatal under the opposite condition.

In cases in which curative measures are out **of** the question, the hæmorrhages and the erosions may be **relieved by** suitable applications, rest in bed, defecation in the recumbent posture, etc. Persulphate of iron is perhaps as good an application to the bleeding surface as any other; and weak solutions of **nitrate of silver often have a good** effect upon the erosions. The reduction of **a prolapse of** the second degree is by no means as simple a **matter as that of the first. When the** sphincter is tight and the **tumor** œdematous, **it may be nearly** impossible; and in **old** cases where the opposite condition of the sphincter obtains, it may be equally difficult to keep the parts within the body after placing them there. The latter may, however, generally be accomplished by the means already enumerated, and the reduction in obstinate cases may generally be obtained through the influence of anæsthesia. The dangers which may attend an attempt at reduction by taxis are well illustrated in the following case.

CASE X.—*Complete prolapse of the rectum; rupture of the bowel during reduction.*[1] The case was that of a woman, aged forty-six years, who about twelve years before, a short time after a difficult labor, had begun to suffer from prolapse which came down daily at the time of defecation, and was easily reducible. She was seen by the doctor at a time when the tumor had been down nearly twenty-four hours and had resisted all the efforts of herself and female friends at replacement. She had passed a restless night and was much fatigued by her journey in an old cart, but had experienced no bad symptoms referable to the stomach or bowels. The doctor found at the anus a tumor larger than the fist, round, red, and covered with bloody mucous.

The prolapse was directly continuous with the margin of the anus in such a manner as to render the introduction of a sound between them impossible. At the extremity of the tumor there was a rounded aperture which admitted the finger without obstacle. To accomplish the reduction the woman was placed on the bed with the thighs separated; the tumor was seized in the palms of the two hands and the ends of the fingers, and a gentle circular compression was exercised in order to diminish its volume and cause it to go up by an operation similar to the taxis. The resistance being great, a few moments were allowed for rest, and after a quarter of an hour the same manœuvre was repeated after having enveloped the tumor in a cold cloth. "After a few moments I felt," says the narrator, "during a violent effort of the patient, the tumor distend under my fingers, and at the same time I heard a noise similar to that made by tearing parchment. At the same time the tumor suddenly disappeared of itself, and syncope, nausea, and a marked change in the expression of the face supervened.

When the patient came to herself she complained of severe colic. I then found outside of the anus a loop of intestine which I easily replaced, and on introducing the finger into the rectum I recognized at a considerable height an irregular longitudinal rent the extent of which I was unable to determine. I placed a tampon of lint over the anus and kept it in place with a T bandage and compress. I sent the patient to her home, ordering that nothing be disarranged. As the case was very serious, I requested a neighboring *confrère* to come and aid me with his advice. At our arrival, six hours after the accident, I found the patient sitting by

---

[1] Condensed from report by Dr. Roché, Révue Méd.-Chirurg., 1853.

the corner of the fire, without the dressings. Between the separated thighs were exposed, in the midst of the ashes, the large and a considerable part of the small intestines, distended with gas, cold, and in several spots livid. The face was Hippocratic, the pulse thready and much accelerated, the voice feeble; and to this was joined colic and continual vomiting. After having placed the woman in bed and raised the intestines, the mass was replaced within the body, the former dressing was applied, and the woman died in a few hours."

Two questions may arise in this connection. Should reduction be tried when the tumor is inflamed; and should it be tried in case of a circular slough? In answering the first question, the distinction must be made between a prolapse which is merely strangulated and one which is inflamed. The appearances may be much the same, but an old prolapse in an old person when found in this condition is much more apt to be inflamed than strangulated, for the sphincter muscle in such cases has generally lost the power of forcible constriction. The danger in returning an inflamed prolapse into the body is that the inflammation may extend and cause general and fatal peritonitis; and as a rule it is safer not to employ the taxis in such a case, but to put the patient in bed and treat it by local applications and rest till the acute symptoms have disappeared.

In answer to the second question, Mollière[1] recommends extirpation of the prolapsed portion rather than its reduction when there is a circular slough, on the ground that no matter how radical such a step may appear at first sight, it is better than leaving the case to nature. For a circular slough means inevitably a cicatricial stricture; and if the prolapse be extensive, a stricture situated high up in the rectum or sigmoid fluxure beyond the reach of art. As preferable to this he recommends the complete ablation of the tumor with all the dangers which attend such a step. These dangers are easily understood to be hæmorrhage, hernia of the intestines through the incision, and peritonitis. Each may be avoided where the surgeon is prepared beforehand for their occurrence, and Mollière relates one case where the operation was performed by himself *with the hot iron*, but the patient "died on the eighth day from the effects of the chloroform" so that he was unable to decide on the value of the operation.

Excision with the surgeon's eyes open to the fact that he is dealing with peritoneum may perhaps be done with success under such circumstances. At all events it is a very different matter from excision of this variety of prolapse under the impression that it is the one previously described, and contains no peritoneum, as the following case will show. Van Buren[2] says: "I have reliable information of a case in which the

---

[1] Op. cit., p. 240.
[2] Op. cit., p. 60.

removal of a 'compete prolapse' of long standing, in a child, was quite recently undertaken by a hospital surgeon of mature years. The protest of a junior colleague led the operator to pass some deep sutures, in deference to a fear expressed as to the probability of intestinal protrusion, but he was confident that the tumor consisted of mucous membrane alone, and proceeded to remove it. Notwithstanding the deep sutures, protrusion of several coils of small intestine did occur, and the child died, in collapse, within twenty-four hours."

In this form of the disease, the surgeon may find it better after mature deliberation not to attempt a radical cure, but to confine his efforts solely to palliation. The following case illustrates the danger of attempted removal of a part of the mass in an old and extensive prolapse.

CASE XI.—"The patient was an elderly man who had a prolapsus as big as a cocoa-nut always coming down, and rendering his life a burden. He had already been operated upon twice by a hospital surgeon, but in vain. The patient was then sent to me, and, formidable as the case looked, I determined to undertake it. I applied the clamp deeply in three different directions. There was a great deal of bleeding and I had to apply the cautery over and over again before I could stop it; and then, just as I was finishing the operation, a most untoward event occurred— severe vomiting, as the result of the anæsthetic, took place. The prolapsus was forced still further down; and before I and my assistants could return the parts, the violent action of the abdominal muscles was such that the weakened coat of the bowel gave way, and a knuckle of small intestine actually protruded through the rent thus made. I carefully returned this as soon as the vomiting ceased, and anxiously waited the result. Our house-surgeon, Mr. Newmarch, watched the patient with great care and treated him with great skill, keeping him constantly under the influence of opium, and locking up his bowels for several days. The result was not a single bad symptom of any kind. On the first action of the bowels there was no protrusion, nor afterwards; and as soon as the man was fairly recovered I removed three longitudinal folds of skin from the anus, so as further to tighten the parts. The man was completely cured. Now, the lesson this case teaches is this—not to employ an agent which could cause vomiting; because, of course, in such a terribly severe case as this it is absolutely necessary to clamp deeply, and thus weaken the bowel. It was a most unlooked-for accident, not likely to occur again; in fact, it is hardly reasonable to expect to meet with another such a case for operation. I have, however, been called to cases as bad or worse, but where no operation could be recommended."[1]

Dr. Kleberg has utilized the elastic ligature in operating upon severe cases of prolapse: and, it may be, that if the mass has to be removed at

---

[1] Henry Smith, Lancet, Mar. 15th, 1880.

all, the method he describes is the preferable one. The operation is performed as follows.[1]

CASE XII.—*Operation.* On the previous day a dose of castor oil was given, and on the morning before the operation an enema of luke-warm water was administered high up the bowel. Immediately before, a glass of wine and one grain of opium were given. After the patient had pressed down the gut as far as he could he was placed on the operating table in the lateral position with the pelvis raised and shoulders turned downward. Chloroform was then administered. In two cases Kleberg has operated without chloroform because the patients were in such a miserable condition that he was afraid to narcotize them thoroughly, and an incomplete narcosis has all the dangers of profound anæsthesia and none of its advantages. After the chloroform, he says, " I carefully examined about the rectum at the junction of the skin and mucous membrane in order to discover the sphincter ani—a procedure that was more difficult than one would think, because it had become so stretched and atrophied that I could only make it out by feeling under the fingers the coarser fibres running across the longitudinal axis of the bowel. Of anything like the normal muscle there was nothing to be discovered.

An assistant, at this point, surrounded with all the fingers the prolapsus from above, the points of the fingers being directed towards the free end of the prolapsus, and pressed as hard as possible into the gut at a point perhaps half an inch below the supposed sphincter. Immediately in front of the ends of the assistant's fingers I then placed a good, fresh, unfenestrated drainage tube of rubber, one and one half lines in diameter, around the prolapsus, and drew it only as tight as seemed necessary to stop the circulation. The elastic ligature was brought to the necessary tension by means of an easily-untied slip-knot of silk thrown under it.

The assistant now had both hands free; and from this time on the operation was performed under the carbolic spray. A few lines beneath the ligature I now made a longitudinal incision two inches long through the prolapsed gut, and in this way opened the sac formed by the drawing down of the peritoneum. Then I seized the elastic ligature with the forceps and fixed it firmy. It was thus an easy matter to push back into the peritoneal cavity a protruding loop of intestine without the slightest bleeding taking place into the wound or any air entering the peritoneal cavity; because the elastic pressure follows so rapidly all the movements that no opening can exist anywhere.

After I had convinced myself that the peritoneal sac was empty, and that no invagination of the intestine was present, but, on the other hand, only that part of the gut which was to be removed lay in front of the ligature, I thrust the largest size Luer's pocket trocar through the pro-

---

[1] Ueber die Anwendung der elastischen Ligatur zur Operation sehr schwerer Fälle von Prolapsus Recti. Arch. für Klin. Chirurg., vol. xxiv., p. 840.

lapsus, immediately below the elastic ligature, from before backwards, and passed through the canula two elastic drainage tubes of one and one-half lines in diameter, and, after removing the canula, tied them as tightly as possible, one on the right side, the other on the left. These knots were secured against slipping by means of the knot of silk. The first provision against hæmorrhage—the elastic ligature applied after Esmarch's plan—was then removed and the prolapsus cut off with the scissors one inch in front of the permanent ligatures. After a few minutes' time, during which I kneaded the parts which still remained and lay above the ligatures thoroughly, and as far as possible removed the fluids from them; I covered the parts around the stump with cotton, and soaked that part of the prolapse which still remained above the ligature with a solution of chloride of zinc, dried it, squeezed the soft parts once more, thoroughly applied the chloride of zinc again, and then covered the whole with dry cotton-batting, giving the patient instructions to remove this as soon as it became moist and to replace it with dry, and to give the air all possible access to the parts."

No fever followed the operation, and the pain was bearable, with the aid of an occasional opiate. On the next day the parts had so far shrunk as to leave a concavity at the anus where before there had been a bulging. There was no bleeding, no peritoneal irritation, and only slight tenesmus. On the fourth day the first ligature cut out, and the second on the fifth. The rectum was irrigated twice a day with water and permanganate of potash, and on the seventh day a dose of castor oil was followed by a large evacuation while the patient was on his back without pain or hæmorrhage. The passage, however, was involuntary. On the fourteenth day the wound was healed, the general condition of the patient excellent, and the evacuations regular but still involuntary. The sphincter at this time began to be appreciable, and there was no protrusion of the bowel, the patient going about and wearing a bandage. One month later he had control of solid fæces, but there was still a slight discharge of mucus; and after another month he was entirely well.

In this case the prolapse was about a foot in length and six inches in diameter. The mucous membrane was spongy, bleeding, excoriated, and ulcerated. The patient had been sick for two years, had been bedridden for two months, and was waxy pale.

Another case by the same surgeon and the same method ended fatally, but can hardly be considered a fair test of the dangers of the operation, on account of the exceedingly bad condition of the patient.

Third and fourth varieties.—These two forms of invagination will be described together because of the fact that they differ from each other not at all in their nature but only in extent and location. It will be observed that the word prolapse is now dropped and invagination substituted which more aptly expresses the condition. The essential difference between the disease now to be considered and the forms already described consists in the fact that while in the latter the bowel begins to slip **down from** its lowest portion at the anus, in the former the lowest portion **at** the anus remains in its proper position and the bowel from above **is** telescoped within it. Under these circumstances it is evident, as is shown in Fig. 39, that the **affected** portion of the bowel must consist of three different and distinct **cylinders**, an outer one which contains the other two, and two included portions, **one of which is** the entering and the other the returning **bowel**.

When the **upper** part of the rectum **becomes** invaginated in this way within the **lower**, the included portion will appear at the **anus** as in the cases of prolapse already described, **and a** distinct sulcus may be felt by the finger **between** the extruded portion **and the mucous** membrane which is continuous with that of the anus. **The bottom of this** sulcus or the point at which the entering portion becomes directly continuous with that into which it enters may also be felt by the finger if it is low enough down; if not, it may be detected by **the** aid of a soft catheter. This **is what** is understood by the third variety of prolapse. When a portion **of** the bowel still further removed from the anus has become invaginated into that immediately below, the included portion may or may not descend sufficiently near **to the anus** to be felt by rectal touch, and the sulcus may not be apparent. **This constitutes** the fourth variety or what is now generally known as **intussusception**. **It is** evident that between a case of prolapse in which **all the coats of the rectum** appear through the anus, and in which **a sulcus can be felt by the** finger passed around the protruded portion; and a **case** in which **the ileum is** telescoped through the ilio-cæcal valve and appears at the anus, **the difference is one** of degree and not of kind.

Of this condition there are many degrees, and **almost** any portion of the bowel from the duodenum to the rectum may become invaginated into the portion next below. The cæcum itself may be so loosened from its attachments as to follow the same course, and the orifice of the appendix vermiformis may be detected at the anus by the side of the orifice of the included bowel.

In 763 cases of invagination collected by Bulteau,[1] 220 were of the small intestine; 151 of the large; and 392 ileo-cæcal.

The mesentery of the two included portions is drawn in with them, and by its attachment and traction gives to them a curve the concavity of which is towards the point of attachment of the mesentery. For this **reason the lower orifice** of the included portion is not found in the axis of the containing portion, but turned **toward** some portion of its circumference, and is, therefore, often **difficult to** detect by a digital examination.

---

[1] De l'occlusion intestinale **au point de** vue du diagnostic et du traitement. Thèse de Paris, 1878.

The immediate effect of an invagination is to interfere with the passage of fæces, but seldom to entirely prevent their passage, for the fæces do pass and in considerable quantity, forced down through the constriction by the contraction of the healthy bowel above.

Another immediate effect which is due to constriction of the bloodvessels in the included mesentery and in the walls of the included portion, is the transudation of serum and consequent swelling of the intestinal walls. By this means the serous surfaces become dark-colored, and the mucous surfaces become infiltrated; blood is effused between the mucous surfaces of the outer and middle layers, and lymph between the serous surfaces of the middle and internal layers, and after a time these become completely agglutinated.

If the constriction be sufficiently severe, the included portions soon become gangrenous and slough away, the lumen of the bowel is again established, and a circular cicatrix is left. This is nature's method of cure, and though life is by it saved for a time, in the end the cicatrix thus formed may become a stricture which shall be more surely fatal than the condition from which it arose. The invaginated portion is at first of necessity short; but as the case advances, it may reach to several feet, and in one case[1] there is a reason to believe that about four yards of intestine came away, piece by piece, *per anum*.

The disease is twice as common in males as in females, and is greatly more common in children than in adults. In adults the trouble will generally be found to involve the small intestine; in children, the large. An invagination of the small into the large intestine begins generally at the ileo-cæcal valve, which with the vermiform appendix is carried up the ascending, and along the transverse colon, till it may finally reach the anus and protrude through it, the valve all the time remaining the lowest portion. In these cases only the inner tube is made of small intestine, the middle and the outer consisting of the large.

Strangulation is much more frequent where the outer layer is composed of the small than where it is composed of the large intestine; because of the greater tightness of the constriction. In the latter case the congestion may be only moderate in degree and the condition may last many weeks without gangrene or ulceration. This condition is known as chronic intussusception.

If sloughing occur at all, it may happen at any time after the first week, generally, however, it occurs within three weeks, though it may be delayed for a much longer time. In one case[2] the separation of fragments of intestine extended over an interval of three years.

In about one-half of the reported cases a favorable termination has followed spontaneous separation, in the remainder death has occurred

---

[1] Peacock: Path. Trans., vol. xv.
[2] Peacock, loc. cit.

after a longer or shorter interval. Several pathological changes may occur. The peritonitis which serves to unite the serous surfaces of the contained portions may become general and cause death. The ensheathing portion may become **ulcerated** and perforated, allowing of the extravasation of fæces. **The ulceration may** perhaps be due to the lateral pressure of the **end of the contained** portion against **the side of the** cylinder which contains it.[1] Separation by sloughing leaves the upper end of the **ensheathing** portion united **with** the lower **end of** the healthy bowel, and **results in** complete amputation of the contained portion. Extravasation **may** also occur from a deficiency in this union at the time when separation occurs.

The **causes of** invagination are **not as yet perfectly understood.** It is easy to understand how in the effort **which the intestine** makes to relieve itself of a polypus or other **tumor by** its vermicular action, **not only the growth** itself may be extruded, but also the portion of the bowel **to which it is** attached; and polypus is one of the recognized causes of this condition. But in the great majority of cases no such palpable cause is **to be detected.** Except in the case of a tumor it is probably always an accident **of** sudden occurrence dependent upon some violent action in that part **of** the bowel. A collection of gas causing an undue dilatation in one part of the intestine, combined with a violent movement of the abdominal muscles, and a peristaltic movement in the portion just above that which is distended, might, it is easily understood, cause the accident. So, also, might any interference with, or undue violence in, the rhythmic action of natural peristalsis, by which the bowel in successive portions is first shortened and dilated by contraction of the longitudinal fibres, and then narrowed and elongated by the contraction of the circular fibres. Since the wave of peristaltic action is constantly passing from above downwards, it may easily happen that a narrowed portion may under unfavorable circumstances be caught in a dilated portion just below, and, once engaged, the exaggeration of the condition becomes natural and easily understood. It is to such explanations as this that we have to look in the absence of any palpable cause.

Symptoms.—An invagination will cause a very different train of symptoms, according to the part of **the** bowel affected and the intensity of the constriction. **As** a rule, the symptoms are more acute and severe in invagination of the small intestine, and **are** more chronic in the large, because the constriction is more intense in the former than in the latter; but an invagination **of** the small intestine may approach in symptoms and chronicity to one of **the** large, and *vice versa.*

Wherever the constriction be located, its first symptom is generally a sharp attack of pain **in** the abdomen, coming on suddenly, and often in the midst **of perfect health.** There **is** nothing characteristic in this

---

[1] Aitken: Pract. of Med., vol. ii.

pain. It may pass off after a few hours and again return; it may or may not be accompanied by vomiting at the start; it is sometimes relievable by direct pressure, and it is not at first accompanied by any tenderness of the abdomen.

Change in the character of the evacuations is also a symptom common to the disease in any part. After the onset there will still be a discharge of the contents of the bowel below the constriction, and a certain amount of fæces may still leak through the invagination. Instead of the natural passages, however, the appearance of bloody stools is a very common occurrence, the blood coming, as has already been explained, from the congested and swollen mucous membrane of the outer and middle portions. There is also present at times a dysenteric discharge and a good deal of tenesmus.

By careful manual examination, a tumor can generally be discovered in the abdomen, which may be characteristic enough to form a basis for the diagnosis; but this may be concealed by the presence of much fat, or by a general distention of the abdomen with gas. The tumor is cylindrical, and may be movable under the hand from its own peristaltic action, or it may be seen to change its position from day to day as the invagination gradually advances, and more and more of the bowel becomes involved.

The other symptoms depend in great measure upon the severity of the strangulation, and, as has been said, are more marked when the small intestine is implicated. In such cases, the symptoms rapidly increase in severity. There may or may not be considerable febrile action; the abdomen soon becomes tender to the touch; there is almost complete obstruction, or else only the passage of bloody mucus; the patient rapidly sinks, and the history ends either in death or in the sloughing of the included part. The latter is shown by a re-establishment of the calibre of the bowel, and, therefore of the passages; by an abatement of all the worst symptoms, and finally by the appearance of larger or smaller pieces of gangrenous intestine in the passages.

The existence and the early appearance of fæcal vomiting have been given as points in favor of the diagnosis of intussusception of the small rather than of the large intestine, but they point rather towards complete obstruction than to the particular seat of the obstruction.

In invagination of the large intestine, the general history of the case is that of a more chronic trouble. The pain is less severe and the paroxysms separated by longer intervals; the fæcal evacuations are larger, and the dysenteric symptoms are more pronounced; vomiting is variable, and after a time often stercoraceous. This state may continue for several weeks before death results from gradual exhaustion or from the supervention of acute strangulation. The history of a case of chronic invagination may at any time be cut short by the occurrence of a general acute perionitis, and this is particularly apt to happen at the time of the separation of the slough.

Diagnosis.—In **any case in which** the invaginated portion descends near enough to the **anus to be felt by** digital examination, the diagnosis is easy to the surgeon **of ordinary care** and intelligence who has studied the symptoms which **infallibly point in** the direction of intestinal occlusion. But **when such an examination has** been made with a negative result, beyond the **fact that** occlusion exists the surgeon may **be** completely at a loss. Under **such** circumstances **the** differential diagnosis rests between the following conditions: 1. **Invagination;** 2. Volvulus; 3. Stricture; **4.** Concealed internal hernia; **5. Pressure from** without the bowel by tumors etc.; 6. Obstruction from **foreign bodies, as calculi,** indurated fæces, etc.; **7.** Peritonitis from perforation. It may **be as well to state at** once that in these cases the differential diagnosis will often **be impossible,** and then go on to throw what light upon the question modern science has made available. It is a good plan to divide all cases of intestinal **obstruction** into the acute and the chronic. An acute case will generally be either an invagination, a volvulus, or an internal hernia. Duplay[1] also has called attention to **the fact that a** peritonitis from perforation may cause **all** the symptoms of **an acute occlusion** and has given the chief points in the diagnosis **of that affection.** In peritonitis the vomiting seldom becomes fæcal but remains bilious **to the end;** the constipation is less marked and the patient **generally passes gas and liquid** fæces or small quantities of solid matter; **the tympanites** is also **less marked, and the** coils of intestine are less pronounced; the pain begins **with** great severity **at** one point and extends over the **whole** abdomen (the same thing may **happen in acute** obstruction, but **in** such cases the other symptoms—fæcal **vomiting, absolute** constipation, absence of the passage **of** gas *per anum*—are all equally severe, while in peritonitis they do not correspond in severity **with** the intensity of the pain); the temperature is elevated in peritonitis and normal or even less than normal in obstruction.

Having then excluded peritonitis from perforation, the diagnosis in any acute case will rest between invagination, volvulus, and internal hernia. Invagination is indicated by the signs of *partial* occlusion, by the moderate tympanites, by the bloody stools mixed with mucus, the tenesmus, and the presence of the tumor. The diagnosis between volvulus and internal hernia will generally be impossible except as the history may point to antecedent peritonitis, or to a hernia which has ceased to come down; or as the careful exploration of the abdomen by palpation and of the pelvis by rectal and vaginal touch may show the existence of an induration or resistance limited to one point.

In other words, **in any acute case of** occlusion the existence of invagi-

---

[1] Duplay: Du Traitement Chirurgical de l'Occlusion Intestinal. Arch. Génl. de Méd., Dec., 1879.

nation may be decided by the presence or absence of its peculiar symptoms, and if excluded the diagnosis rests either with volvulus or internal hernia, but with which it may be impossible to decide.

In a case of chronic intestinal occlusion, the diagnosis rests between invagination, occlusion by the pressure of solid or fluid tumors outside the bowel, stricture of the intestine, abnormal adhesions of the bowel, and obstruction by foreign bodies within the bowel, such as biliary calculi, indurated fæces, tumors, etc. The easiest of these to diagnosticate is that which comes from the pressure of a tumor without the bowel. Chronic invagination may be made out by the symptoms already given. For the symptoms of stricture, we must refer the reader to the chapter on that subject, and these symptoms are much the same whether the obstruction be due to a narrowing of the calibre of the bowel by a deposit in its wall, or to the presence of a foreign body, or abnormal adhesions of the peritoneum which cause acute flexures and obstructions in its calibre.

It will thus be seen that the differential diagnosis is shrouded in difficulty, and that the difficulty is rather greater in a case of chronic than of acute obstruction. A well-marked case of invagination, whether acute or chronic, is, however, the easiest of all the forms of occlusion to distinguish, and the diagnosis can generally be made with sufficient approach to certainty to guide the surgeon in the selection of his plan of treatment.

Treatment.—It is evident that the treatment of the conditions we have been describing must differ in every particular from that of those previously described. When the invagination has occurred in the rectum, that is, when the upper part of the rectum has become telescoped into the lower, and has appeared as a prolapsed mass outside of the anus, the case may still be relievable by the methods of reduction and taxis. The mass must be replaced by a process exactly the reverse of the one by which it came down, the most dependent portion being first carried into the body, and the entanglement unfolded in this way. In a child, with the assistance of anæsthesia, the inverted position, and gentle manipulation with the fingers or possibly a soft bougie, this may sometimes be accomplished where the point of constriction is low down near the anus. Prall[1] reports a case where replacement was successfully accomplished by manipulation with the tube of a stomach-pump, though the mass could only just be felt in the rectum.

In cases, whether of adults or children, where the constriction is still higher in the intestine, and manipulation with the hand or bougie is out of the question, various other mechanical means may be tried with a prospect of success. These consist in applying indirect pressure to the invaginated portion, and to the constricting part by means of copi-

---

[1] Brit. Med. Journ., July 31st, 1880.

ous injections of water or air, **but it** should be understood that **they are only applicable to cases affecting the large intestine alone, and the lower down** in the large **intestine the** constriction may be, the better **is the** prospect of their **success. In cases of this** kind, the mechanical treatment may be **assisted by the previous** administration of opium and belladonna in full doses, **the one to quiet** peristalsis, **the** other to relax the unstriped muscular **fibres of the intestine.** To **these** means may be added **the reversal of** position **and** anæsthesia, **and then** the copious injection of large quantities of warm **fluid, or of air by means** of a bellows, may in a few cases be successful.

The following case illustrates the method **of treatment by** injection, and what, under favorable circumstances, **may be accomplished by it.**[1]

Case XIII.—A well-nourished infant, **seven months old, was in** perfect health till noon of the day of **attack, when** she **suddenly screamed,** and immediately afterward became pale, cold, and **collapsed. She was** put into a warm bath, after which she lay quietly in the nurse's **arms** for an hour and a half, the bowels acting slightly once or twice. **At 3 P.M.,** the child **had** become warmer, and was sleeping quietly, occasionally, however, waking **up with a scream,** and drawing up her legs with an expression of **severe pain. There** was occasional vomiting, and **at 6** P.M., two passages **of bloody mucus. At** 11 P.M., a distinct but ill-defined oval tumor, **about an inch and a half in its** longest diameter, could be felt through the **parietes, at a spot two inches to the** left of the umbilicus. A considerable quantity (perhaps **a drachm) of dark** blood came away, and it was determined to distend **the large intestine with thin** gruel. The child was put thoroughly under **the influence of chloroform,** and placed on the table with the nates well raised **on** a pillow. **The gruel was** slowly injected by **means** of a Higginson's **syringe, the** upper part of the nozzle being pressed firmly against the anus **to** prevent any from **escaping.** After a pint or more had been injected, the abdomen became tense, and **the** distended bowel could be felt like **a** hard **rope an** inch in diameter, across the upper part of the abdomen, almost as far as the right iliac region, and considerable force would have been required to inject any more of the fluid. When the nozzle of the syringe was removed, a portion of the gruel escaped, and soon afterwards a much larger quantity. The child slept well at intervals during the night, took the breast well, and **there** was neither vomiting nor pain. Next morning the **skin** was **a little hot** and the pulse **a** little quick, and one small healthy motion **had been passed.** The tumor which had been felt in the abdomen had disappeared. **At 1 P.M., all** the feverish symptoms had disappeared, and **the child had passed a** copious motion of green color, and there had been **no pain or spasm.** At 4 P.M., there was another large motion of the same character. From this time the child appeared

---

[1] Dr. N. P. Blaker, Brit. Med. Journ., Jan. 11th, 1879.

in perfect health, but the motions retained their unhealthy look for four days longer.

The success of this treatment undoubtedly depended in a great measure upon the speed with which it was adopted before reduction became difficult from strangulation.

Instead of warm gruel the enema may consist of simple water, or of soda-water from a siphon, or of a portion of a seidlitz powder,[1] the idea in the latter case being to gain the distention by the gas as well as by the water. A good formula when it is desired to make use of the pressure of gas is two parts of a solution of bicarbonate of soda, and one of tartaric acid injected separately. There are now many cases recorded in which these means have been successful, and the relief following such a procedure has been instantaneous, but as a rule injections of fluid are more easily managed, the amount of pressure produced by them better gauged, and, therefore, they are safer.

There is much to be said against the practice of trying to relieve the condition of distention by puncture of the intestine, though Broadbent reports a very successful case in which cure was affected by that means. The danger is that fæcal extravasation may occur, and to guard against this he offers the following suggestions: 1, To secure, if possible, absolute freedom from peristalsis by an extra dose of opium. 2, To select, if possible, a coil of intestine which shall contain only gas and not liquid. This will be found (if anywhere) in the jejunum, and therefore above and not below the umbilicus. *An indispensable condition is that scarcely any food shall have been taken during the entire attack.* 3, To pierce the coil exactly at its most convex part. The abdomen should be carefullly watched for some time at every visit, and especially before the operation. In some cases where the walls are thin the outlines of various coils may be traced even in repose; but this will be more distinct when peristalsis is provoked by pressure or manipulation of any kind; it will be seen also which coils shift and which keep the same position when contracting. The spot chosen for puncture should be as near as possible over the centre of a coil which does not roll about, and by preference in the *linea alba*. 4, To exercise great care and patience during the escape of gas. The needle should be held lightly but rather firmly, perpendicular to the abdominal wall, and should not be allowed to follow too freely the rolling of the coil of intestine. As the gas escapes from the coil which has been punctured, it will collapse, and the flow from the needle will cease; very soon, however, the air in the intestine will distribute itself and enter the empty portion, when it will again escape. This may be aided by gentle manipulation and pressure. Should the tube get blocked, aspiration may free it; but it is safer to drive a little air through the tube into the bowel

---

[1] Case, Dr. Morton, Practitioner, July, 1875.

than to exert powerful suction which may draw the mucous membrane against the point of the needle.

Dr. Broadbent, in spite of the rules for its use which he has so carefully laid down, believes **that puncture can** relieve obstruction only very exceptionally. His own **experience** leads him to recommend it as a palliative, and he suggests **that it may be a** useful preliminary to inflation, manipulation, **suspension** in the inverted position, etc., in the treatment of intussusception.

The chief hope of relieving an **invagination, however, lies in** prompt and efficient surgical interference by opening the **abdomen.** The propriety of such **a** course has in the last few years been **the** subject of much argument. In its favor have been adduced the rarity of ultimate recovery from the disease even after sloughing **of** the included portion **and** temporary relief; the fact that when the large intestine is affected **the** bowel may remain in a comparatively healthy state for weeks; and above all the actual saving of life which has now sufficiently often followed **the** performance of the operation to attest its undoubted value. Against **the** operation still stand, however, the difficulty of positive diagnosis, especially early in the disease; **the** speedy formation of such adhesions as will prevent reduction **even after the** abdomen has been opened, and the early supervention **of gangrene which renders** reduction improper; and the comparative **frequence of spontaneous** recovery by sloughing.

At the present **time it is admitted that** in cases of acute or chronic invagination, **where the** diagnosis **is reasonably** certain, and where the means of relief which have been enumerated **have been tried** and failed, the abdomen **should** be opened. The discussion **at present has** changed its bearings to the question of abdominal section **where the diagnosis** as to the form of obstruction cannot be **arrived** at. **The surgeon having** arrived at this conclusion, no time is to be lost; for success, **if the** operation be successful, will depend more than anything else **upon the time** at which the operation is done.

The operation of laparotomy or opening the abdominal cavity is to be performed as follows. The incision should be about five inches long, in the *linea alba* above the umbilicus. The tissues should be divided slowly and all bleeding should be stopped before the peritoneum is opened on a director to an extent equalling the opening in the skin. The seat of the obstruction is **to be** sought for by first noticing the condition of the cæcum. **If this be** flaccid, the obstruction **is** in the small intestine, if it be distended it is in the large. **If** the **cæcum** be found undistended the hand is to be passed gradually along the **small** intestine, till the obstruction **is** encountered; **if the** opposite condition obtains, the ascending, transverse, and descending **colon are to** be successively examined.

When the invagination **has been** found, it should be unfolded as Hutchinson suggests, rather by expressing the included portion out of its sheath from below upwards than by traction upon it from above. If the bowel should be found perforated, or gangrenous in any part so that perforation seems probable, an artificial anus is to be formed by stitching the bowel to the lowest part of the abdominal wall.

CHAPTER VIII.

NON-MALIGNANT GROWTHS OF THE RECTUM AND ANUS.

Polypus.—Definition.—Hypertrophy of Villi.—Characteristics.—Villous **Tumor.** —Adenomatous Polypus.—Fibrous Polypus.—Structure; Characteristics.— Symptoms of Polypus.—Diagnosis.—Diagnosis from Malignant Disease.— Treatment.—Vegetations.—Definition.— Description. — Microscopic Appearances.—Relation to Syphilis.—Symptoms of Vegetations.—Diagnosis.—Treatment.—Condylomata.—Distinction between Condylomata and Vegetations.— Description.—Syphilitic and Non-syphilitic Condylomata.—Benign Fungus. —Gummata.—**Rarity and** Literature.—Ano-rectal Syphiloma.—Definition of Fournier. — **Fibromata.** — **Lipomata.** — Characteristics. — Enchondromata.— Cysts.—Dermoid **Growths.—Characters.** — Pilo-Nidal Sinus. — Hydatids.— Fœtal Inclusions.—**Spina Bifida.—Congenital** Cysts.

UNDER this head will be included polypus, vegetations, condylomata, benign fungus, gummata, ano-rectal syphiloma, **fibromata,** lipomata, enchondromata, **and** the various forms of **cysts.**

Polypus.—A polypus may be defined as a benign tumor composed of **one or** more of the normal elements of the wall of the **rectum; an** hypertrophy either of the mucous membrane or of the **submucous connective** tissue. Those which are composed of the elements of the mucous membrane are known and generally spoken of as "soft" polypi; while those into which the submucous connective tissue enters are known as the "hard" or fibrous. In many works the former class are spoken of as the polypi of childhood and the latter as those of adult age—a classification of little practical value.

The mucous membrane, as has been shown, is composed of villi, of the follicles of Lieberkuhn or tubular glands, and of occasional closed or solitary follicles. **A polypus** composed of **an** hypertrophy of the villi is well represented in **Fig. 41.**

A polypus **of this variety may reach** the size of a pigeon's egg, it is **soft to** the feel, **and has a** shaggy or **cauliflower** surface. On section the **cut surface is** of grayish-red color, **the substance of** the growth homogeneous, and the fluid which may be forced **from it** by pressure will be found **to** be full of cylindrical epithelium. A microscopic examination shows it to be composed of long fine papillæ bifurcated at their extremities and covered by cylindrical epithelium.[1]

Although the polypi are generally small, Dr. Goodsall has reported a case from St. Mark's Hospital,[2] in which the tumor attained the size of an orange. It was rough and tuberculated on the surface, and was attached to the rectal wall by a pedicle long enough to permit of its extrusion from the anus without pain. It was attended by a frequent, copious, watery discharge, but never by any very free hæmorrhage at one time, and the patient showed no emaciation.

Villous polypus (granular papilloma, Gosselin; villous tumor, Cur-

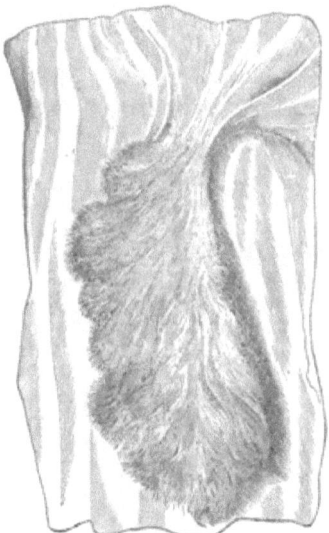

FIG. 41.—Rectal Polypus (Esmarch).

ling; villous polypi, Esmarch; "peculiar bleeding tumor," Quain).—Figs. 42 and 43.

It is a question whether this form of growth should be classified with the polypi already described, or with the warty growths, whose description is to follow. It consists of an hypertrophy both of the villi and of the follicles of Lieberkuhn, with a centre of connective tissue and generous vascular supply. According to the description given by Dr. A. Clark[3] of a specimen in the London Hospital Museum, the tumor is "essentially an outgrowth of dense areolar tissue, permeated by blood-vessels, and assum-

---

[1] Lücke: Die Geschwülste. Handbuch der allgemeinen und speciellen Chirurgie. Pitha u. Billroth, p. 250.
[2] Lancet, May 21st, 1881, p. 828.
[3] Curling, p. 85.

ing a papillary form, the papillæ being flattened and curled so as to represent hollow cylinders, and being clothed with layers of epithelium, the free layers being cylindrical."

These tumors are very rare; they have the feel of a large warty polypus with cauliflower surface; are of red color; bleed easily; are of relatively slow growth, existing in Gowland's case several years. They adhere to the wall of the rectum by a pedicle, sometimes composed chiefly of mucous membrane, and at others large, short, and fleshy.

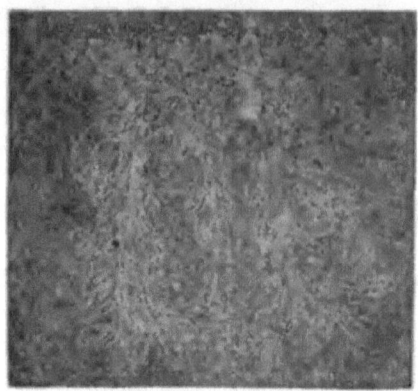

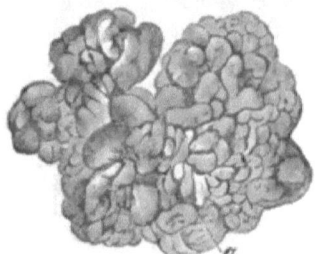

Fig. 42.—Villous Polypus (Bryant).

The pedicle may be absent (Curling); and the growth will vary in structure according to the proportion of its different elements. It may reach the size of an orange[1]; it is found only in adults or in old persons, and the symptoms are the same as those caused by other polypi; viz., discharge and hæmorrhage: but the hæmorrhage is not a constant symptom, and varies greatly in frequency and amount in different cases.

The adenomatous polypi, or those developed from the glands of the mucous membrane, are well shown in Fig. 44.

---

[1] Syme: Diseases of the Rectum, 2d ed., p. 82.

These may be due either to an hypertrophy of the follicles of Lieberkuhn or to an hypertrophy of the closed follicles. They occur most frequently in young persons; are generally of the size of a small plum, rarely reach that of a pear, and yet Esmarch reports one weighing four pounds.[1] They are very vascular tumors, and, therefore, of reddish color; they are sometimes smooth on the surface, but oftener mammillated, like a strawberry, and are attached by a pedicle, most often to the posterior wall, but occasionally to the sides of the rectum, and at a point generally within reach of the finger, but sometimes higher up. They may indeed occur anywhere along the large intestine as high up as the ileo-cæcal valve.

The pedicle is generally large and short, and not long and slender as

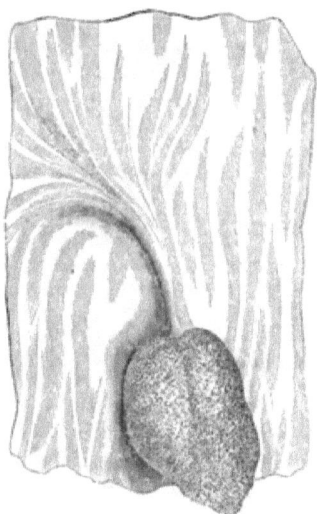

FIG. 44.—Glandular polypus (Esmarch).

in the case of the fibrous polypi soon to be described; but there are frequent exceptions to this rule, and these tumors will sometimes be spontaneously expelled by rupture of the slender pedicle in defecation.

The pedicle is also sometimes double (Smith). It consists of mucous membrane covering the vessels, which carry the blood to the tumor, and return it again—an artery and generally two veins, but when the tumor is very large, sometimes two arteries and a collection of veins.

Polypi which consist of an hypertrophy of the closed follicles of the rectum are often found in considerable numbers. Fochier[2] removed sev-

---

[1] Op. cit., p. 176–177.
[2] Molliére, p. 362. Note.

eral hundred of them from a patient aged eighteen, and Richet[1] from sixty to a hundred in a man aged twenty-one. Van Buren[2] speaks of the same condition, adopting Broca's name of "polyadenomata." To this variety of polypus belong also certain cysts (closed follicles), distended by viscid and transparent fluid; and Bathurst Woodman has reported one such case in which the cyst was lined by a membrane similar to peritoneum.

On section, these adenomatous polypi are found to contain much viscid fluid, full of cylindrical epithelium and rudimentary glandular tubes. Under the microscope a vascular stroma of connective tissue will be

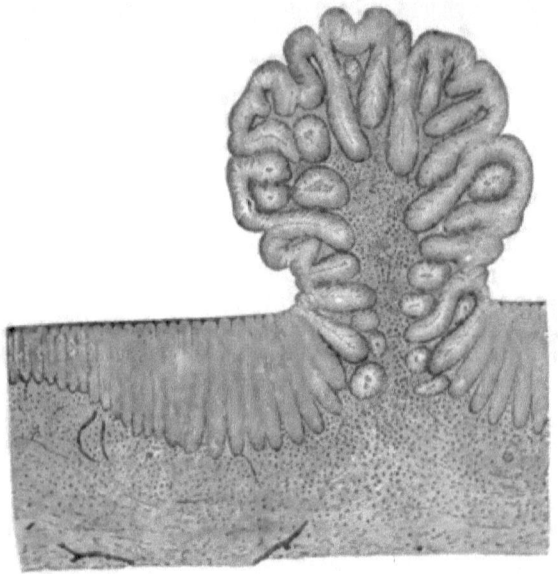

Fig. 45.—Vertical section of glandular polypus (Esmarch).

found, in which there are enlarged glandular tubes sometimes branching at their extremities; and also cystoid spaces filled with reddish viscid fluid (Esmarch).

The microscopic appearances of a section of such a polypus are shown in Fig. 45.

The hard or fibrous polypus (sarcomatous polypus, Esmarch) which is composed primarily of the elements of the submucous connective tissue, is much rarer than the soft variety, and is most commonly found in adults, where it may be isolated or multiple. It is chiefly composed of fibrous

---

[1] Traité Prat. d'Anat. Méd.-Chirurg. 4th ed., Paris, 1873.
[2] Op. cit., p. 103.

tissue, and resembles the uterine fibroid; but it may contain both muscular and glandular elements. When the glandular elements are filled with fluid which resembles glue, these **tumors have been** know as colloid, and when cysts are found filled with jelly-like **substance,** the name myxoma has also been applied.

These hard or fibrous polypi vary greatly in their degrees of hardness to the feel, according to their turgescence and their composition. They **may creak under the knife on section,** and look very much like hypertrophied and œdematous skin, or they may resemble the better-known nasal polypus in their consistence.

The connective-tissue fibres are generally irregularly disposed, **and cross each other in every direction, though** a regular stratification, such as is seen in uterine myxomata, **may be present (Esmarch). When seen in the rectum** before removal, the surface **is red from their** vascularity, but after removal, they are pale, and generally smooth, though sometimes uneven and irregular in surface, **and** covered with hypertrophied papillæ. The mucous membrane is generally **easily** stripped off, though if there has been local inflammatory irritation, it may **be firmly** attached. The vascular supply is abundant, and distributed **both to the** substance **and surface** of the tumor. **This accounts for their rapid development.**

The pedicle is generally very slight, **and is formed mechanically by the traction of the growth on** the mucous **membrane beneath which it is located. It is composed, as in the** soft variety, **simply of** mucous **membrane and blood-vessels. There may,** however, in a **case where the pedicle has been formed** by traction upon and prolapse **of all the coats of the** bowel by a **tumor** located primarily above the reflection **of the peritoneum, be a** peritoneal *cul-de-sac* within the pedicle.

An hypertrophy and increased vascularity of the mucous **membrane at the** attachment **of** the pedicle has been noted in certain cases.

If left to its natural course, the pedicle gradually becomes **longer and more slender, and finally ruptures** in the act of defecation, **and in this way a patient may relieve himself of the growth.**

These tumors are benign in character, and when once removed, do not generally return at the same point. They may, however, recur, if not at the same point, at one very near it, and the same patient may be relieved of a succession of them.

Symptoms.—A rectal polypus may exist for many years, and give no sign of its presence. The two chief symptoms which it is apt to excite are hæmorrhage and discharge. The hæmorrhage may be **a daily occurrence, or may be present only at** long intervals, and it **may vary in amount from a few drops to a** quantity which shall cause **grave disturbance and alarm. When the** mucous membrane covering the **tumor has once become** ulcerated, **the hæmorrhage will** be frequent, **and the discharge will be more or less fœtid. The** vessels are apt **to bleed freely when opened, because of their being imbedded** in **fibrous tissue, and of** their inability to contract. When the tumor is so high and the pedicle so short as to be beyond the grasp of the sphincter, there is no suffering, but after prolapse once begins to take place, the suffering may be very severe. The sphincter may become dilated and relaxed, or the pedicle may be firmly grasped by it after the act of defecation, and a cure may result from the strangulation thus caused.

The discharge from the rectum which a polypus may cause is sometimes extreme in amount and constant, escaping not only at the time of defecation, but at frequent intervals between, and being of an excessively fœtid character. This discharge may by its irritating qualities cause secondary congestion of the rectal mucous membrane, erosions around the anus, vegetations, constant diarrhœa, and tenesmus; and joined with the loss of blood the condition of the patient may be easily mistaken for that of chronic dysentery or even malignant disease.

There are several points worthy of attention in examining a patient for this disease. It is a good plan, as suggested by Chassaignac, to first administer an enema of water before making the examination that the polypus may float freely in the distended rectum. The finger is, in the vast majority of cases, all that is necessary for the examination; and as Mollière suggests, the examination should be made from above downwards and not, as is usually the case, from below upwards. In the former case, by passing the finger up along the anterior wall and withdrawing it along the posterior, the tumor may easily be caught in the descent after the pedicle has been put upon the stretch, while, in the latter case, it may easily be carried up the bowel and escape detection altogether.

Diagnosis.—Hæmorrhage from the rectum in a child, with or without pain on defecation, generally means polypus; and it often means the same in an adult, though it will oftener indicate hæmorrhoids. The secondary symptoms which seem to point to dysentery: the erosions and vegetations, must never cause the original disease to be overlooked. There is in fact but little difficulty in the diagnosis of a polypus in the vast majority of cases; but once in a while, where the attachment is broad and the pedicle not well marked, the question of benign or malignant growth may arise and be difficult to solve except by the subsequent history and development of the case.

In the chapter on cancer, attention will be called to the fact that the distinction between epithelioma and a benign polypus of the adenoid variety cannot always be made by the microscopic examination; and we here emphasize the fact that the diagnosis must rest rather upon the clinical history and gross appearances than upon histological investigation of the growth when removed. In children, malignant disease is so rare that the chances are greatly in favor of benignity. Malignant growths, moreover, do not tend to spontaneous extrusion and are not pedunculated, and the presence of a pedicle is therefore greatly in favor of benignity. But given an adult with an adenoid polypus which has ulcerated and which is not pedunculated, and the diagnosis between it and malignant disease may be impossible, either by the microscope or the clinical history; for the ulcerated and bleeding tumor may cause a wasting and cachexia which strongly resembles cancer. A soft polypus may also be mistaken for an internal hæmorrhoid when no pedicle is present, but the point of attachment is different in the two cases.

Treatment.—The treatment of polypi is generally a simple matter,

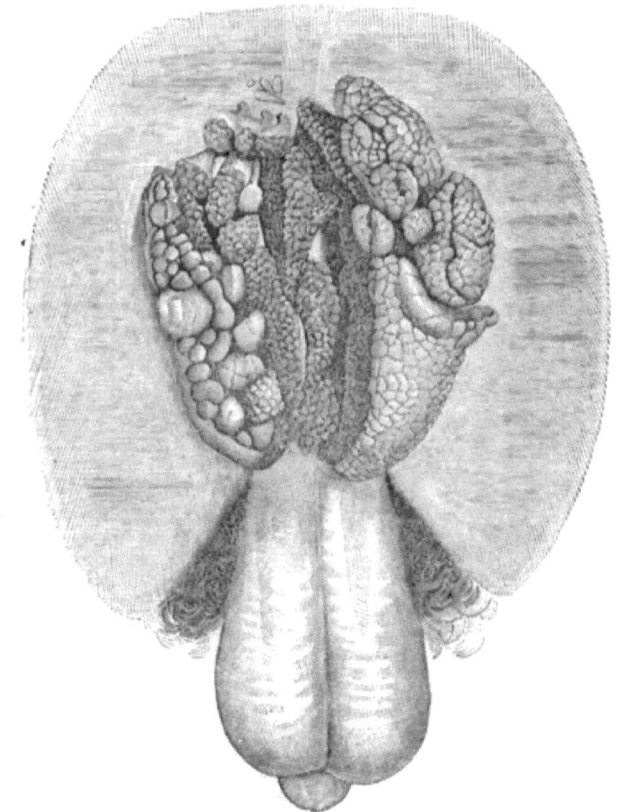

FIG. 46.—Vegetations (Esmarch).

and consists in their extirpation, after which they rarely return. There are two dangers to be considered; the first is that the pedicle, when a pedicle exists, many contain large vessels; the other is that it may contain peritoneum. The extirpation of a polypus, which has come down from its attachment in the sigmoid flexure, has been followed by death from wounding the peritoneum, at the hands of no less a surgeon than Broca. Where the pedicle is long and slender, the polypus may generally be twisted off by simple torsion without danger. It is generally safer, however, first to apply a ligature, and then cut away the tumor. Should there be no pedicle, the mass must be extirpated as any tumor would be, and the hæmorrhage which occurs must be treated upon general surgical principles.

Vegetations.—These growths, known also by the names of warts and papillomata, may be defined histologically as an hypertrophy of the papillary layer of the skin and of the papillary layer only. They are composed of the connective tissue, the epithelial covering, and the blood-vessels which, in their natural quantities, form the papillæ of the derma.

The gross appearances of these warty growths are represented in Fig. 46.

Under the influence of any of the exciting causes which will soon be mentioned, little tumors resembling ordinary warts appear, and grow rapidly till they reach two or three millimetres in size. The extremity of the tumor shows a decided tendency to branching and bifurcation, and when there are many of them their branching extremities may fuse together and form a large flat tumor, which will be attached to the skin, however, by numerous little pedicles, so that, if shaved off, the skin will not be wounded except in numerous small points where the pedicles have had each its independent attachment.

When the wart is isolated it is dry, but when several are united they become macerated in the secretion of the part which decomposes between them and gives rise to inflammatory phenomena. The tumor then becomes moist and fœtid, and all the adjacent parts become irritated. According to the size of the growths, the condition of the patient, the abundance of the secretions, and the irritation to which they are originally due, these vegetations take on various shapes, and have been described as cockscombs, cauliflower excrescences, etc., etc., but the elementary structure of them all is the same—an hypertrophy and branching of the papillæ of the derma.

On placing a longitudinal section of one of these warts under the microscope, the following structures will be seen. In the centre, a frame-work of connective tissue composed of a prolongation of the papillary bodies of the derma, in the centre of this a vascular loop; the whole covered by one or more layers of epithelium, the form and size of which are variable, and depend apparently on several conditions, such as the moisture and dryness of the parts, and the amount of pressure to which the growths are subject. When the connective tissue is abundant and the epithelial layer relatively thin, the vegetations are dry and hard. When the conditions are reversed, they are moist. When the vascular network is greatly developed, the tumors are red and turgescent, and bleed easily.

The growth occurs from the cells of the proliferating zone between the summit of the papilla and the epithelial covering. The intercellular substance of the connective tissue becomes less abundant, while the cellular elements increase, and mingle above with the epithelial layer and below with the connective tissue. Similar proliferating zones may be seen on the lateral surfaces of the ramifying warts and, through their medium, the ramifications develop at the extremity of the wart, while on the level with the proliferating zones, the capillary loops grow and develop by which the afferent and efferent vessels communicate (Rindfleisch, Mollière).

These vegetations were formerly considered as proof positive of the existence of syphilis, and even of sodomy, and were treated as such. Mollière[1] relates how, at the time of Dionysius, there was a special hospital at Rome for the treatment of these growths; and Dionysius himself tells how the surgeons spared neither the iron nor the fire, and were not moved to pity by the cries of the patients, inasmuch as this disease was the result of unnatural intercourse between man and man.

The same false idea has lasted until the present time, and is even now far from unpopular; and yet the independence of these growths upon syphilis would seem to be beyond question, except to the extent that any syphilitic sore in this neighborhood may, by the irritation of its discharge, cause their production. They owe their growth, in the first place, as pointed out by Diday,[2] to a special predisposition to the formation of warty growths on various parts of the body in the individual, and this predisposition is assisted by the presence of any irritation of the part. Thus the discharge from a gonorrhœa or a leucorrhœa, or any disease of the rectum or genitals, may cause them to grow, and they may appear in persons apparently perfectly healthy and cleanly. Pregnancy has an undoubted influence upon their production, and they sometimes disappear spontaneously after delivery. From what has been said, it is evident that these growths are neither contagious nor inoculable, and that anti-syphilitic treatment can be of no avail.

Symptoms.—These vegetations may occur at any age from infancy to adult life, though they generally belong to the latter period. They may vary in size and quantity from a single enlarged papilla at the verge of the anus to a mass such as is represented in the plate, and which weighs as much as a pound. The symptoms, in any case, will vary with their size, number, location, and the amount of the secretion. When they grow from one side of the intergluteal fold, and are large enough to press with their moistened surface upon the corresponding point of the opposite side, a second patch may be developed at the point of contact. The irritation from any other source would have the same effect. The development of the growths may be slow or rapid, and when the tumors

---

[1] Op. cit., p. 506.
[2] Exposition critique et pratique des nouvelles doctrines sur la syphilis. Paris, 1858.

are of large size, the patient is constantly troubled by the feeling of a foreign body, by a sanious and foul-smelling discharge, and by fresh erosions and superficial ulcers in the adjacent parts. Great pain in defecation may be produced by a small wart situated just at the verge of the anus, and such a little tumor may give rise to all the characteristic symptoms of a painful fissure, including a slight discharge, and an occasional drop or two of blood. They are not very infrequent on the line of junction of the mucous and cutaneous surfaces, just within the verge of the anus. They may, also, spring entirely from the mucous membrane, above the sphincter, though they are generally confined to the first inch of the canal, and, in such cases, give rise to a much more aggravated train of symptoms, and to much difficulty of diagnosis. There they are generally smaller and harder than when on the cutaneous surface, and cause a serous discharge, which may be so profuse as to escape from the anus between the acts of defecation, and cause much suffering from pruritus and rectal tenesmus.

On examination in such a case, the mucous membrane will be found dry and glistening, as a rule, though sometimes there may be a more or less extensive proctitis; and the little hard, tender, warty excrescence, which is the cause of all the grave train of symptoms and of so much suffering, may easily escape detection. The only treatment for such a condition is to seize the little tumor with the toothed forceps, and excise the mucous membrane to which it is attached. It may, however, return many times.[1]

Diagnosis.—The dignosis of these growths is not generally difficult, though care is necessary when they are small and located within the grasp of the sphincters. The mistake most commonly made is to consider them as syphilitic condylomata; and, indeed, they may not always be easily distinguishable from the raised mucous patch or flat condyloma which is a manifestation of true syphilis. A careful examination of a raised mucous patch can scarcely fail, however, to show the difference between its general character and that of a cauliflower growth which has sprung up from the surface like a shrub, and is attached to it by numerous little pedicles. The two may exist simultaneously, the wart being caused by the irritation of the discharge from the other. There is little danger of mistaking these vegetations for malignant growths, though they have been known to assume a semi-malignant epithelial character, and to return frequently after removal.

Treatment.—The surest, most rapid, and in every way most satisfactory way of curing these vegetations, is by simple excision, with the knife or scissors. The ligature is often inapplicable, and cauterization is not always easy to limit in its action. The growths may, however,

---

[1] Des **Verrues de l'intestin rectum. Rognetta, Gaz.** méd. de Paris, June, 1835.

often be induced to dry and shrink up by applications of powdered alum or tannin, and by washing with astringent lotions, such as Labaraque's solution.

Condylomata.—The term condyloma has been applied to many different growths around the anus, as well as to the raised mucous patch already spoken of, and to the remains of external hæmorrhoids. It will be used here to refer to the non-syphilitic growths of skin frequently seen around the anus, which are attached by a broad base, are pinkish in color, soft, fleshy, glistening, moist, and irregular in shape, flattened where two are pressed together, or where one is subjected to the pressure of the buttocks, and generally giving out a slight secretion.

They generally have one of the radiating folds of the anus as their point of departure, and they differ from the class of vegetations last described, in that they consist of an hypertophy of the whole thickness of the skin, and not alone of the papillæ. The epithelial element in them is not as marked as in the warts, and the blood-vessels are also less developed. They are merely the result of a localized chronic inflammation and thickening of the skin, and often follow an external hæmorrhoid or any local irritation such as has been spoken of in connection with vegetations. They are generally isolated and few in number; but it may happen that after the irritation to which they owe their origin has ceased, the growth may continue, becoming harder and more movable, and resembling a true fibroma. Such a hard tumor may, under sufficient irritation, take on an ulcerative and suppurative action, its size all the while increasing, until a foul, painful, indurated mass results which strongly resembles malignant disease. Paget[1] once said that without considering these growths as absolutely and always syphilitic, they are so rare without it, that, as yet, he had not seen a case. They are a very common accompaniment of any ulcerative process within the rectum, and hence of stricture, and many a stricture has been untruly stamped as syphilitic because the discharge from the anus had caused a development of these fleshy tags. They are indeed common in syphilis of this part, but they are not syphilitic.

These condylomatous tumors occasionally reach a large size, as in a case recently reported by Dr. Barnes.[2] The tumor in his case was the size of an ordinary orange, and had been protruded from the anus during labor. It proved to be a dense growth attached to the margin of the anus, the rest of the anal circumference being surrounded by piles more or less indurated. At one point, the tumor was greenish, as if about to sphacelate. It was removed by galvano-cautery. It had a broad base, and Dr. Barnes looked upon it as an outgrowth from a hæmorrhoidal tumor. Dr. Goodhart reported it as, for the most part, composed

---

[1] Med. T. and Gaz., vol. i., 1865, p. 279.
[2] Br. Med. Jour., April 12th, 1879.

of loose fibro-cellular tissue, covered by a tough and altered mucous membrane; the deep parts were, however, cavernous in structure. He was of opinion that it originated in some chronic overgrowth of connective tissue round a pile.

The diagnosis of these growths is generally easy. They can scarcely be mistaken for aught except a syphilitic gummy deposit or malignant disease, and they are not apt to be confounded with either. I have seen malignant deposit, however, mistaken for simple condyloma, and treated by mercurials, ablation, and the hot iron, it is needless to say without benefit.

The necessity for distinguishing between the syphilitic and non-syphilitic condylomata around the anus has already been referred to.

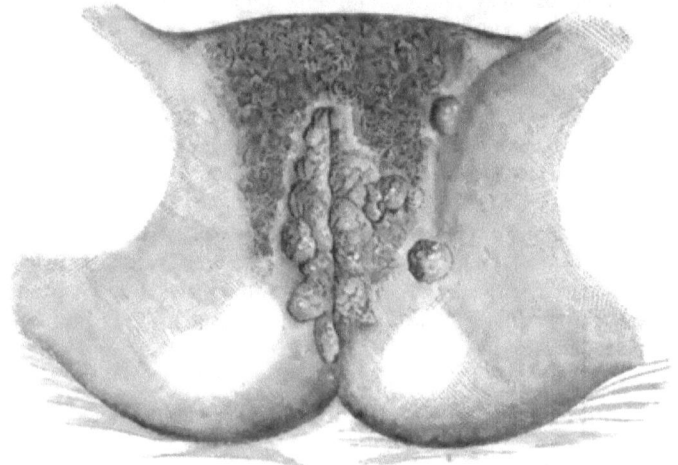

FIG. 47.—Condyloma lata or vegetating condyloma (Bumstead and Taylor).

There is a variety of mucous patch situated upon the skin near the anus which is often spoken of as condyloma lata or vegetating condyloma.

The syphilitic condyloma first manifests itself as a red spot and by a slight effusion beneath the epidermis, which is soon rubbed off by friction, exposing a raw surface, generally covered by a grayish pellicle. This surface is subsequently elevated by an upward growth, and by branching of the papillæ, with formation of connective tissue, and dilatation of the blood-vessels. Where this development of the papillæ has reached a considerable extent, the cauliflower appearance is the result, and what was at first a simple mucous patch may become a large pedunculated wart surrounded by other vegetations which have sprung up around the original lesion, and which are due to the irritation of its presence (Bumstead and Taylor, Keyes, Bäumler).

It may be impossible to distinguish this **form of** syphilis from the simple vegetation already described, except by the history, the fact of its infectiousness, and the **results of** treatment. Under the microscope, both are **composed of an** hypertrophy of the **papillæ of** the derma. It ought not, however, to be difficult to distinguish between this syphilitic **mucous patch and the simple** hypertrophy of the skin, such as is seen **at the site of an old external pile,** to which **we** here **limit** the name **of condyloma.**

This loose and undefined use by the word condyloma is much to be regretted, but is so common as to make any change out of the question. **It is used here to denote only one form of growth,** the simple non-**syphilitic hypertrophy of the whole skin.** What is usually called the **syphilitic condyloma is here referred to as the raised** or vegetating **mucous patch.**

The only treatment necessary in cases of condylomata is their simple **excision,** after which there will generally be no return.

Benign fungus.—Under this title Mollière[1] describes **a** granular condition **of** the mucous membrane of the lower end of the rectum occasionally **seen in** children as **a** result of prolapse. It is composed of soft, friable, vascular tissue identical with the granulations of a cicatrizing wound. The surface of **the mass is** red and uneven, the base is marked by dilated **veins.** After **defecation** the tumor may remain prolapsed, **but it** is easily **reducible, and when** prolapsed is not painful, which is a distinguishing **mark between it and** polypus. The hæmorrhage attending this form of growth is always **abundant** and may cause much wasting. On account **of this** hæmorrhage **the** growth is **best treated** by cauterization and astringents.

Gummata.—These also may affect either anus or rectum, though their **rarity in** the latter may best be judged by the statement of Fournier[2] that **he has never seen one, and only admits their** existence on the **testimony of** Verneuil who has seen one. However, their presence, *a fortiori* probable, **has** been demonstrated by other observers than Verneuil. Esmarch[3] **admits it;** Zeissl[4] **reports a case in a** male, **and** Zappula[5] another; Mollière[6] has seen one starting at the anus and extending into the ischio**rectal fossa; and** Fournier[7] himself met **one** in a young woman starting in the left buttock and secondarily involving the anus and then the rectum.

---

[1] Op. cit., p. 524.
[2] Lésions tertiaires de l'Anus et du **Rectum, Paris, 1875, p. 8.**
[3] Op. cit.
[4] Vrtljschr. f. Dermatol. u. Syph., 1876, H. ii.
[5] Ann. Univ. de Med., Milan, ccxiii., 1870.
[6] Op. cit., p. 645.
[7] Op. cit.

Ano-rectal syphiloma.—This affection is defined by Fournier[1] as "an infiltration of the rectal walls by a neoplasm, whose initial structure is still undetermined, but susceptible of degenerating into retractile fibrous tissue and of constituting in this way more or less extensive intestinal strictures." He speaks of its also as "hyperplastic rectitis becoming later a fibro-sclerous rectitis," and as identical or at least analogous to other lesions of the same order developed in different viscera, as the liver or testicle. He particularly emphasizes the fact that this process begins in the submucous layers, and that the mucous membrane is only secondarily destroyed, being at first entirely free from ulceration or cicatrices. Its point of predilection is the rectal pouch, but it may be found below. He has never seen it above. Sometimes only two or three centimetres of the wall are involved, but when it begins at the anus it may reach seven or eight centimetres up. It forms a cylinder around the whole circumference of the bowel. In the initial stage the rectum is only stiffened and thickened but not contracted. When the infiltration is limited to the vicinity of the anus, it is not uniformly diffused around its circumference, but forms irregular masses which are at first covered by healthy tissue. These are painless unless inflamed, but are liable to erosion and ulceration. The disease is more common in females than in males—eight to one.

Unfortunately the specific character of this ulceration cannot be proved under the microscope, there being nothing distinctive in its structure. The theory advanced by Fournier has held its own, however, and has gained adherents. Duplay[2] adopts it, and Van Buren has distinctly recognized this form of disease, and has also "seen it disappear under anti-syphilitic treatment," though Fournier says distinctly the anti-syphilitic treatment exercises no curative influence on confirmed syphilitic retraction, and this he explains on the ground that the contraction is less a syphilitic lesion than the ultimate consequence of a syphilitic lesion, just as a cicatrix is the ultimate consequence of a wound.

The remaining tumors which occur in this part of the body are very rare, so rare as to be rather curiosities than otherwise; and yet, as they may be met with at any time, it will not be a waste of time to enumerate them and say a few words concerning each in turn.

Fibromata.—True fibrous tumors may develop outside of the anus. Curling[3] gives a description of one such case removed by Mr. Hovel, of Clapton, which had been growing for seven years and weighed upwards of half a pound. It was composed of fibrous tissue arranged in several lobes, was pendulous and attached to the margin of the anus by a narrow neck. The surface was ulcerated from friction. He remarks that they

---

[1] Lésions Tertiaires de l'Anus et du Rectum, Paris, 1875.
[2] Le Progrès Méd., Nov. 30th, 1876.
[3] Op. cit., p. 188.

seldom exceed the size of a chestnut, and that their surface is generally irregularly lobulated.

Lipomata.—Of these fatty tumors there are only a few scattered cases in literature from which to derive a general knowledge of their characteristics in this part of the body. Esmarch[1] speaks of two cases, one observed by Weiss, the other by Bose. The former occurred in the surgical clinic at Prag, its size was that of a plum, and it had caused an invagination of the sigmoid flexure into the rectum and a prolapse nearly four inches in length. After extirpation of the tumor and ligature of the pedicle, the prolapse was reduced and the invagination overcome by forced injections. The second case was somewhat similar and occurred in Langenbeck's clinic. Mollière[2] gives two cases in full. One from Cl. Bernard[3] in a woman eighty-three years of age, who complained of obstinate constipation and dyspepsia, and a sensation as if of the weight of a foreign body in the rectum. By making a digital examination upon herself she could feel the tumor, and she soon succeeded in evacuating it. It weighed twenty grammes, was about the size of a pigeon's egg, was composed entirely of fat, and had a distinct and slender pedicle. The other case,[4] reported by Castilain, occurred in a man aged forty-three, who complained of the same symptoms of constipation and dyspepsia, and this also was expelled spontaneously by the straining of the patient. The doctor at first supposed the mass to be a ball of hardened fæces, but a closer examination proved it to be an ovoidal tumor measuring twelve centimetres in length by six in thickness. The consistence was firm, and the section reddish in color. The tumor showed numerous lobules and was enveloped in a resisting envelope. At one end there was a distinct pedicle two or three centimetres long, and slender. Spencer Wells[5] has also reported a large lobulated fatty tumor weighing two pounds which he removed from the recto-vaginal septum.

Fatty tumors may also occur in the region around the anus and encroach upon it to a greater or less extent. Molk[6] in his well-known thesis gives several such examples. They may be divided into the pedunculated and non-pedunculated. The former occur especially in children, and are easily removed by knife, scissors, or galvano-cautery wire, and generally without great danger. The non-pedunculated variety is much rarer. Molk relates one, in a still-born child, which filled the pelvis, and de-

---

[1] Op. cit., p. 154.

[2] Op. cit., p. 525 et seq.

[3] Azefou, Bull. de la Soc. anatomique, séance du Mars 26, 1875.

[4] Gaz. hébdomadaire, Mai, 1870, p. 318, et Bull. Méd. du Nord de la France, Mars, 1870.

[5] Trans. London Path. Soc., vol. xvi., p. 277.

[6] Des tumeurs congénitales de l'extremité inférieur du tronc. Thèse de Strasbourg, 1868, No. 106.

scended to the calves of the legs. Robert[1] has recorded another in which the tumor sprang from the ischio-rectal fossa and was at first mistaken for a perineal hernia. It occurred in a riding master, forty-five years of age, and measured ten centimetres by seven. The operation at first consisted in cutting down upon the tumor layer by layer as in the case of a hernia, but as soon as its true nature was evident it was followed into the ischio-rectal fossa and extirpated. The patient was well in a fortnight.

Virchow[2] has made a study of these intestinal fatty tumors from which the following general facts may be derived. The fatty tissue of which they are composed is apt to undergo inflammatory changes by which the general appearance of the tumor is changed, so that when it appears at the anus it may seem like a hard fleshy tumor of dark-red color on section. Another result of the irritation to which they are exposed is the formation of a hard crust on their surface which may finally become cartilaginous and cause them to be confounded with fæcal calculi. Instead of an inflammatory hardening, a central softening may occur, and a cavity be formed containing free liquid fat. Cretaceous masses may also be found in the centre of the tumors.

In general, these tumors are attached high up the bowel, and hence the pedicle may contain peritoneum. They are very apt to cause invagination, as in Esmarch's case, and this coincidence should always be borne in mind when one is found presenting at the anus.

Enchondroma.—Cartilaginous tumors of the rectum proper are of exceeding rarity, and when found they are generally the result of a secondary change in a tumor primarily glandular, and do not therefore present the well-known characteristics of the typical enchondroma. M. Dolbeau has reported[3] a case of enchondroma of the lower part of the rectum, removed from a young man aged twenty-seven. The tumor was the size of a hazel-nut, was hard and movable, and located at the entrance of the anus, where it caused no pain except when a sound or syringe was used. Around the tumor, the mucous membrane was eroded. The microscopic examination showed a predominance of the fibro-cartilaginous element with glandular *culs-de-sac*, in the proportion of one to four. M. Dolbeau did not believe that the tumor was developed from the glands of the rectum, and Robin thought that the glandular elements of the tumor were of new formation.

Cysts.—Cysts in the neighborhood of the rectum and anus may be of many varieties. Of the dermoid, there are several recorded examples. At a meeting of the London Pathological Society, May 18th, 1880, Dr. Port[4] showed a tumor he had removed from the rectum of a girl aged

---

[1] Lipome de l'anus simulant une hernie périnéale. Annales de thérapeutique, Oct., 1844.

[2] Pathologie des Tumeurs. Translation par Aronssohn, vol. i., Chap. 14.

[3] Bull. de la Soc. Anat., second series, t. v., p. 6.

[4] Brit. Med. Jour., May 29th, 1880, p. 811.

sixteen. It was mainly composed of fibrous tissue inclosed in an integument like ordinary skin, covered with long hair, and containing abundant involuntary fibre like that seen in the normal cutis. Growing upon it also was a well-developed canine tooth. The author refers to a somewhat similar case, recently reported in Germany, in which the tumor contained not only a tooth but brain substance.

Danzell[1] reports a case in a woman, aged twenty-five years, in whom a lock of brown hair, the size of the finger, protruded from the anus occasionally after defecation. In the front wall of the rectum, about two and a half inches from the anus, a hard tumor could be felt about the size of a small apple. This was extirpated by introducing the whole hand into the rectum after Simon's method, death following some months after from localized peritonitis.

The hair growing from this tumor was from twelve to eighteen centimetres long. The tumor itself, when extirpated, measured 4.5 cm. in length, 4 cm. in breadth, and 3.5 cm. in thickness, and the microscopic examination showed the usual cyst-wall and contents.

Perrin[2] gives an account of three cases of these tumors, which may be briefly extracted.

CASE XIV.—Woman, aged thirty years. First noticed small tumor at point of coccyx a few months after confinement. Tumor round, elastic, well defined, firmly adherent to point of coccyx, painless to the touch, but more sensitive at menstrual epochs, and when the patient was in sitting posture. At this time it was the size of a small nut, but a year later it had increased considerably, and extended from the anus to the sacrum; it gave a sense of fluctuation to the touch, and was unattached to the skin. Defecation painful. The sac of the tumor was extirpated after its steatomatous contents were emptied without difficulty. It was adherent by fibrous tissue to the point of the coccyx, but not elsewhere. The examination after removal showed it to be about the size of a hen's egg, with the large extremity turned toward the anus. It was composed of an envelope and contents. The envelope was composed of two distinct layers; the outer, fibrous and elastic, and showing the elements of cellular tissue under the microscope; the inner, thin, transparent, and resembling a very thin layer of cartilage. Under the microscope this transparent layer was composed of flattened, transparent, polygonal epithelial cells about one-fortieth mm. in diameter.

The contents of the sac consisted of whitish matter, disposed in layers at the circumference, but mingled in a tallowy mass in the centre; seen under the microscope to be composed of epithelial cells filled with fatty matter. Cure.

---

[1] Geschwulst mit Haaren im Rectum. Arch. für Clin. Chirurg., 1874, p. 442.
[2] De la Glande coccygienne et des tumeurs dont elle peut être le siége. Strasbourg, 1860, Thèse No. 536.

CASE XV.—Woman, aged twenty-seven years. This tumor had been growing for five years. It first appeared as a small tubercle about one-third of an inch in size, very hard and painless, at the left side of the coccyx. For the first three years it was painless, but during the latter two had caused more uneasiness when struck or pressed upon. After a time the pain was increased, and became continuous with remissions and exacerbations, and the size began to increase, while the surrounding parts took on an inflammatory action. The pain followed the course of the sciatic nerve on the side of the tumor, and after a while it became impossible to lie on the back or to walk. At this time the tumor had increased to the size of a child's fist, and rested on the left sacro-sciatic ligament. The skin and subcutaneous tissue over it were healthy and not adherent. The tumor itself was hard and somewhat elastic, and adherent to the subjacent parts.

The tumor having been completely separated by enucleation and dissection from surrounding parts, was cut away with curved scissors, care being taken to cut the osseous portion as much as possible in a longitudinal direction. The excised portion presented a fibrous shell, like that of a cyst, containing in its upper part a caseous, grayish substance which increased in consistence in proportion as it neared the base, where it was of fibrous hardness and appearance, then became fibro-cartilaginous, and, at the base, where it was adherent to the bony outgrowth from the coccyx, it was almost cartilaginous. The interior of the tumor was perforated with spaces inclosing a liquid matter resembling pus. Cure.

CASE XVI.—Man, aged twenty-four years. Fibrous cyst, size of a pigeon's egg, filled with liquid contents. Cure.

Mollière also reports one case of his own, in a young girl in whom the tumor, the size of a small almond, was covered by healthy skin.

From these cases, the general characters of these tumors may be deduced. The are generally soft, pasty, and indolent, covered by healthy skin, to which they are not adherent, and firmly attached to the sacrum or coccyx. They occur most frequently in adults, and seldom attain any size larger than that of a hen's egg. They grow slowly for a longer or shorter time, until an inflammatory action is excited, when acute symptoms supervene, and they demand attention. They may contain sebaceous matter, hair, or teeth, and may be located either within the rectum, which is very rare, or in the ano-coccygeal region, which is more common.

While speaking of tumors containing hair, etc., it may be well to refer to an affection which Dr. Hodges,[1] of Boston, has described under the name of "pilo-nidal sinus" (pilus, a hair; nidus, a nest), and which has for some time been known in French literature by the name of the posterior umbilicus. The affection is simply a ball of hair and dirt in a

---

[1] Boston Med. and Surg. Journal, Nov. 18th, 1880.

sinus between the anus and the tip of the coccyx. **The sinus** is a deep, symmetrical, somewhat conical dimple of congenital origin, representing an imperfect union **of the** lateral halves of the **body,** involving the integument alone, in which, as life advances, **short hairs** and other particles accumulate. These, by their irritation, cause a purulent discharge from the fistulous opening of the cavity, and when **the case comes under the observation of the surgeon,** it is usually mistaken for fistula-in-ano. The hair being removed, the sinus closes by granulation.

This sinus is never found in children, never in men who do not have a large amount of hair about the nates, **and so** rarely in women that the records of the Massachusetts General Hospital include but a single instance, **and** in this patient there **was, for a** female, an unusual growth of hair. For the development of the affection, there are necessary a congenital coccygeal dimple, an abundant pilous growth (hence adult age, **and almost of necessity** the male sex), and insufficient attention to **cleanliness. The affection** is, therefore, met with in persons of the lower **class,** and in hospital, rather than private practice.

Hydatids.—The number of hydatid cysts of the **pelvis** which have been reported is by no means inconsiderable. F. Villard[1] has collected thirteen of them in women, and the standard surgical writers mention their occasional occurrence. Bryant mentions removing a "basinful" of secondary cysts from one in this position. These swellings **are to be** recognized by their tense, globular, and elastic feel, and by the fact of their causing no symptoms except those due to pressure, except in cases of suppuration after the death of the entozoon. The cyst has **laminated walls** lined with a granular layer, and **is** usually surrounded by a connective tissue capsule formed from the part in which it is imbedded. It **may be** of any size, and contains a clear, watery, albuminous fluid, in which **may be found parts** of the entozoon.

Fœtal Inclusions.—In these congenital cysts, any fœtal structure may be found. They are not so rare but that **several** very complete studies **have been made of them.** Molk[2] gives numerous examples; **Verneuil**[3] has collected ten cases; and Paul[4] has written exhaustively on the subject, his article being founded **on a study of** twenty-eight cases. That variety which **is** located **in** the sacro-perineal region is the **most frequent of all.** The sac is composed of three layers, cutaneous, fibrous, and serous. The skin is thinned from distention, is violet or bluish in color from congestion, and an inflammation or a spontaneous rupture may cause perforation of the sac, and the escape of the fluid contents. The

---

[1] Considérations cliniques sur les Kystes **hydatiques du petit bassin chez la femme.** Annales de Gynécologie, 1878, p. 101.

[2] Surgery, p. 152, Amer. ed.

[3] Arch. Gén. **de** Méd., 1855.

[4] Etude pour servir à l'histoire des monstrosités parasitaires de **l'inclusion fœtal** situé dans **la region sacro-perinéale.** Arch. Génl. de Méd., t. xx., **1862.**

fibrous layer may be more or less resistant. It is sometimes composed of a simple hypertrophy of connective tissue, at others, it is aponeurotic in character. When the sac communicates with the spinal canal, this fibrous layer is a direct extension of the dura mater of the cord. The serous layer is smooth, and covered by pavement epithelium, and to one side of it the included fœtus will be found attached. This may also be a continuation of the arachnoid of the cord.

These cysts contain a serous fluid and fœtal contents in **the form** of an irrgular mass, hard and soft in spots. Any and every part of a fœtus may be discovered in this mass. The tumor is ovoidal in shape, resembling an egg when small, or the scrotum when larger. The size is generally equal to that of the head of the fœtus which bears it, but sometimes equals that of the head at term, and may be larger. The tumor may be bilocular, its contents generally give fluctuation, and are irreducible except where there is a communication with the spinal canal. There is no pain unless inflammation has supervened. The diagnosis is generally made by discovering a hard mass of fœtal elements in the **midst** of a serous cyst. When the cyst communicates with the spinal canal, the differential diagnosis between it and a spina bifida may be impossible.

Such a cyst may **cause death by** obstructing labor, or by the development of a gangrenous **inflammation after** birth. As a rule, operations for their removal have **not resulted successfully** when undertaken during the first three years of life. **One operation practised** at a later date has, however, been crowned with success.

Spina Bifida.—Concerning this variety of cyst little **need be said** except as regards its diagnosis. It should be borne in mind that a **tumor** due to a deficiency of the spinal bones may be entirely within the **pelvis**, in which case it would present great difficulties in diagnosis. **Such a case** is the following.[1]

CASE XVII.—Woman, aged 36, single. The patient stated that ten years before, she detected a swelling as large as a goose egg in the right iliac region, her attention having been called to it by shooting pains through the abdomen starting from this point. The size of the tumor increased slowly, had once caused retention of urine, and now caused œdema of the right leg. The patient was cachetic and emaciated. . . .

The abdomen was uniformly enlarged and tympanitic. On making a vaginal examination, the cervix uteri could be scarcely reached, situated as it was above the pubes, while a mass was felt behind in the *cul-de-sac*, extending to the right, apparently an ovarian cyst. But from a digital examination in the rectum it was evident that the rectum was pushed forward by a large, soft, fluctuating tumor behind it, which filled up the hollow of the sacrum to within a short distance of the anus. . . .

---

[1] Emmet: Prin. and Prac. of Gynæcology, 1st ed., p. 773.

The patient was placed under ether, and a fine trocar was introduced into the sac, about three inches beyond the anus, by which an ounce or more of its contents were aspirated by Dieulafoy's pump. This fluid was serous in character, perfectly clear and limpid, resembling hysterical urine. It contained no albumen, and the microscope revealed nothing more than a few oil globules, which had, beyond question, been attached to the instrument before its introduction.

Autopsy nine and a half hours after death. On opening the abdomen, the colon was so much distended as to fill the whole cavity, and reached to a level of the fourth rib, being filled with flatus and fæces. . . . A cyst which contained some three quarts of fluid was found behind and to the right of the rectum, filling completely the cavity of the pelvis, and extending up to a line with the second lumbar vertebra. . . . The rectum was greatly constricted in its upper portion. . . . In attempting to discover the attachments of the cyst in the hollow of the sacrum it was ruptured. The sacrum was removed, and a *spina bifida* found, the three lower bones of the sacrum being deficient on the right side. A funnel-shaped opening communicated directly with the spinal canal, from which projected portions of the cauda equina an inch or more in length. . . . Although the posterior portion of the bones were wanting, no external bulging of the sac could take place posteriorly in consequence of the dense ligamentous structures bridging it over.

The diagnosis of spina bifida can generally be made by the reducibility of the tumor, the signs of pressure on the brain and spinal cord which are produced by pressure on the tumor, the fluctuation at the fontanelles, and the chemical character of the fluid which may be withdrawn for the purpose of diagnosis. The fluid of a *spina bifida* contains both sugar and urea as does that of the cerebro-spinal canal, and though both these substances may be found in cysts entirely independent of the cerebro-spinal canal, they will always be found in spina bifida.

There still remains a class of congenital cysts which are neither connected with the spinal canal (spina bifida), nor parasitical (containing fœtal remains). These are often of large size at the time of birth, and may consist of a single cyst or be multilocular. They are generally attached by a pedicle near the tip of the coccyx, though the cyst or cysts may have prolongations in the perineum or the ischio-rectal fossæ. The cyst-wall in these cases is fibrous, and when many cysts are present it sends prolongations between them. The integument covering it is thin and generally marked by large veins. The cyst is filled with a yellowish, tenacious, gelatinous fluid, transparent to light as is a hydrocele. It will be seen at once that the great difficulty in diagnosis lies between this form of cyst and a *spina bifida*, and though the diagnosis may not always be possible, it will generally turn upon the presence or absence of the signs of communication with the spinal canal when pressure is made upon the tumor.

The treatment of these growths is by extirpation. Injections of iodine, etc., have in them the element of danger from prolonged and extensive suppuration. When extirpation is attempted it should be complete; and where the cyst is multilocular it should be followed into the perineum and ischio-rectal fossæ if necessary, in order that no parts of it may remain to undergo subsequent development.[1] These cystic formations, unless of sufficient size to cause death during labor, are not incompatible with life.

---

[1] Buneau: Bull. de la Soc. Méd. de la Suisse romande (Mollière).

CHAPTER IX.

NON-MALIGNANT ULCERATION.[1]

Varieties.—Simple Ulcers.—Generally due to Traumatism.—Various Forms of Injury to which Rectum is Subject.—Sodomy.—Injury of Rectum in Labor.—Ulcers due to Surgical Interference.—Fissure or Irritable Ulcer.—Nothing Distinctive in the Ulcerative Process.—Characteristics of Irritable Ulcer.—Theories concerning this Form of Ulcer.—Description.—Herpes.—Tubercular Ulceration.—Distinction between True Tubercular Ulcer and a Simple Ulcer in a Tuberculous Person.—Description of Each —Scrofulous Ulceration.—Esthiomène.—Rodent Ulcer.—Dysentery.—A Cause of Stricture.—Venereal Ulceration.—Gonorrhœa.—Chancroids.—Chancroidal Stricture.—Discussion. —**True** Chancre.—Secondary and Tertiary Syphilitic Ulcerations.—Diagnosis of Syphilitic Ulcers.—Ano-rectal Syphiloma as a Cause of Ulceration.—Ulceration Secondary to Stricture.—Gangrene.—Symptoms of Ulceration.—Gravity of the Disease.—Diagnosis.—Treatment.—General and Local Measures.—**Treatment of Fissure.—Fissure Complicated with Polypus.—Treatment by Rest, Fluid Diet and Incision of the Sphincter.**—Local Applications.

THE many different varieties of non-malignant ulcers which are met with at the anus and within the rectum may best be classified, from the stand-point of etiology, into the following groups: 1. Simple. 2. Tubercular. 3. Scrofulous. 4. Dysenteric. 5. Venereal. 6. Those due to stricture. 7. Those due to gangrene around the rectum.

Simple Ulcers.—These are almost always of traumatic origin, and the most frequent traumatism to which the rectum is subject is, perhaps, that arising from the presence and passage of hardened fæces. From this cause alone, or from this, combined with their extrusion from the anus, the surface of projecting hæmorrhoidal tumors may become ulcerated for a considerable extent; and, by this means, a fissure is often produced within the grasp of the sphincter. The latter I have known to happen on the first evacuation of the bowels after an operation for hæmorrhoids (the bowels having been confined by medicine for several days), rendering necessary the usual operation for its cure at a subsequent time. Another frequent cause of direct injury is the presence of foreign

---

[1] A part of this chapter and of the following one, on cancer, originally appeared in the American Journal of the Medical Sciences, Oct., 1880; April, 1881.—AUTHOR.

bodies, either fish-bones, date-stones, etc., which have been swallowed, or larger substances which have been intentionally introduced *per anum*. The presence of such substances may exite extensive ulceration which will lead to subsequent stricture.

An infrequent cause of direct violence to the rectum, and of subsequent ulceration due to the direct injury, and independent of any venereal disease, is sodomy, either attempted or accomplished. Burgeon[1] describes the rectum of an idiot, who for a considerable time had practised this vice, as much dilated and infundibuliform in shape, the mucous membrane as blackish, swollen, and ulcerated in spots; and the submucous and muscular layers as hypertrophied to four or five lines in thickness. It is doubtful whether passive pederasty should be included among the causes of stricture, as the injury done does not generally reach to this extent; and, indeed, the anus is not often dilated to any such extent as in this case. Ligg[2] describes a deaf-mute, thirty-five or forty years of age, the victim of this habit, whose anus offered no trace of traumatism, and was well closed, being marked only by the absence of the radiating folds. The mucous membrane of the rectum also was normal. This absence of the radiating folds, together with the presence of spermatozoa in the rectum or in the mucous discharge from it, are given as the best medico-legal proofs of the vice.

An injury to which women alone are subject, and which is believed by many to go far towards accounting for the greater frequency of ulceration and stricture in them than in men, is bruising of the rectal wall between the head of the fœtus and the sacrum in parturition. Most of the standard authors mention such cases.

An ulcer of the rectum is a not infrequent result of surgical interference with diseases of this part. Although in certain subjects a wound made by the surgeon may refuse to heal under the best of treatment, ulceration from this cause will generally be found to be due to careless or ignorant manipulation, rather than to the unfortunate constitutional state of the patient. Two cases occur to me now: one of a large ulcer, with hard and elevated edges, looking much like a true chancre, which resulted from the persistent application of caustics to a simple fissure; and another, of three separate ulcers which marked the former site of three internal hæmorrhoids which had been removed by ligatures. The patient suffered only slight discomfort from the operation, and was allowed to go to his business on the following day—a thing which may sometimes be done with apparent impunity, but which should never be countenanced by the operator.

The application of nitric acid to prolapse is said to have been followed by disastrous ulceration and stricture, but such need not be the case; nor

---

[1] Bull. de la Soc. Anat., 1830, p. 80.
[2] Corr. Bl. f. schweiz. Aerzte, No. 3, p. 71, Feb. 1st, 1879.

is any such use of the acid necessary to effect a **cure** in any case where its use is indicated at all. Prolapse is not, however, a rare cause of stricture, due to the strangulation and sloughing of the prolapsed portion, and to the subsequent cicatrization.

Irritable Ulcer, or Fissure.—An injury due to any of the causes already mentioned may, in certain persons, and when located at the **verge of the anus**, assume the characteristics of an affection which **has been elevated into a separate class, and is known** as fissure, or irritable ulcer. The irritable ulcer differs in **no respect from** other simple ulcers in the same locality, except in the **fact of its** irritability. There is nothing peculiar in the ulcer itself. It may be due to a slight rent in the mucous membrane from hard fæces; to a congenital narrowness of the anal orifice and a naturally over-powerful sphincter,[1] to the irritation of a leucorrhœal discharge in women; to an herpetic vesicle, or to the venereal sore which it so strongly resembles—the soft **chancre**. Any sore which is fairly in the grasp of the external sphincter is apt to become an irritable or painful one; and a fissure may be painless at one time and painful at another in the same person, or painless in one person and painful in another.

For this reason Gosselin[2] has divided these ulcers into two distinct varieties, the tolerant and intolerant—a classification which Mollière[3] still further improves by suggesting the words tolerable and intolerable. An ulcer associated with contracture, spasm, irritability, and sometimes with actual hypertrophy of the sphincter is what is known as an **irritable one;** and without this condition of the muscle it will not properly come under this classification.

This **contracture of the muscle may** be temporary or permanent, and is due to the irritation of the sensitive nerve filaments on the surface of the ulcer by the passage of fæces, and to the reflex action excited thereby; and to many slighter causes such as laughing, coughing, sneezing, or position. It may even come on spontaneously in persons of a highly nervous organization, **or with** such slight **provocation as to appear to be** spontaneous.

There are two well-known theories regarding the causation of this little sore. According to Boyer,[4] the foundation of the trouble is a spasm of **the sphincter muscle, and** the **fissure is** merely a secondary lesion due to **the passage of fæces** through the spasmodically contracted anus. Trousseau,[5] on the **other hand, reverses** the relation, and very properly, holding that the fissure exists first, **and that the spasm of the sphincter and**

---

[1] Sarremone, Thèse de Strasbourg, 1861, No. 555, Mollière, p. 134.
[2] Dict. de Méd. et de Chirurg. Prat., art. Anus.
[3] Op. cit., p. 149.
[4] Traité des Maladies Chirurg., T. x., p. 105.
[5] Clin. Méd., T. iii., art. Fissure.

the resulting pain are reflex, being specially apt to occur in **persons** of neuralgic tendency, and being in many cases merely the local manifestations of a general nervous state.

Although these **ulcers** are generally stated to be due to an actual laceration of the mucous membrane, **or to** its abrasion from some irritation, they not unfrequently **originate** within the sinuses of Morgagni and a true fissure **may be entirely concealed** from view within one of these pouches, as in the following instructive case reported by **Dr.** Vance[1] which for brevity I will slightly condense.

CASE XVI.—A lady, aged 18, had suffered for more than a year from all the symptoms of fissure, had been frequently examined to no purpose, **and** was reduced to a very miserable state. On examination the integumentary folds were congested, thickened, and œdematous, doubtless as a result of constant scratching, but there was no trace of **anything like** a fissure. The lining membrane was searched with the utmost care, **but no** lesion of any sort was revealed except slight hypertrophy of the sphincter. A second painstaking review of every part of the rectum gave the same **result**, and the author was about to abandon the hope of finding any local **lesion**, when as a matter of form, for there was no evidence of disease **about** them, he determined to pass a probe into each of the pouches. The probe could not be forced **into** the first one, and with the second he fared no better, but **with the third, after an** ineffectual attempt, the probe passed into **the sacculus.**

No sooner had the **probe entered**, however, than the patient screamed with pain, and **there** was a spasmodic retraction of the levator ani and sphincter muscles **and the** part was forcibly **withdrawn from view.** The site **of** the sacculus felt as if a buck-shot **had been imbedded in** the tissues, so hard and swollen was the part. A small probe-pointed tenotome was carefully passed along the canal, and as soon as the sensitive point was touched, the handle was brought down and the edge of the knife made to sever the inner wall of the sacculus and expose the diseased point. This done, the cause of the suffering was revealed. On the left side of **the** anus, and at a point where there had been no unusual sensibility, **an** indurated ulcer had formed within one of the little pouches. When the sacculus was opened and the ulcer exposed, it seemed very much like an ordinary fissure of the anus, but before cutting it open there was no evidence whatever, save the symptoms the patient complained of, to indicate the existence of such a lesion.

These ulcers **are generally situated at** the posterior commissure, but may be found **anywhere on the anal circumference.** They are generally single, but when of **venereal** origin **there** may be two or three. They are more common in women than in men, because constipation is more common in the former and **because the** skin **is finer.** They are confined to no age

[1] **Med.** and Surg. Reporter, Aug. 14th, 1880.

and are by no means relatively rare in infants. **They are** generally oval in shape with **their long axis vertical,** and involve both skin and mucous membrane, being situated just at the junction **of the two.** In some cases they have the appearance of a simple erosion, in others **of** on old ulcer with grayish base and indurated edges which has involved the whole thickness of the mucous membrane and extended fairly down to the muscle beneath. In the majority of cases they are not attended by suppuration or the discharge of pus. They may exist for years without gaining in surface or depth. Allingham[1] has pointed out how commonly they are attended **by** small polypi situated at their upper end or on the opposite side of the rectum; and they will often be found in conjunction with **hæmorrhoids** and condylomatous tags, the dragging upon which in the act of defecation has seemed to me in some cases to account mechanically for a slight tearing **of the mucous membrane.**

An eruption of herpes around the anus, similar to what is seen on the lips, may result after rupture of the primary vesicles in numerous small superficial ulcers of a reddish color and secreting a little pus. These may coalesce at their edges and form a serpiginous sore. They are apt to be accompanied by similar eruptions on other parts of the body, and must be carefully distinguished both from mucous patches and soft chancres. The ulcerations which result from acute and chronic eczema and from pruritus present no special characteristics. They are generally due to the injury inflicted by the nails of the sufferer.

From what has been said of the etiology of these simple ulcers it is **plain** that they must present many variations in appearance; yet the diagnosis of each from the other, and of the whole class from those which are to follow, will not generally be found difficult if proper attention is given to the history, the appearance of the lesion, and its course. The disease is generally of a healthy type, and tends to self-limitation and spontaneous cure rather than to increase. The ulcerative action is generally superficial, and tends to extend on the surface rather than in depth. It is generally surrounded by the signs of reparative action, and **with proper care will** undergo cicatrization **which** when extensive will result in stricture.

Tubercular Ulcers.—There are two varieties of ulceration met with in persons of the tubercular diathesis; one due to the actual deposit and softening of tubercle, the other a simple ulceration containing no tubercular deposit, but modified in its course by the patient's general condition of malnutrition. The former may properly be called **tubercular ulceration,** and the latter is better known as the ulceration of the tuberculous. The former is very rare. It may occur in the rectal pouch or indeed anywhere along the course of the alimentary canal, but its favorite site is at

---

[1] Op. cit., p. 192.

the verge of the anus where it may exist before any general manifestation of tuberculosis.

The characters by which such an ulcer may be recognized are its pale-red surface covered with a small quantity of serum but devoid of healthy pus and appearing as if varnished; the absence of all surrounding inflammation and of the granulations which exist in a healthy sore; its tendency to spread in depth rather than on the surface; the absence of any marked pain; the regular outline ending abruptly in healthy skin; and above all its chronicity and the utter failure of all remedies to affect its steady course. The diagnosis may be confirmed by the microscope[1] and the disease is analogous to tuberculosis of the larynx which, however, has been studied much more thoroughly.

Whether such an ulcer is ever a cause of stricture is doubtful, it being doubtful whether a truly tubercular ulceration in this place ever heals, or, in other words, results in the formation of contractile tissue. It is exceedingly difficult to induce them to take on a healthy reparative action; and if cicatrization begins, the process is generally incomplete, and the cicatrix easily breaks down. Sands,[2] however, relates a case of stricture in a boy aged eighteen due to tubercular deposit, both in the rectum and peritoneum, for which he performed colotomy, the deposit being on the anterior wall at the level of the pubic symphysis, and the rectum being so nearly occluded as not to allow of the passage either of an instrument or an injection. On autopsy, a portion of the small intestine seven feet long, was also found to be so narrowed as to admit of the passage only of a full-sized bougie, but the narrowing in both cases seems to have been due rather to the encroachment of the tubercular mass than to cicatrization and subsequent contraction.

A tubercular ulcer starting in the wall of the rectum may end in perforation and fistula (fistula with large internal opening), and, as a matter of course, the usual operation in such a case would be followed only by disappointment. Such an ulcer has also been known to cause sudden death from hæmorrhage in a child, aged four years, the subject of acute general tuberculosis.[3]

---

[1] In the excellent monograph of Péan et Malassez, Etude clinique sur les Ulcérations anales, Paris, 1872, there may be found the history of a case of this kind with the microscopic report and drawing of Cornil. Gosselin also gives a clinical lecture on a similar case in the Gaz. Méd. de Paris, Mar. 27th, 1880, calling attention to the main points in the diagnosis and treatment; and Allingham speaks of cases in which the diagnosis was confirmed by Paget, and remarks parenthetically that the disease is not as rare as is generally supposed. Other literature on the subject may be found in Habershon, On the Diseases of the Abdomen, London, 1862, p. 302 et seq.; in Mollière, Traité des Maladies du Rectum et de l'anus, Paris, 1877; Spillmann, De la tuberculization du tube digestif (Thèse d'agrégation en Médecine, 1878); and Lionville, Bull. Soc. Anat., 1874.

[2] N. Y. Med. Journ., April, 1865; continued in December number of same year

[3] Ashby, Trans. Manchester Med. Soc., Brit. Med. Journ., July 31st, 1880.

The treatment is, therefore, only palliative, though Mollière[1] propounds the interesting question whether, if such an ulcer were completely extirpated or destroyed, before general symptoms of tuberculosis had shown themselves, it might not be possible to prevent the general manifestation of the disease, as may be done in cases of tubercular testis. He bases the question on a case in which such an ulcer existed nearly four years before any other sign of tuberculosis was apparent.

The other variety of so-called tubercular ulcer is a simple sore in a phthisical patient, modified in its course and characteristics by the general condition. It may result from any of the causes already mentioned, and any of the varieties already described may, under the proper conditions, assume its characteristics. It may occur either within the rectum or at the anus, and may vary in size from a mere spot a quarter of an inch in diameter to a sore covering the whole lower part of the rectum. It may extend in depth as well as on the surface; may perforate and cause abscess and fistula, or be attended by thickening of the wall without decrease in calibre. It is often accompanied by numerous polypoid growths; it is generally painful, and the discharge is purulent. It neither extends rapidly nor heals easily, and yet it is surrounded by a healthy reparative action, and, unlike the true tubercular sore, it may be induced to heal, and is one of the causes of grave stricture. The process is essentially a chronic one, and several of the cases of "chronic ulceration of the rectum" reported by Curling come properly under this category. It may easily be distinguished from true tubercle, but may readily be confounded with some of the varieties which are to follow.

Scrofula.—Allied to the class of ulcers last named are those in which the scrofulous taint manifests itself, as it may do either in follicular ulcers of the rectum and large intestine, in lupus or *esthiomène*, and in rodent ulcer. The last two affect primarily the anus and perineum.

Follicular ulceration is due to a chronic inflammation and fatty degeneration of the follicles of the rectum. These which, when first affected, appear as small caseous nodules, break and leave small, deeply excavated ulcers, which, being multiple, may coalesce and leave larger ones of the chronic variety, capable of subsequent healing with the formation of cicatricial tissue.

They may perforate the bowel or form fistulæ of the blind internal variety when low down, or cause peritonitis when higher up. They may be only one of many manifestations of the scrofulous tendency in the same patient, and they frequently co-exist with pulmonary disease.

Under the title of *esthiomène* (lupus exedens of the ano-vulvar region) a number of phagedenic ulcerations, complicated with more or less hypertrophy of the nature of elephantiasis, have probably been described; but, in spite of the confusion of statement, this would seem to be a rare manifes-

---

[1] Op. cit., p. 651.

tation of scrofula, which may precede any others in its development. It commonly starts from the external organs of generation in the female, and invades the anus, rectum, and vagina secondarily. It is almost never seen in men. Its favorite starting-point is in the perineum, and instead of being superficial, it may be perforating and produce great loss of tissue, turning the rectum and vagina into one cavity. At this stage other ulcers are apt to appear in the rectum and colon, causing diarrhœa and sometimes peritonitis; but whether these are of the variety just described as follicular, or are due to further deposits of lupus, has not yet been positively decided.

The ulcer is irregular in outline, with a granular base of a violet-red color; and there is a slight sanious discharge. The edges are but little elevated, and are not undermined, and there is more or less hypertrophy of the surrounding tissue which, in some cases, is exceedingly well marked. The ulcer may cicatrize in part, the cicatrix being thin and white, at the same time that the ulcerative process is extending in the opposite direction. At a little distance from the ulcer there is often a pathognomonic appearance of slight, reddish, hard nodules of tubercular lupus, separated from the primary sore by healthy skin. With this amount of disease the constitutional disturbance is often not sufficient to confine the patient in the house.

The diagnosis is not generally difficult, though the disease may be confounded with cancer, phagedenic chancroid, and with elephantiasis with secondary ulceration. It is best distinguished from cancer by the cicatricial bands which it leaves behind in its ineffectual attempts at healing; and from chancroid by the surrounding tubercles which in lupus develop in the thickness of the derma, and ulcerate secondarily; while the ulcers which sometimes surround a chancroid are ulcerous from the first, being due to secondary inoculation. Van Buren advances the theory that most of these ulcerations are due to the grafting of the syphilitic poison upon the scrofulous diathesis in women of improper lives. The duration of the disease is indefinite, and it seldom leads to fatal results. It is best treated by destructive cauterization and reclage.[1]

Rodent Ulcer is very closely allied to epithelioma, and may, in fact, be considered one of its varieties; but it is distinguished from it clinically by the fact that it does not infiltrate surrounding tissue, does not involve the lymphatics, and does not become generalized. It is the same disease met with upon the face, and is exceedingly rare at the anus, being seen only twice in four thousand consecutive cases at St. Mark's Hospital.

According to the classical description of Allingham, it is found by preference at the verge of the anus, and extending from this point up-

---

[1] See also Huguier, Mém. Acad. de Méd., 1849; **Harday**, Scrofule et Scrofulides, p. 80; and Péan et Malassez, op. cit.

wards into the rectum. It is irregular in shape, and its edges end abruptly in healthy tissue. Its surface is red and dry; it destroys superficially, attacking mucous membrane rather than skin, and undergoes rapid but only partial cicatrization under proper local and constitutional treatment. It never entirely heals, and is not to be included among the causes of stricture. It is at first generally mistaken for a late syphilitic manifestation, but is distinguishable from it by the powerlessness of all treatment to prevent its steady progress. It is one of the most painful of all the ulcerative affections of this part, and ends fatally, unless some other disease cuts short the history. It is best treated by complete excision, and this, in one case of Allingham's, secured immunity for a period of four years during which the patient was under observation.

Dysentery.—In dysenteric ulceration, the diseased portion of the lower bowel becomes infiltrated with fibrinous exudation, and, as a result of the compression which this exercises, is necrosed and sloughs. When the slough is cast off, there results a loss of substance, and if this is superficial, the membrane may regain its former state; but, if deep, the usual callous cicatrix is produced in its place, and stricture is the result.

The ulcers resulting from this proces vary much in size, location, and appearance. They may be minute circles, but are generally large, and, though their favorite site is the rectum or sigmoid flexure, they may be found anywhere in the large intestine. They may extend so as to coalesce and leave only islands of mucous membrane between the extensive patches. The process usually involves only the mucous coat, but may extend in depth, and result in perforation and its attendant evils. The coats of the bowel may become sinuous abscesses, so that, on dividing the prominent portion of mucous membrane between two ulcers, several drachms of pus may escape (Habershon). Although all the symptoms of dysentery may result from ulceration due to other causes, as in Annandale's case,[1] there is no doubt that in this country the disease is one of the causes of chronic ulceration and stricture, and Habershon concludes that the disease is more common in our climate than is generally supposed.

In the Medical and Surgical History of the War of the Rebellion,[2] Dr. Woodward remarks that stricture resulting from dysenteric ulceration seems to have been much rarer than might have been supposed, and that no case has been reported at the Surgeon-General's office, either during the war or since; that the Army Medical Museum does not contain a single specimen; nor has he found in the American medical journals any case substantiated by *post-mortem* examination which this condition is reported to have followed a flux contracted during the Civil War. In the *Amer. Journal of the Medical Sciences*, for April, 1881,

---

[1] Brit. Med. Journ., 1872, p. 681.
[2] Part ii., vol. i., Med. Hist.

I published a case which I then believed came under that category, and the subsequent history of which has only the more convinced me of the correctness of the diagnosis.

Venereal Ulcers.—Gonorrhœa of the rectum has already been spoken of under the head of **proctitis**. Without attempting to decide upon the specific character of the inflammation which may follow the contact of of gonorrhœal virus, it may be well to call attention to the severity of that inflammation and to the fact that it may cause ulceration and, probably, subsequent stricture. During the height of the process, the rectum is hot, red, swollen and granular, and there is an abundant purulent discharge issuing from the anus, from time to time in clots. The irritation of this may cause erosions and fissures which may reach a considerable size; or a previously existing fissure may become inoculated in this way and spread in extent.

Chancroids.—One of the most frequent of all the superficial ulcerations at the anus is the soft chancre. It is said by Péan and Malassez to have constituted nearly one-half of all the ulcerations in this region examined at the Lourcine in 1868. It is much more common in females than in males, constituting one in nine cases of chancroids in the former and one in four hundred and forty-five in the latter.[1] To account for this greater relative frequency only two things are necessary: the frequency of accidental contact of the male organ in coition and the facility of auto-inoculation which is due to the proximity of the vulva and vagina.

These ulcers are seen either on the skin around the anal orifice, or just within the canal, and show a decided tendency not to pass above the upper border of the internal sphincter. So marked is this trait that their existence in the rectum proper has been denied, and the mucous membrane supposed to furnish no suitable ground for their inoculation. They may be single or multiple, may be situated at any point in the anal circumference, or may completely surround it. In one case of my own, the anus was completely surrounded by a group of these sores, and the ulceration extended from the posterior commissure backwards in the intergluteal fold its whole length, as far as the base of the sacrum, being superficial, however, in the whole of its course. In such a case the pain is apt to be severe; a careful examination is impossible without ether, and there is often free hæmorrhage. The bleeding at the time of defecation was the chief cause of alarm to the patient in the case mentioned. These sores have the same characteristics as the soft chancre in other parts of the body. The class of women in whom they occur is always an aid to the diagnosis, and if suspicion as to their nature exists, the test of auto-inoculation may always be tried.

Sores of this variety tend to spontaneous cure with cleanliness, and, if

---

[1] Fournier: Dict. de Méd. et Chirg. Prat. Art. Chancre, p. 72.

necessary, with judicious cauterization; and no matter how completely they may have involved the anus or the skin around it, they seldom leave any traces of their former existence. On the other hand, the cure may be delayed even for months, and the sore may assume a chronic type, due either to the existence of other disease in the rectum, as hæmorrhoids, or to a syphilitic or scrofulous taint in the patient. They may be complicated by a chronic œdema of the surrounding parts, and resemble the lupus exedens already mentioned, or by the gangrenous process known as phagedæna, generally of the chronic variety, and advancing in one place while healing in another.

And now we come to the debatable ground upon which so much has been said and written, and about which much still remains to be learned. Do these soft chancres ever cause stricture of the rectum, and are they the most common cause of those grave strictures so often met in women who have had syphilis, and which are generally known as syphilitic? In the light of our present knowledge, and yet subject to such modifications of opinion as future experience may teach, we shall answer yes to the first of these questions, and no to the second.

That a soft chancre may extend into the rectum and cause great destruction of tissue, cicatrize, and leave stricture, is beyond doubt. Van Buren[1] says, "I have certainly seen this in several cases, but only in women;" Bumstead and Taylor[2] speak in the same way; Mollière[3] says, "Nevertheless, the soft chancre of the rectum does exist, and has even been seen to assume frightful proportions in this deep region;" and Bridge's[4] case is generally considered as conclusive, though its authority rests much more upon the well-known character of the men who pronounced judgment upon it than upon its history as it stands recorded; for there is at least a strong suspicion of syphilis, and there is no account of the crucial test of auto-inoculation.

Dr. Mason's[5] paper to prove the chancroidal nature of this kind of ulceration and stricture has this great advantage over the similar one of Gosselin,[6] that he leaves the reader in no doubt as to what he means by chancroid, and unhesitatingly adopts the dualistic theory. That this is not the case in the latter article, the reader may readily convince himself by a careful perusal; and, for my own part, I am unable to see where in this justly-celebrated article the non-syphilitic nature of the affection in question is taught, for the author leaves us in absolute ignorance as to which of the two at present well-known varieties of "chancre" is, in his

---

[1] Op. cit., p. 243.
[2] "Venereal Dis.," Phila., 1879.
[3] Op. cit., p. 677.
[4] Arch. of Dermatology, Jan., 1876.
[5] Amer. Journ. of the Med. Sci. Jan., 1873.
[6] Arch. Génl. de Méd., 1854.

opinion, the primary cause of the stricture; and it is rather by inference than otherwise that his "chancre" is interpreted to mean chancroid.

The idea left on the mind of the reader is not that the disease is not syphilitic, but that it is neither a primary, secondary, nor tertiary manifestation of syphilis, as such are generally understood, but something developed in the neighborhood of the primary sore.

Gosselin, though he comes nearer to it than had ever been done before, just missed enunciating the chancroidal nature of these strictures, though Bassereau had distinguished between the two chancres two years before. What he does assert is, that they are not to be considered as manifestations of constitutional syphilis, but that they are of local character, "due to a special modification of the vitality of the tissues contaminated by the virus of the chancre, comparable to the lengthening and hypertrophy of the prepuce with contraction of its orifice, which follows a chancre on its under surface, in which the disease is evidently neither an œdema, nor a specific induration, nor a constitutional affection, but a local lesion, due to the presence of the chancres, and consecutive to the inflammation which they have caused." In the same class of lesions, he places hypertrophy of the labia, condylomata, and other vegetations.

The weight of the evidence, then, is decidedly in favor of the occasional causation of stricture by the chancroid. But that all of the many so-called syphilitic strictures are not due to this cause is rendered certain by the fact that many of them occur in women above the suspicion either of a chancre or a chancroid, and many more are developed late in the course of true syphilis, but are not preceded by any ulceration, chancroidal or otherwise, at the anus, and have their starting-point well above the sphincter muscle. Of the true nature of these we shall speak later.

Chancre.—True chancre at the anus is not very uncommon. Though Péan and Malassez saw only one case at the Lourcine in 1868, they explain the fact by the slight local disturbance which the sore causes—so slight that the sufferers do not seek treatment. They give the proportion in this place as compared to chancres in other parts of the body as one in sixty-eight, and as much more frequent in women than in men (one in thirteen in the former, to one in one hundred and seventy-seven in the latter). These are about the same figures reached by Jullien. In the female, a sore in this locality is easily accounted for by accidental inoculation; in the male, it means sodomy. They are most likely to be mistaken for simple fissures, but have a hard, raised outline and indurated base, are less painful, and devoid of the healthy surface of the former. In any case of suspicion, constitutional treatment should be delayed till the diagnosis is completed by the appearance of general symptoms.

True chancre within the rectum is very rare indeed. Ricord,

Fournier, and Vidal de Cassis each report a single case, and in the latter the induration is said to have been so great as to cause stricture.[1] Mollière carefully analyzes the evidence on this point up to date, and concludes that though a true chancre may exist within the rectum, it never causes stricture, for the reason that it does not produce any great amount of ulceration, and that the induration tends to spontaneous resolution, or, at least, rapidly yields to the influence of mercury. The difficulties surrounding the diagnosis of such a sore are manifest. Its mere appearance would scarce be conclusive, and in women, the absence of any other sore which might cause secondary symptoms would need to be absolutely proved—a very difficult thing to do.

Secondary and Tertiary Syphilis.—One of the secondary manifestations of syphilis is to be looked for at the anus and rectum—the mucous patch, not an infrequent sign in the former locality, and one liable to assume ulcerative action from local irritation or inoculation with the virus of the chancroid. Generally, however, they are devoid of symptoms, and disappear spontaneously without treatment, or simply with cleanliness and the use of an astringent wash. That the mucous patch may appear in the rectal pouch also is rendered probable from analogy with the fauces, and such cases have been reported;[2] but as they never form cicatrices, they must be counted out of the etiology of stricture.

Tertiary syphilis.—Well marked cases of tertiary syphilitic ulceration in the rectum such as are seen in the mouth and throat are seldom mentioned; and yet that they may exist and may cause extensive destruction is not only probable from analogy, but clinically true. Smith[3] says, "I am strongly impressed with the view that stricture of the rectum is produced either directly by the specific ulceration in the part affected, or by contact of the discharge from the surrounding parts."—A sentence of which the the last clause weakens the first, for the question is not whether ulceration may be set up in the rectum of a syphilitic person by the irritation of a discharge from the surrounding parts, but whether there is such a thing as true tertiary syphilitic ulceration of the rectum.

Curling[4] describes a case presented by the late Mr. Avery at a meeting of the London Pathological Society,[5] which he says clearly showed the connection of the lesion with syphilis. "Immediately within the anus, which was surrounded by a circle of vegetations, the ulcer commenced extending three inches upwards, and occupying the whole of the internal surface of the rectum to that extent. The edges were rough and uneven above, and below soft and rounded, the whole surface was smooth, exhibiting the muscular fibres of the intestine quite bare. The

---

[1] Van Buren.
[2] Mollière, p. 641.
[3] Diseases of the Rectum.
[4] Diseases of the Rectum, p. 112.
[5] Trans. Path. Soc., vol. i., p. 94.

patient died with numerous indelible marks of syphilitic eruption on the limbs and trunk."

Paget[1] also describes a case very fully and gives the main points by which syphilitic ulcers may be distinguished from tubercular; he says "The whole mucous membrane is destroyed except one small patch which is thickened and opaque. The exposed submucous surface has a lowly-tuberculated, undulating, uneven appearance, and is thickened by infiltration. In the early stages the tissue is soft, as it is from recent inflammatory effusion or œdema; but as the infiltration organizes it hardens, becoming callous, with fusion of the mucous and submucous coats, and then contracts and thus brings about the stricture. The affection commonly extends from the anus, as if by continuity with the excrescence (condylomata), to about five inches up the rectum; but it is rarely so marked in the first inch of the rectum as it is higher up."

In the case spoken of, there were also ulcers in the colon which, as the patient died of phthisis, had to be carefully distinguished from tubercular disease. He says, "On the mucous membrane of all parts of the colon there are ulcers of regular round or oval shape, from one-sixth to two-thirds of an inch in diameter, with clean, sharply cut, scarcely thickened edges, surrounded by healthy or only too vascular mucous membrane. Their bases are for the most part level, flat, or with low granulations resting on the submucous tissue, nowhere penetrating to the muscular coat, with no marked subjacent thickening or hardening. On some of them are ramifying blood-vessels; on some few there is at the centre of the base a small island of mucous membrane, giving to the ulcer an evident likeness to the annular syphilitic ulcer of the skin." In a few places they had coalesced so that the annular shape was less distinct. In the colon they were continuous with those in the rectum which Paget conjectures to have been originally of the same shape.

The diagnostic marks are thus given: "These ulcers were limited to the large intestine and decrease in size and number from the rectum upwards—conditions which I think are never observed in tubercular disease. There is not a trace of tubercle, i. e., of circumscribed, crude, or softening tuberculous deposit, in the submucous or any other tissue of the intestine, none in a Peyer's patch, or at the base or edge of any ulcer, or in the subperitoneal tissue below an ulcer. The shape and other characteristics of the ulcers are quite unlike those of intestinal tuberculosis; they are regular; with sharp, even, well-defined edges, with level bases; they are not excavating, nor do they extend through the submucous tissue; their edges are nowhere eroded or undermined, sinuous, thickened, or brawny or infiltrated; the subjacent and intervening structures appear healthy except at the rectum. These ulcers are not grouped, and where by extension or coalescence they have lost their first shapes they have acquired one

---

[1] Med. Times and Gaz., 1, 1865, p. 279.

altogether irregular, and have in no instance even tended toward that girdle-like shape, encircling **the canal of the intestine,** which is so characteristic in the large coalesced tuberculous ulcer. Thus by negative as well as by positive characters, these ulcers are clearly distinguished from the tuberculous, and, as I have said, there is no other form of intestinal ulcer to which they bear **even a** remote resemblance."

I have seen two cases of ulceration in syphilitic women where I could find no more satisfactory explanation of the cause than the presence of this constitutional state. In both the disease began well within the rec**tum and not at the anus,** which is rare **but** which proves that they were **not an upward extension of** a chancroidal ulcer at the anus; and in both it began **as an** ulcer **of** the mucous membrane, **and** was not at all similar to what has been described as ano-rectal syphiloma. In one it coincided **with a late** syphilitic eruption, but though the eruption promptly yielded to general treatment, the rectal disease did not.

A strong argument in favor of the syphilitic origin of many cases of ulceration and stricture is found in the fact that a large proportion of them all, nearly one-half, occur in persons with an undoubted syphilitic history.

Both Smith and Paget remark on the occurrence of large condylomatous tags of skin around the anus in these cases as a sign of value in the diagnosis of syphilis; and the former remarks that he has more than once made the diagnosis of syphilitic stricture from their presence alone. As a sign of ulceration and probable stricture they are of value; but they can hardly be said to point to the character of the ulceration.

The ano-rectal syphiloma of Fournier (see non-malignant growths) is not primarily an ulceration, but like the gummata it leads to ulceration, and according to him it is the most common cause of that form of stric**ture** which is called syphilitic and which we have spoken of in connection with the chancroid. It is primarily an infiltration of the wall of the rectum by a new deposit of peculiar, doughy, inelastic feel, covered by shiny, livid integument, which is prone to break down into ulceration; and it causes stricture, not by a process of ulceration and subsequent cicatrization, but by an actual blocking up of the outlet of the canal.

Stricture.—Not only is ulceration a common cause of stricture, but any form of stricture is liable by its obstructive action to set up ulceration in the wall above.

At first there is dilatation of the rectal pouch and hypertrophy of its walls, due to the effort to overcome **the obstruction.** In this way the coats may become **double their** natural thickness. **Next** an ulcerative action is set up in the mucous membrane, probably due to the irritation and traumatism of **fæces.** Beginning as a simple congestion, it advances to complete destruction of the tissue over the whole circumference of the **bowel, and sometimes for several** inches above the stricture. As a result of this process the muscular layer may be entirely denuded, and even perforated at a point high above the original disease.[1]

Gangrene.—The gangrene which sometimes follows the continued fevers and is particularly liable to affect the female genitals, and the more severe forms of abscess **in this region** may by their extensive sloughing end in subsequent **deformity and** stricture. The following case[2] shows the extent of the ravages which may be caused in this way.

CASE XVIII.—Colored woman, aged eighteen years, stated that six days before she had been taken in labor **at** full term, and was delivered of her first child after an easy labor of less than twelve hours. She was left after delivery in a soiled condition upon the filthy bed for three or four days, when she experienced some uneasiness and felt some pimples upon the vulva. On examination on admission to the hospital, the labia were found swollen, black, and sloughing; and escaping between **them** was a purulent discharge of intensely fœtid odor, mixed with the **urine** which constantly trickled away. With this local condition these was **as**sociated a slight fever, and small, quick pulse. Eight days after admission, the whole vulva and vagina, which had separated at its junction with the uterus, were thrown off, **leaving** a deep excavation, five inches from above downwards, **two and a** half inches across, and three inches in depth. The greater portion of **the back of** the cavity was filled with a globular body, red and bleeding **when touched**, which was taken for the bladder. In the lower portion **of the cavity a remnant** of the posterior wall of the rectum which had suffered in the general destruction could be seen. The slough which **came away** was nearly eight inches **in** length and two or three in thickness.

This disease is not to be confounded with the idiopathic gangrenous cellulitis already spoken of under the head of abscess, and **which is also, when** recovery takes place, very apt to result in subsequent deformity **and** stricture.

Symptoms.—The symptoms of what is known as the irritable ulcer or fissure are so well marked as to render its diagnosis in most cases easy. The chief is the peculiar pain which may be constant, but is always increased by defecation. The act of defecation itself may not be notably painful, but after the act, sometimes almost immediately, sometimes after a short interval, the characteristic suffering begins and may last in mild cases an hour or two, or in severe ones nearly all of the twenty-four hours. The pain is described by the sufferers as dull gnawing and aching rather than lancinating, **and** with it there will often be associated **neur**algic pain in the loins **and down the** thighs.

As a result of this suffering, at first periodic and later constant, a very

---

[1] See Mollière, p. 294; Gosselin, loc. cit.; Lancereaux, Bull. de la Soc. Anat., 1859; Malassez, Dict. Encyc., p. 728.

[2] Dr. Sparkman, Trans. South Carolina Med. Ass., 1879.

miserable general condition is often developed. The sufferer soon learns to dread the act of defecation and to postpone it as long as possible, till a state of chronic constipation is produced which is overcome at long intervals by purgatives; and in this way the whole digestive apparatus is thrown out of order. In women also there is apt to be reflex irritation of the bladder with tenesmus; and in men there may be spasmodic stricture of the urethra. In women, also, it is not uncommon to find uterine trouble combined with that at the anus. It is sometimes a matter of amazement to the physician to see how long a woman will suffer from a simple sore of this kind, and to what a condition of invalidism she will allow herself to be reduced before she will seek for aid. The struggle between feminine modesty and the desire for relief may last for many years before common sense finally gains the victory.

It will sometimes be found that as great suffering may be caused by a simple erosion at the anus as by more extensive and deeper ulceration, and indeed the amount of pain is not at all indicative of the depth or extent of the sore. The element upon which the pain directly depends is probably the exposure of nerve-filaments. Moreover, the susceptibility to pain varies greatly in different people, and a woman of high nervous organization may be completely invalided by a sore which would not prevent a laboring man from attending to his daily avocations.

It must be remembered in this connection that all fissures or ulcers in this part are not painful, that many heal spontaneously, and many more exist for years without causing any particular trouble.

Ulceration within the rectum is also attended by a certain train of symptoms which render its existence extremely probable, and which in themselves are sufficient to denote the presence of an ulcerative process, though throwing little light upon its nature. These have been so well described by Allingham that we cannot do better than give them in his own words.

"In the majority of these cases, the earliest symptom is morning diarrhœa, and that of a peculiar character, in my opinion, quite indicative of the disease [ulceration], and can only be confounded with cancer. The patient will tell you that the instant he gets out of bed he feels a most urgent desire to go to stool; he does so, but the result is not satisfactory. What he passes is generally wind, a little loose motion, and some discharge resembling 'coffee-grounds' both in color and consistency; occasionally the discharge is like the 'white of an unboiled egg' or 'a jelly-fish;' more rarely there is matter. The patient in all probability has tenesmus, and does not feel relieved; there is something of a burning and uncomfortable sensation, but not actual pain; before he is dressed very likely he has again to seek the closet; this time he passes more motion, often lumpy, and occasionally smeared with blood. It also may happen that after breakfast, taking hot tea or coffee, the bowels will again act; after this, he feels all right, and goes about his business for the rest of the day, only perhaps being occasionally reminded by a disagreeable sensation that **he** has something wrong with his bowel. . . . After this condition has lasted for some months, more or less, **as** influenced by the seat of the ulceration and the rapidity of its extension, the patient begins to have **more** burning pain after an evacuation, there is also greater straining and an increase in the quantity of discharge from the bowel; there is **now not so much** jelly-like matter, **but** more pus— more of the coffee-ground discharge and blood. The pain suffered is not very acute, but very wearying, described as like a dull **toothache,** and it is induced now by much standing about or walking. At this stage of the complaint, the diarrhœa comes on in the evening as well as the morning, and the patient's health begins to give way, only triflingly so perhaps, but he is dyspeptic, loses his appetite, and has pain in the rectum during the night, which disturbs his rest; he also has wandering and apparently anomalous pains in the back, hips, down the legs, and sometimes **in** the penis."

We need scarcely call attention to the extreme gravity of this condition, or to the certainty with which, if untreated, and sometimes indeed in spite of the best of **treatment, it** will end either fatally, or in stricture which will require **the gravest** surgical procedures for its relief. The picture is unfortunately **a familiar one** to every general practitioner, and a case of severe or **extensive ulceration** of the rectum, is perhaps one which calls for as much skill in **treatment** and yields as poor results as anything in the range of surgery.

Diagnosis.—The diagnosis of the existence of ulceration is generally easy with sufficient care. A small ulcer **within the grasp of the** external sphincter, or patially concealed within one of the sacculi, may easily escape a cursory examination, but no ulceration within four inches of the anus is beyond the reach of actual touch and vision, and none need, therefore, escape detection when the examination is properly conducted. In many cases the diagnosis is plain, the sphincter will be found destroyed, and the rectum and vagina will present one common cavity of foul appearance, from which issues a fœtid purulent discharge. In other cases, by a careful and gentle pulling apart of the lips of the anus and a gentle straining down on the part of the patient, a small ulcer within the grasp of the sphincter, or at least its lower edge, will be brought into view without the use of the speculum or ether. In others, a digital examination will reveal an eroded painful spot within the rectum, and when the finger is withdrawn, it will be found stained with blood. In all such cases the diagnosis is easy; in others, there is but one way to make a diagnosis, and the secret of success will be found in the two words—ether and the speculum. This **is the way,** I am sorry to say, which is least often followed by the general practitioner. It is much easier to give a lady a diarrhœa mixture and trust in Providence for a cure than to gain her consent to take ether and be thoroughly examined, and for this reason many a case of curable disease has been allowed to reach an incurable stage before its existence has been certainly determined. The existence of a chronic diarrhœa, or of a discharge of any kind from the rectum, is always a good and sufficient reason for a thorough physical examination, and with ether, a dilated sphincter, and a good speculum, no one need be in doubt as to the existence of ulceration in the lower part of the rectum.

The existence of ulceration being decided, its nature remains to be determined. We have already, in speaking of the different varieties, given some of the chief points in the differential diagnosis, and to these we must again refer the reader. In every case, the history must be taken into account, as well as the appearance of the lesion. Of the many varieties we have mentioned, some may almost certainly be excluded from their great rarity. Amongst these are the true chancre, the tubercular deposit, lupus, and rodent ulcer. In the majority of cases, after excluding syphilis, the ulcer will be of the simple variety first described, modified more or less by the general condition of the patient, or it will be malignant.

Treatment.—In speaking of the treatment of ulceration of the rectum and anus, we will first deal with the simplest form, the irritable ulcer, and then with the more severe, postponing the question of stricture which is the most frequent result of severe ulceration to a separate chapter.

The treatment of fissures at the anus should in the first place be preventive in those persons in whom the skin of the part is sensitive and liable to cracks and small sores; and for such there is nothing better than the daily washing of the part with cold water and a soft sponge, and the avoidance of anything which may tend to irritate it, such as printed or rough paper after defecation.

When fissures really exist, but before the sphincter has become irritable, they may often be cured by a nightly application of Goulard's liniment on a pledget of lint, or by gently touching the surface with a solution of nitrate of silver to coat the sore (gr. v. or x.— ℥ i.). Allingham strongly recommends the following ointment for use in such cases, to be applied several times during the day.

℞ Hyd. subchlor....................................gr. iv.
Pulv. opii..gr. ij.
Ext. belladonnæ..................................gr. ij.
Ungt. sambuci.................. ....℥ i.

The occasional light application of the solid stick of nitrate of silver will sometimes effect a cure, but cauterization should be used with great caution. An ointment of the oxide of mercury (ʒ ss.— ℥ i.) will sometimes prove effectual and I have myself been very well satisfied with the results obtained by the occasional passage of simple hard bougie well oiled, and allowed to remain a few minutes within the anus.

With these local measures must always be combined the greatest possible amount of rest, and the daily administration of a mild laxative to insure a soft evacuation. If there is already considerable pain after defecation, it is a good plan to have the bowel emptied before going to bed at night, and to administer an opium suppository or enema after the motion, by which means a quiet night may often be obtained. An ointment of ext. belladonna may also be used for the same purpose.

By such means as these, varied according to the requirements of each case, a fissure or superficial ulcer of the anus may generally in children and often in adults be cured before it reaches the stage at which it may properly be called irritable. But after this stage has once been reached, there is no longer any tendency on the part of nature toward spontaneous cure; and except the rest and laxatives which are important adjuvants to all treatment, they may as well be abandoned for more active measures.

The method of cure which at the present time has succeeded all others in these cases and which is so invariably successful as to leave little to be desired consists in temporarily paralyzing the sphincter muscle by stretching it, the patient being under ether. This is an outgrowth of the original operation of Boyer,[1] which consisted in completely dividing the muscle with the knife. Syme saw that this was unnecessary, and substituted for it the division of those fibres of the muscle which formed the base of the ulcer, an operation equally effectual and in every way preferable to the other, involving no danger of permanent loss of power of the muscle, inasmuch as its fibres are not completely divided. Dumarquay[2] also proposed another substitute which he believed would succeed where other measures failed and which consists in a subcutaneous section of the muscle by passing the knife first between the mucous membrane and the muscle and then cutting till the muscle gave way very much, as the tendo Achillis may be felt to do when similarly operated upon. The objections to this procedure are the occasional occurrence of suppuration in spite of the greatest care; and the risk of a concealed hæmorrhage which may be none the less severe and infiltrate the parts with blood.

The operation of stretching was originally performed by Recamier and as performed by him consisted rather in a thorough kneading of the muscle with the fingers than in stretching; and this was once again improved upon by Maisonneuve,[3] who brought it to essentially its present condition. This operation has been already described.

In fissures complicated with polypi, the polypus must always be removed at the time of the operation; and in women suffering from the union of uterine and vesical trouble with painful ulcer, the uterus must

---

[1] Traité des Maladies Chirurg., etc., t. x., Paris, 1831.
[2] Arch. Génl. de Méd., 1846.
[3] Clin. Chirurg., t. ii., p. 1864.

be treated as well as the ulcer, or the operation on the latter will be apt to fail.

In cases where the patient refuses to take ether, the operation of drawing a sharp knife through the ulcer and muscular fibres directly beneath it may sometimes be performed quickly, and with only momentary pain. It is customary to use a fenestrated speculum in such an operation, but it may easily be dispensed with when a straight, blunt-pointed knife is used. The knife should be very sharp, and the operation must be skilfully done, but when properly done it is usually successful.

It is not necessary to cut entirely through the sphincter, and yet those fibres of it which form the base of the ulcer should be fairly divided, for it is by putting an end to the contractions of these fibres that the operation works its cure. The operation should always be extensive enough to produce a certain amount of relaxation of the muscle.

The most frequent cause of failure in any of the procedures commonly employed for the cure of fissure is the presence of a small polypus or an external hæmorrhoidal tag in connection with the sore. These should always be searched for with great care, hence with a speculum, and should always be removed when found. Otherwise neither stretching nor division of the sphincter will be of much avail.

NOTE.—Kjellberg (Nordiskt Med. Arkiv, Bd. VIII., Heft 4) has called attention to the comparative frequency with which fissure is met with in children, which he believes to be much greater than is generally supposed. In 9,098 children brought to the Polyklinik of Stockholm, it was found 128 times; 60 of the cases were boys and 68 girls. The majority were under one year of age and 73 under four months. The symptoms resemble those in the adult, but are less severe, and the treatment is the same, care being taken to remove anything which may act as a cause of the trouble, such as constipation, worms, rectal catarrh, etc.

The treatment of ulceration within the rectum is a much more difficult matter than the treatment of that at the anus, and yet in principle they are the same. In both we give the ulcer rest, and try to assist nature in her own methods by avoiding anything which shall interfere with the process of repair. The treatment of ulcer of the rectum may therefore be summed up in two words, rest in bed and fluid diet. I do not think I exaggerate when I say that these alone will cure most cases that are curable, and that without them no treatment is likely to be of much avail.

The rest in bed must be absolute, and is not such rest as is usually considered by ladies to be compatible with a morning bath, a rather elaborate toilet while standing before the mirror and walking round the room, and a final sitting down to comparative quiet in an easy chair or on a lounge for a part of the day till the reverse of the performance is repeated. Rest in these cases means rest in bed for weeks at a time, and the line should be drawn on exercise at just what is necessary for the use of the commode which is brought into the room and placed by the patient's bed when necessary. After considerable experience I have found it easier to begin right in these cases than to waste a couple of months while the patient is half resting, and then have to come to it in the end; and have again and again been surprised to see how quickly reparative action will begin in the one case, and how long it may be delayed in the other. An hour's walking and standing around the sick-room will undo more than the other twenty-three can gain.

This point being carried to the surgeon's satisfaction, the milk-diet need not be so absolute; but may be varied with soups and easily digested solids, as bread and crackers; care being taken to secure soft and unirritating passages. With such diet as this, it will sometimes happen that a movement of the bowels every two or three days will be all that nature requires, and, as long as such a condition causes no uneasiness, I am not accustomed to interfere with it by laxatives.

In cases where it is well borne, cod-liver may be administered both as food and laxative, often with excellent effect upon the general condition and the local trouble.

In the way of local applications suppositories answer the best purpose. The menstruum should be of some substance which may be easily dissolved at the temperature of the body; and in the way of drugs I have had more satisfaction with bismuth and iodoform than with anything else. The practice of introducing local remedies in this form has many advantages over that of applying them by means of a speculum, because a speculum examination of an ulcerated rectum, repeated two or three times a week, is apt to do more harm by its mere introduction than the remedies will do good. The utmost gentleness must be used in all cases, and the greatest care is necessary to keep from irritating the part. I have also found it well to mix about the tenth of a grain of morphine with the suppository, and administer this at night and morning. It certainly ministers to the local rest of the part, and it renders rest in bed much more endurable in persons of a nervous tendency.

Certain good results may be gained by applications to the ulcerated spot by means of enemata, and when the disease is situated high up, the amount of fluid injected should be large. Three pints of water may be thrown into the upper part of the rectum, the sigmoid flexure, and the lower part of the colon, if the proper means be adopted, without causing any uneasiness at the time or any subsequent desire for an evacuation. Long, flexible, soft-rubber tubes may now be obtained from any of the surgical instrument-makers, which are suitable for this purpose. The tube should be small and the opening in it just large enough to hold securely the smallest end-piece of an ordinary Davidson's syringe. The injection should be given with the patient on the side, and given slowly. The drug from which the best results may be expected when used in this way is the nitrate of silver, and the solution should vary in strength from twenty to forty grains to three pints of water. This plan of treatment has recently been very successfully employed in cases of dysenteric ulceration. Dr. Mackenzie[1] reports five cases of cure by it, and in one of them, where the disease had lasted two years and a half, the cure followed a single injection.

The knife may serve a good purpose under several circumstances. Where the sore is of small dimensions and well-limited in outline, even though it be above the external sphincter, it is sometimes of advantage to draw the knife across the muscular fibres which form its base, and secure rest for it in this way. The operation is one of delicacy, but is also one which may assist greatly in the cure.

In cases of more extensive disease above the sphincter and at its level, where the latter by its action causes constant pain and suffering (and indeed ulceration of the rectum is seldom very painful unless the sphincter is involved, and in advanced cases where it has been entirely destroyed, may be almost painless), I am in the habit of freely dividing that muscle in the median line posteriorly by a single incision through all its fibres. In this way relief is given to suffering, more perfect rest is obtained than is otherwise possible, and a way is opened for such further local treatment as may be necessary.

The operation may be followed by incontinence, though it is not apt to be if the incision is in the median line, so that the nerves are not implicated, and if the internal sphincter be not involved in the incision. The operation is preferable to that of stretching the muscle simply, because its effect is more permanent; and, indeed, is a substitute for colotomy in the same class of cases. Of this operation I shall say more in the next chapter when speaking of the most frequent secondary effect of ulceration—stricture.

---

[1] On the Treatment of Chronic Dysentery by Voluminous Enemata of Nitrate of Silver. The Lancet, April 22d, 29th, 1882.

CHAPTER X.

NON-MALIGNANT STRICTURE OF THE RECTUM.

Stricture due to Changes in the Rectal Wall and to Pressure **from Without.**—Spasmodic Stricture.—General Division into Venereal and Non-Venereal Strictures and into Fibrous and Cicatricial.—Frequence of Syphilis in Connection with Stricture.—Non-Venereal Strictures.—Congenital, Dysenteric, Traumatic, Varieties.—Stricture from Hypertrophy of Valves.—Pathological Anatomy.—Changes in Rectal Wall above and below the Stricture.—Changes in Parts around the Stricture.—Symptoms.—Value of Flattened Passages as Symptom.—Signs **of** Obstruction.—Obstruction with Stricture of Considerable Calibre.—Diagnosis.—**Dangers to be** Avoided in Examination.—Difficulty when Disease **is Situated high up in** the Bowel.—Use of Bougie for Diagnosis.—Treatment.—Advisability **of** Anti-Syphilitic Medication.—Palliative Treatment.—Medicinal **Treatment of** Threatened Obstruction.—Surgical Measures.—Dilatation, **Gradual or Sudden.**—Rules for Gradual Dilatation.—Divulsion, **Dangers of,** and Methods **of** Performing.—Treatment by Free Division.—Description **of** Operation.—Collection of Cases.—Results of this Treatment.—Comparison with Colotomy.—Cases from **Author's** Practice.—Knife for Operation.—Excision of Non-Malignant Stricture.—Colotomy.—Restrictions to **the** Operation.—General Considerations **Regarding** it.—Treatment of Stricture High Up.

A STRICTURE of the rectum may be due either to a change in the wall of the bowel or to pressure from without. A tumor of any kind in the pelvis will not infrequently press upon the rectum so as to obstruct its calibre. An abscess in the ischio-rectal fossa may be accompanied by an amount of inflammatory deposit around the rectum sufficient to obstruct it; and a pelvic inflammation in women may be accompanied by an exudation which in the form of bands across the bowel shall partially close it, and at the same time lead to compensatory muscular hypertrophy of the rectal wall. Medical literature is full of cases of this nature, and here it is only necessary to refer to them as a not infrequent cause of obstruction both of the rectum and of other parts of the canal.

Much has been written in times past upon the question of spasmodic stricture of the rectum, but at present the condition is looked upon by the best authorities with great doubt, if not with absolute unbelief. Spasmodic contraction or stricture of the external sphincter is not an unusual condition, and cases of it from my own practice and that of others will be reported further on; but spasmodic stricture of the canal above this point has always been a matter of belief and assertion rather than of demonstration.

Allingham upholds its existence in connection with organic stricture, as a complication of the latter, and gives the following case as proof. He says: "There are, no doubt, many cases of stricture in which there is very little deposit and much spasm; and there are, on the other hand, cases where much obstruction exists, but very little spasm. A patient under my care at St. Mark's had a stricture so tight that I could not make the point of my little finger enter it; on putting her under the full influence of chloroform, I could get two fingers through without difficulty."

This case, if it be admitted, as it generally will be on so good authority, actually proves more than has ever been proved before with regard to this question, and is about the only one which really proves anything. I have already referred to the difficulty which often exists in passing a rectal bougie from the natural conformation of the parts. It is upon this difficulty that nearly all the arguments for and the supposed cases of spasmodic stricture rest. When the bougie cannot be passed, a spasmodic stricture is supposed to be the cause. When, after numerous trials, by a lucky manipulation an entrance is effected, the spasm has been overcome. To this may be reduced nearly all the reported cases of this affection which from time to time have appeared in the writings of those who have devoted attention to the subject.

Mollière,[1] with his usual happy style, has gone very nearly to the bottom of this question. He says that at a not very remote period there flourished by the side of Ashton, Curling, and the surgeons of St. Mark's Hospital certain specialists as expert in finding strictures in the rectum, as are our laryngologists in discovering polypi in the larynx. These estimable practitioners gave themselves up to the daily exercise of dilatation by bougies, and to facilitate the practice, one of them had invented a pair of pants of a special pattern, dressed in which novel livery his patients came daily to have a sound introduced into the anus.

Another anecdote is repeated by several authors which illustrates the ease with which patients may deceive themselves or be deceived by others in this matter.

A lady went to consult a rectologist for some reason or other which is not stated, and a sound was introduced into her anus. Her husband learning this, rushed to the house of the scoundrel in a violent rage and armed with a whip. Half an hour later he returned disconsolate. He had found out that, like his wife, he had a stricture of the rectum, and, like her, he had submitted to catheterization.

This whole question of spasmodic stricture has been very ably dis-

---

[1] Loc. cit., p. 320.

cussed by Van Buren,[1] and if the reader wishes to follow it further, he can scarcely do better than to consult that article. Uncomplicated spasmodic stricture of the **rectum is a thing** whose existence is not admitted by the best authorities, and which will seldom be found by a skilful examiner. It is perhaps too much to say that it never exists, but a well-marked case of it within easy reach of the finger, which could be plainly detected by an ordinary examination, and which disappeared under chloroform, is what those who do not believe in its existence are calmly waiting to see.

The changes in the wall of the rectum which may cause stricture, independent of malignant disease which will be considered later, may be divided into the two general classes of venereal and non-venereal, and each of these may be again divided into the cicatricial and fibrous.

Venereal Stricture.—In the chapter on ulceration stricture has been frequently referred to as a not infrequent consequence of that process, and the various forms of ulceration which by subsequent cicatrization were capable of producing this result have been mentioned. In a general way it may be said that any ulcer which destroys even the thickness of the mucous membrane to any extent will, when healed, leave a cicatrix, and if such a cicatrix be at all extensive it will by its contraction cause subsequent diminution in the rectal calibre.

It has been shown that many of the more severe forms of rectal ulceration are of venereal origin. The venereal sores capable of producing a stricture are the chancroid, and the later syphilitic ulcers. We shall leave out of consideration the true chancre, and the mucous patch, for the reason that their influence in the causation of stricture is still rather a matter of surmise than of proof, and the same thing may be said regarding gonorrhœa of the rectum.

For a description of these ulcerative venereal processes the reader may again refer to the chapter on ulceration; but there is a class of venereal strictures which are syphilitic, but are not primarily ulcerative and therefore not cicatrical. In this class are to be placed the gummata, the ano-rectal syphiloma which differs from gummy deposit rather clinically than microscopically, both of which have already been described; and a third late manifestation of constitutional syphilis, which is an inflammation of the rectal wall. This inflammatory change may involve a large portion of the rectum. It begins in the muscular fibre, the interstitial tissue of which becomes filled with round cells which ultimately form a connective tissue, and this connective tissue by its hardening and consolidation finally causes the complete destruction of the muscular element. This is not to be confounded with the ano-rectal syphiloma in which

---

[1] On Phantom Stricture and Other Obscure Forms of Rectal Disease. Amer. Journ. of the Med. Sci., Oct., 1879.

there is an actual deposit of large masses of new material in the rectal wall—masses which it may be very difficult to distinguish from cancer.

In these various ways venereal disease and especially syphilis may result in rectal stricture, and this accounts for the fact that in about fifty per cent of all cases of stricture there is a syphilitic history.

Non-venereal Rectal Stricture.—The non-venereal strictures may be classified as congenital, dysenteric, and traumatic.

The congenital narrowing of the rectum which is sometimes seen has been already described in speaking of the malformations of this part. There is also another form of obstruction of the rectal calibre which is supposed to be due to an hypertrophy of the folds of mucous membrane which are normally present in every one.

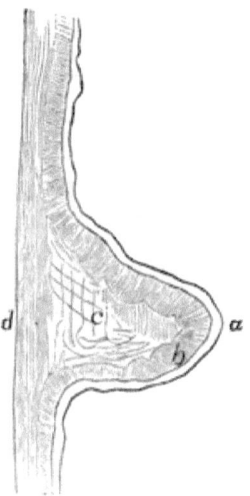

FIG. 48.—Longitudinal section of stricture of the rectum at the *plica recti inferior* (Kohlrausch). *a*, Mucous membrane. *b*, Circular muscular layer entering into the fold of the stricture. *c*, Cellular tissue. *d*, Longitudinal muscular layer passing over the stricture.

Quain,[1] under the head of impaction of fæces, describes the case of a man, aged forty years, who died with a large accumulation which was evidently due to the presence of two crescent-shaped shelves of mucous membrane projecting into the rectum, one attached opposite the prostate and the other about four inches higher. Each of these was more than an inch in breadth, and into each the circular muscular fibres fully entered, while even the longitudinal layer dipped slightly inward at their bases. Kohlrausch also describes an analogous case, in which he made an autopsy on a criminal who had been executed. (Fig. 48.) He found an enormous dila-

---

[1] Diseases of the Rectum, London, 1854, p. 273.

tation of the rectum above the spot at which he locates the plica transversalis. At that point he discovered an undoubted stricture which, from its hardness and extent, he at first considered cancerous. It presented, however, nearly the same anatomical condition as the one just described; the mucous membrane was sound and formed a considerable duplicature; the circular muscular fibre entered into this duplicature and formed a hard, hypertrophied, muscular ring several lines in thickness. The longitudinal fibres passed directly over the affected spot in this case, however, and were not unusually thick or firm, and the space left between the outer and inner muscular layers by the bending inward of the latter was filled with connective tissue. A stricture was in this way formed without degeneration of the mucous membrane—a condition, however, which led to no less serious results. Such a state furnishes in itself the ground for constant aggravation, for the longitudinal fibres passing entirely over the fold must, by each contraction and by the necessary increase in their normal function, augment the substance of the fold more and more, and thus decrease the lumen of the gut. Nélaton, indeed, has written that valvular retractions of the rectum are most often only an hypertrophy of his superior sphincter, and that the projection formed by it into the cavity of the intestine is the point at which foreign bodies are most frequently arrested, as well as that at which invaginations in young children generally begin; and in all these points he is borne out by Velpeau.[1] Sappey[2] says "at the level of this band most of the organic contractions of the rectum are situated; its study, therefore, offers no less interest in a pathological than in a physiological stand-point." This idea of the pathological relations of the mucous folds and muscular bands in the causation of organic strictures may be traced through the works of Arnold, Tanchou, Hyrtl, and Houston; and has its foundation in the fact that, as these folds are the most subject to injuries, so they may be the most frequent starting-point of those contractions of the rectum which are due to injuries, especially those from foreign bodies introduced *per anum* or swallowed, and from masses of hardened fæces, intestinal concretions, etc.

Dysenteric stricture and ulceration have also been already described. Stricture due to this cause is, perhaps, more often multiple than when due to any other.

The last cause to be enumerated is the simple traumatism which may result in stricture, either by causing ulceration and cicatrization or by exciting a chronic inflammation in the submucous connective tissue. Amongst these traumatisms may be enumerated operations upon hæmorrhoids, applications of strong acids, the performance of some surgical operations, foreign bodies, kicks and falls, and the injury produced by the head of the child at birth.

---

[1] Velpeau, Anat. Chir., 3d ed., 1837, p. xxxix.
[2] Anat. Descript., t. iv., p. 222.

Pathological Anatomy.—In studying the pathological anatomy of stricture, there are several points to be observed, for changes will be found not only at the stricture itself, but both above and below it, and in the surrounding parts.

From what has been said already, it will be inferred that a stricture which is not the direct result of a deposit of new material in the rectal wall will be composed either of cicatricial tissue, such as is found in other parts of the body, or else of connective tissue which is firm and dense, and creaks under the knife on section. All the connective tissue in the rectum at the diseased point, whether submucous, subperitoneal, or intermuscular, will be found to have increased in quantity; and this accounts for the increased thickness of the rectal wall. The mucous membrane at the seat of stricture will generally be found destroyed, and replaced by granulation tissue on this fibrous base, which bleeds easily when scraped.

Above the constriction a process occurs which will be found to be almost constant. This begins by a dilatation of the bowel and an hypertrophy of the muscular layer, with, at first, a thickening of the mucous membrane. Later, the mucous membrane, due, probably, to the irritation of retained fæces, will show all the stages of ulceration, from simple congestion in some points to a complete destruction in others, and an exposure of the muscular tissue beneath. This ulcerative process may extend for several inches up the bowel. The wall of the bowel above the stricture may be as thin as paper in spots, and at such points perforation is apt to take place. In a case reported by Goodhart,[1] the changes of which we are speaking had gone on to actual gangrene, extending in spots along the transverse and descending colon, and were undoubtedly due to the intensity of the inflammatory action caused by the retained irritant matters. The bowel is also generally distended with gas and fæces, and the latter are more often fluid than solid, though fæcal tumors, with their well-known characteristics, will sometimes be met.

The dilatation above the stricture may reach an enormous size, and may ultimately result in a *cul-de-sac* or pouch which will fill a large portion of the abdomen, and dip down below the point of constriction, and an ulceration in this pouch may result in its perforation and the establishment of a fistulous outlet for the fæces. Such an opening may be into the rectum, either above or below the stricture, or into the ischiorectal fossa, with the necessary result of abscess. An opening may also be made into the bladder in either sex, and in females, into any part of the genital tract.

As showing what efforts nature is capable of making to overcome the occlusion caused by stricture, the following account of the post-mortem appearances found in the body of Talma, the tragedian, is of great interest. The whole history of the case may be found in Quain.[2]

---

[1] Med. Times and Gaz., Feb. 28th, 1880.
[2] Op. cit., p. 190.

In the examination of the body the intestines were all found largely distended with air and fæcal matter. . . . The pelvis was filled with an enormous sac—the upper part of the rectum largely dilated. When the sac was raised a circular **narrowing of** the gut was discovered. This was the stricture. It was **at the distance of** six inches from the anus, and proved, upon close **examination, to** be wholly impervious. It was, in fact, a solid fibrous **cord, but** on the surface irregular, and having **the** appearance of a purse, drawn tightly **and** puckered, with the strings tied around it. The **great** dilatation of **the** bowel at **its lower end,** dipped down below the level of the stricture **in the form** of a dependent sac, in which was an opening about an inch in diameter, and from this opening issued a fluid, the same as that diffused through the abdomen. The rectum below the stricture was no more than the size of a child's intestine, and upon it, close to the stricture, was an ulcerated surface with a narrow opening, to which the edges of the aperture above the stricture had been adherent. A new communication, but an imperfect one, had **thus** been established between the two parts of the gut—severed one from the other by the stricture. But the connection had given way, doubtless in consequence of the violence **of** the expulsive efforts, and thus the contents of the bowel **had escaped a short** time before death.

The cellular tissue **in the ischio-rectal** fossæ around a stricture may also become hard **and lardaceous, as a result** of chronic inflammation; and this change may extend **to some distance** from the original starting-point along the sacrum, **as high as the promontory, and into** the subperitoneal tissue of the iliac fossæ.

Abscess is always liable to occur in the neighborhood **of the stricture,** probably from lowered vitality in the parts, and this accounts for **the** relative frequency of fistulæ in this disease. These may be both **numerous and** extensive, and may make communications between the rectum **and any** of the adjacent organs. For this reason a fistula should always lead **the** surgeon to think of stricture and to examine for it.

Allingham has also called attention to the frequent existence of a low form of peritonitis in connection with stricture, an inflammation marked by tympanites, vomiting, and pain, especially on walking or moving, and attended by thickening of the peritoneum and old and recent adhesions.

Below the stricture the rectum may sometimes be found unchanged from its normal condition, but it will generally be ulcerated as it is above, or else there will be hæmorrhoidal tumors, excoriations, and vegetations and condylomatous tags of larger or smaller size. These condylomatous growths are the result simply of irritation of the discharge from the process above.

Most strictures are located in the lower part of the rectum, and hence their presence is easily detected in the majority of cases. They are far more frequent in females than in males, because many of the causes which produce them operate chiefly in females. Age has little influence upon their frequency after the period of adult life. A stricture may or may not involve the whole circumference of the bowel; and the contraction may be so slight as not to be apparent till the bowel is distended with the speculum, when a falciform band may spring out from one side. In more extensive disease, there is still usually a passage for the fæces, but this may be very slight. The most extensive disease will be found to be due generally either to syphilitic deposit, syphilitic sclerosis, or dysentery; and in such cases the calibre of the bowel may be lessened for a space of several inches.

Symptoms.—Where stricture is the result of ulceration, the signs of ulceration will at first mask those of the stricture, and the patient will complain of pain, discharge from the anus, excoriations, and warty growths, together with the failure of the general health, gastric and intestinal disturbance, and wandering pains.

The one sign of a stricture is the obstruction, and this may show itself in several ways, generally at first by alternate attacks of constipation and diarrhœa. The constipation is mechanical, and is due to the accumulation of fæces above the constriction. The diarrhœa is secondary to the accumulation, which, in time, begins to act as a foreign body, setting up a catarrhal inflammation, as a result of which sufficient fluid is poured into the bowel to soften the hardened mass, and large quantities are discharged, only to be followed by a fresh accumulation.

It has often been asserted that a well-marked lessening of the rectal calibre must, in the nature of things, produce a change in the shape of the fæces, but this is not quite true. The flattened, tape-like stool is a sign of value when present, and should always lead to careful exploration, but it may not be present even in the worst cases of stricture, and it may exist without stricture; in the latter case generally being due to an irregular spasmodic action of the sphincters, or to pressure from without the bowel. This point, to which attention was called by White[1] as long ago as 1815, has again recently been made the subject of discussion. In an able article on "Annular Stricture of the Intestine; its Diagnosis and Treatment," in the *British Medical Journal* for May 31st, 1879, Mr. Stephen Mackenzie wrote: "The fact that full-sized, properly formed fæces are occasionally passed, of course shows that there can be no

---

[1] "With regard to the lessened diameter of the fæces, just noticed, which must necessarily be the case whenever a permanently contracted state of the gut takes place; yet it has happened in some instances where that change had been observed, that, in a more advanced period of the disease, fæces of a natural size had occasionally passed. The knowledge of this circumstance I consider of some importance, inasmuch as, if properly attended to, it will prevent the practitioner from hastily concluding there is no stricture merely from an examination of the evacuations, when symptoms may otherwise indicate the presence of the disease."
—"Observations on Stricture and other Affections occasioning a Contraction in the Lower Part of the Intestinal Canal, etc.," Bath, 1815.

organic stricture." Under criticism, he withdrew the statement in the issue of the same journal for May 15th, 1880, with the explanation that it was founded on his personal observation, which had since been supplemented and corrected by that of others.

In a case which I recently saw in consultation with Dr. De Long, of Brooklyn, I had a long-wished-for opportunity to observe, in the presence of a number of physicians, the actual mechanism by which tape-like stools are produced. The woman suffered from a stricture one inch above the anus, which was of sufficient calibre to admit the ends of two fingers easily. She had never noticed any deformity of the fæces. While under the influence of ether, and after the sphincter had been very thoroughly dilated, an O'Beirne's tube was passed through the rectum, which was empty, into the sigmoid flexure, which was full. After resting there a few moments, it provoked a movement of the bowels. The stricture was instantly crowded down into view, appearing at the anus, and taking the place of the anus, which, owing to the complete dilatation, ceased to have any action, and was simply a patulous ring. Through the stricture there came a long, tape-like evacuation, the mould which gave it its peculiar form being the stricture pressed to the surface of the perineum, and greatly lessened in calibre by folds of mucous membrane, which were crowded into it from above. While remarking to those present on the peculiar mechanism of its production, the straining ceased, the stricture rose, the mucous membrane was relaxed, and a passage of natural formation was the result. This alternation was repeated several times. At each violent effort the stricture was forced down to the anus, the membrane above it was crowded into it so as to greatly lessen its calibre, and a flat passage was the result. When the effort was less violent, there was still a passage, but the stricture having risen to its place, and not being so tightly filled with the mucous membrane, the passage was natural. The lesson to my own mind was this: that a stricture of large calibre might, as a result of straining, cause a passage of very small size; and that, to get this peculiar shape, the stricture must be crowded down so as to actually take the place of the external sphincter, and be the last contracted orifice through which the soft substance is expressed. It is well known that, with the closest stricture high up, the fæces may be re-formed in the rectum below, and be passed normal in size. At the bedside but little importance is to be attached to the statements of patients concerning this matter.

After a stricture has existed for a certain length of time, signs of obstruction will be manifest by abdominal palpation and inspection. The transverse and descending colon can be felt partially distended with masses of fæces, and will be dull on percussion, tender to the touch, somewhat movable, and pitting on firm pressure. After an attack of diarrhœa, or after a brisk purge, these accumulations may disappear, only to form again in a short time. Generally complete obstruction does not occur without ample warning in this way. It is preceded by eructations of fetid gas, the abdomen swells and becomes very tender on pressure, the coils of intestine are visible through the abdominal wall, and their visibly violent peristalsis gives proof of the effort nature is making to overcome the obstacle. After a short time the patient is exhausted, and, unless surgical aid is given, dies. Complete obstruction has been seen to occur very suddenly, forming almost the first intimation of serious disease; and this is more apt to be the case where the stricture is high up in the rectum or at the junction with the sigmoid flexure. It comes on with the usual signs of acute intestinal strangulation—pain, swelling of the abdomen, bloody passages, etc., and it may be caused by some indigestible substance which has been swallowed and refuses to pass the stricture, or merely by hardened fæces or prolapse of the bowel above into the constriction. The following case is one of quite a large class:

"The patient, a middle-aged woman, was admitted into St. Bartholomew's Hospital with symptoms of sudden obstruction. She stated that she had enjoyed good health up to the onset of the attack, nor had she previously been troubled with constipation. The attack commenced suddenly while at work, and was followed by obstinate vomiting and constipation. The symptoms having existed for some days, and the case appearing urgent, while the sudden onset of the symptoms suggested mechanical strangulation, it was deemed advisable to open the abdominal cavity. This being done, Mr. Marsh felt a hard cancerous mass in the walls of the bowel, which caused the obstruction. The bowel was opened above the obstruction, stitched to the sides of the wound, the patient making a good recovery.[1]

There is one important element in the obstruction due to stricture, which must not be forgotten. It will sometimes happen that fatal obstruction will occur even when, on post-mortem examination, the calibre of the stricture is found to be large enough to permit the passage of the finger, showing that the obstruction could not have been due merely to the contraction of the new growth. John Hunter remarked a fact of this sort, as is proved by the following account:

"On introducing the pipe by the anus, it was found to come butt against one side of the upper part of the cavity of the tumor, where there was a bend in the passage; but why a crooked pipe did not pass when attempted to be passed by turning it to all sides, I cannot conceive, or, why a bougie which was slightly bent did not hit the hole, is not easily accounted for; but, what is more extraordinary than either, why a clyster did not pass freely up; or why did not the wind or soft excrementitious matter that did yet lay [sic] pass readily down, while I could pretty readily pass the end of my finger down from the gut above into the

---

[1] Cripps, Cancer of the Rectum, p. 107.

tumor? The folds of the contracted part did not appear after **death** to have been sufficient for an **entire** stoppage of this sort."¹

Notwithstanding the **statement** that the folds of the part did not appear *after death* to have been sufficient to produce the stoppage, it seems that a prolapsed **fold of mucous** membrane is the only thing likely to give rise to it. In **cases of advanced** disease a spasmodic stricture (if such ever occurs) **would seem out** of the question, whereas partial **or** complete invagination in this part is known to be of **frequent** occurrence. As shown by Rokitansky,² the paralysis above the stricture **is also an** undoubted element in **the** production of the occlusion.

Diagnosis.—The first means of diagnosis in stricture **is** the examination with the finger, and as the great majority of strictures are **confined** to the lower portion of the rectum this is in itself generally sufficient. It **is the** best and safest and least painful of all the means of diagnosis **when** properly executed, and yet it may be the immediate cause of death to **the** patient when roughly practised. There is an inborn tendency on the part of many, when the index finger comes in contact with a tight stricture, to bore through the narrow passage which is left and feel what is on the other side—a tendency **to be** struggled against and overcome. If the surgeon has deliberately **determined** to practise divulsion, this is one way to do it, but at present **we are speaking of** diagnosis, and forcible dilatation is not diagnosis, **but a very grave** surgical procedure. The finger should therefore be **passed slowly up to the** stricture, and unless the calibre admits of it **without straining, it should not be** passed further. The condition of the **parts** below may also **be** appreciated, **the** amount of induration estimated, **and a** general idea **formed of the nature** and extent of the disease. In women the vaginal touch will **generally be found of** the greatest value and should never be omitted.

As a rule all can be learned in this way that can be learned in any other where the disease is within reach of the finger, and nothing is to be gained by a painful speculum examination **or** the use of the bougie— means of diagnosis which, however valuable where the stricture cannot be felt by the finger, are of little use for the lower four inches of the rectum.

When a stricture is situated high up in the rectum or in the sigmoid flexure, the confidence in diagnosis which comes from actual contact of the finger with the disease is entirely lost, and there is perhaps nothing in the whole range **of** surgical diagnosis which requires more skill than the detection of stricture in this part, **and** nothing attended with more uncertainty. The symptoms of stricture **of** the upper part of the rectum are not the same **as when** the disease **is** lower down, for the nerve-supply is not the same, nor **is the sphincter muscle** involved. For this reason

---

¹ Hunterian MS. Cases and Dissections, No. 59, in "Descriptive Catalogue," etc., vol. iii., p. 98. From Mayo, op. cit., p. 249.

² "Manual of Path. Anat.," vol. ii., translated by Sieveking.

the patient is much more apt to suppose himself suffering from chronic constipation and dyspepsia than from hæmorrhoids. Pain in the abdomen, not always localized at the left side, pain in the loins and down the legs, obstinate constipation and occasional diarrhœa, are the things usually complained of, and in these there is nothing upon which to base a positive diagnosis. The fæces may never present any peculiarity, for the reason that they are accumulated in the rectal pouch below the obstruction and passed in the natural shape. They are apt to be lumpy and unformed rather than misformed, but they may be streaked with blood or slime which is always a valuable sign and one calling for careful physical exploration.

A stricture in the locality in question must be examined for with the greatest care and gentleness, and the examination will often be negative in its results. The attempt to decide the question by the use of bougies is altogether unsatisfactory and by no means free from danger. It is unsatisfactory because an obstruction will generally be encountered in trying to pass an instrument of any considerable size through this part of the bowel, and the passage of an instrument of small size, which is much easier, proves nothing. It is dangerous because, with the ordinary rubber rectal bougies, a diseased bowel may easily be ruptured with what may seem to the operator to be no more force than is justified in attempting to overcome the natural obstructions to this part of the passage. The bulbous-pointed bougie on the flexible stem appears *à priori* to be the most suitable for the exploration, but it has two objectionable features. It is not at all an easy instrument to pass, and if passed through an obstruction too much force is required for its withdrawal after the abrupt shoulder is in contact with the stricture.

O'Beirne gives the following description of the way to pass his tube: "A gum-elastic catheter of the largest size was inserted into the anus, and passed to the height of about two inches up the rectum, where its further progress was felt to be opposed by strong expulsive efforts, which lasted but a few seconds, then relaxed and again became renewed. By first yielding somewhat to these efforts, and then taking advantage of the succeeding relaxation, the instrument was gradually passed to the height of seven or eight inches. At this point the resistance was sensibly felt to be much greater than at any former, but, instead of allowing it to yield, the instrument was pressed more firmly upward. Having steadily continued this pressure for about one minute, the resistance suddenly gave way, the tube passed upward as if through a narrow ring," etc.

Even with the softest instrument, the moment when the obstruction suddenly gives way, and the instrument passes forward, will be an anxious one for the surgeon, and the life of the patient may be sacrificed to desire for certainty of diagnosis.

A bougie intended for this purpose should always be hollow and the opening at the lower end should be of a size to admit the small tube of a

Davidson's syringe which should be fitted to it before the attempt to pass it is begun. Then with a basin of warm water close at hand the bougie may be introduced and at the first obstruction the bowel should be filled with water until it is moderately distended. In this way the folds of mucous membrane are drawn out of the way by the distention of the whole bowel and one great obstacle is eliminated. The next is the promontory of the sacrum which is much more easily passed by a soft than by a stiff instrument. Without these precautions, and sometimes with them, the inexperienced examiner will find a stricture in the rectum of nineteen persons out of twenty, no matter how healthy they may be; and for this reason it is seldom safe to rest the diagnosis of stricture on the fact that a bougie cannot be made to pass. Moreover a bougie of good size will often pass a stricture small enough to produce great trouble.

In certain cases information may be gained by the use of a long cylindrical speculum with the patient bending over the table or chair and straining down to bring the parts into view. Fortunately, however, we are not limited to either of these means for a diagnosis, for, if the stricture be cancerous and of any size the mass may be felt through the abdominal wall by careful palpation; and if not, and the symptoms warrant it, the sphincter may be stretched or incised sufficiently to allow of introducing the hand into the rectal pouch. Passing the whole hand into the rectal pouch, and then the finger into the sigmoid flexure as far as possible, is a very different affair from trying to pass the whole hand into the flexure, and is free from danger, because the distention by the hand is not carried to the point where danger is located at the reflection of the peritoneum. Though seemingly a much more serious matter, it is really safer than any forcible use of the bougies, and by it the diagnosis may be rendered certain for all that part of the bowel at present under consideration. I know of no other way than this by which a stricture in the sigmoid flexure which cannot be felt by external manipulation can certainly be recognized.

Treatment.—The treatment of stricture of the rectum is both constitutional and local, medicinal and operative. The first question to be answered is as to the advisability of anti-syphilitic medication. In recent cases where syphilis is to be suspected this should never be omitted.

It is well to exercise caution in this matter, however, and the cases in which the patient should be submitted to this form of treatment should be carefully chosen. The practitioner who considers the majority of strictures as syphilitic and indiscriminately uses mercury and iodide of potash will be mistaken about as often as he who looks upon most of his cases as cancerous and therefore incurable. The general condition of a patient with a stricture is never up to normal, and an unnecessary course of medication may do great harm instead of good.

Cicatricial tissue, though the result of specific disease, is beyond the reach of specific treatment, but where the case can be seen early enough, much improvement can be gained by a thorough course of mixed treatment and a gummatous deposit or a syphilitic sclerosis may be checked. Mercury and iodide of potash should both be given, neither being relied upon alone. Mercury in the form of an ointment or the oleate may also be administered by the rectum, and the full constitutional effects of the drug may be gained in a very short time by this method; it is, however, an irritating application and in cases of much ulceration and sensitiveness it may not be well borne.

M. Trelat[1] has seen good effects follow internal medication in cases of ano-rectal syphiloma, though Fournier speaks so positively as to their uselessness. He gives two cases in which the disease was of long standing, but yielded to a considerable degree to the use of mercury and iodide of potash internally, with glycerin applied locally. Van Buren[2] has also seen good effects in a case of this kind from the use of the modified Zittman's decoction, in mild doses, guarded by bismuth, combined with inunctions of the oleate of mercury.

The following case taken from Zappula[3] is worth reproducing entire, proving as it is supposed to do that a syphilitic stricture which is so extensive as to give rise to the diagnosis of malignant disease may be made to completely disappear by specific treatment. The author says: "The patient who is the subject of this case is one of my colleagues and an intimate friend, a man thirty-six years of age and of nervous temperament. The family history is good. The patient has always enjoyed good health with the exception of some attacks of malaria, a gonorrhœa contracted in 1851, and some months after an ulcer in the balano-preputial fold, which was followed by a painful adenitis in the right groin which, however, did not suppurate. The ulcer was of considerable size, lasted about forty days, and ended by healing under the influence of repeated cauterizations. Nothing more is known of the character of that ulceration, and it is impossible to establish any connection between it and the disease under consideration. But it is certain that the patient used in inunctions more than one hundred grammes of mercurial ointment, and that an examination of the former site of the ulcer shows now no trace of its existence.

"The first symptom of the present disease was pain which started from the right side of the anus, extended as far as the tuberosity of the ischium on the corresponding side, or sometimes took an opposite course, but always was confined to the ano-rectal region. The pain was of neuralgic character, intermittent, returning with more or less frequency, but always very severe and accompanied by the phenomena of spasm. Defe-

---

[1] Le Progrès Méd., June 22d, 1878.

[2] On Phantom Stricture, etc. The Amer. Journal of the Medical Sciences, October, 1879.

[3] Annali universali de Medicina, vol. ccxvii., p. 137.

cation became a little less frequent, but was painless except once when there was a sharp pain about the anus. A fissure was suspected, and though it was impossible to discover it, a suitable injection of laudanum and rhatany was administered.

"The pain disappeared from the ischio-rectal fossæ, but symptoms of impaction followed which purgatives in large doses failed to relieve, and which on the contrary led to still more alarming accidents. It was under these circumstances that I first saw the patient on the 24th of September. He had suffered for one month and his condition seemed to be very serious. Three large fæcal tumors occupied the left iliac fossa, the epigastrium, and the right flank. Severe colic starting from the left iliac fossa extended over the whole abdomen and reached to the anus. The abdomen was swollen and painful to the touch, and pain was also caused by pressure in the ano-ischiatic region where, however, no trace of organic disease could be discovered. An examination of the anus led to the discovery of a stricture so tight that only the end of the little finger could be introduced without causing great pain.

"Such was the group of symptoms the patient presented when I first examined him: retraction of the anus and probably of the rectum; absolute necessity of causing the disappearance of the obstacle to the exit of fæces and of exciting intestinal contraction. But it was impossible for me to know whether the contracture was due to ragades located immediately within the anus, to the neuralgic symptoms described above, or to some neoplasm in the lower part of the rectum. Nevertheless I attacked the symptom of contracture by the method of Récamier, and it may be imagined how painful this proceeding was while the state of the sufferer did not permit me to give ether. However, during the operation I discovered an enormous dilatation of the lower portion of the rectum from which escaped a considerable quantity of glairy matter. Twice afterwards I administered large doses of purgatives, but the patient vomited them almost immediately, and the abdominal meteorism increased. Then the vomiting became spontaneous, the fever increased, and the symptoms of strangulation became so intense that the life of the patient seemed to me about to be sacrificed, when again, under the influence of two inunctions of croton oil on the abdomen, there followed a tumultuous expulsion of fæces. More than twenty hard, round, fæcal masses came away and after this relief all went well. But the patient's ease only lasted a few days, for the fæces very soon accumulated afresh, without forming tumors, however; the passages were made with difficulty; and purgatives administered from time to time caused the expulsion of hardened masses mixed with mucus and sometimes with blood. However, the suffering continued, and was especially violent after the administration of purgatives even in small doses; the abdominal pain became more and more severe; the ischio-rectal pain, together with the neuralgia which he had at the commencement, returned and resisted the most powerful local anodynes; but the anal spasm did not return. Inspite of these frightful sufferings there was as yet little loss of flesh.

"But the organism could not long withstand such sufferings and emaciation supervened; there was fever at irregular intervals always preceded by a chill, and a pale-yellowish tint to the skin. An examination of the rectum, which had been delayed on account of the repugnance of the patient, was extremely painful; but instead of finding as before a considerable dilatation of the lower extremity, I found the tissues soft and uneven, giving to the finger the sensation of folds and anfractuosities, in a way that without a speculum examination would have led one to believe in the existence of condylomata and extensive destruction of tissue; but by the aid of that instrument I was able to prove that we had to deal with an hypertrophy of the mucous membrane which was mammillated.

"This condition was found completely surrounding the rectum and reaching as high as the eye could see. The sensation which my finger experienced could not, therefore, be due to a duplicature of the hypertrophied mucous membrane. A sound introduced into the rectum passed freely eleven centimetres, but, arrived at that point, it was arrested by an insurmountable obstacle, and caused great pain. A second examination practised about a fortnight later permitted me to observe a small tumor on the right side of the intestine four centimetres above the anus. This tumor was the size of a hazel-nut, spherical, smooth, somewhat elastic, and indolent even to pressure. It was absolutely immovable and did not seem adherent to the mucous membrane beneath which it lay. But all these details were very difficult to appreciate well on account of the hypertrophy of the mucous membrane and the irregularities of its surface.

"The retraction of the rectum was then an evident fact, revealed not only by the rational symptoms, but by the physical examination and the hypertrophic thickening of the mucous membrane. But the diagnosis of the nature of the constriction still remained doubtful, for the data furnished by direct examination seemed insufficient. We were therefore reduced to making a diagnosis by exclusion, and rejecting successively the valves of mucous membrane, strictures due to ulceration or simple inflammation, excluding also the idea of a spasmodic or venereal stricture, tubercular stricture, polypus, and hæmorrhoids, we were naturally led to the conclusion that we were dealing with a cancer. However, we had no pathognomonic sign on which to base this diagnosis; and the origin and evolution of the disease were not those of cancer, the march of which is slow and rarely takes such an exceptionally rapid course. Thus, hesitating to admit a cancer, I thought of syphilis. But it was necessary to know for certain whether our patient was suffering from syphilis. It was necessary to be able to establish by well-observed facts that a syphilis may remain latent nearly nineteen years without causing any species of manifestation. The emaciation, the coloration of the skin, the daily fever, all seemed to indicate the presence of cancer, and to exclude the idea of syphilis.

"However, the **powerlessness** of art in the presence of a heteroplastic lesion determined **me** to attempt an antisyphilitic treatment which I **commenced** by **administering** large doses of iodide of potash. After twelve days **of this treatment, the** patient experienced relief of all the worst symptoms. **The first to** yield was the ischio-anal pain which for some time had **been** exceedingly severe. The anal tumor diminished little by little, **the** mucous membrane subsided, there were several normal passages, the colic became less frequent **and less severe, and** disappeared finally after **some** violent pain which **the evacuation of a** considerable quantity of **hard** fæcal matter provoked. ·**From that time** the passages were daily and easy, the local symptoms became definitely better. The flesh returned, the fever disappeared, with it disappeared the yellowish tint of the integument, and at the end **of** three months the patient **was** completely cured."

This case is also quoted by Mollière[1] in full, as proof of **what** may be accomplished by anti-syphilitic treatment in syphilitic stricture. He remarks that one such case **seems to** him to pass all comment; and to prove what caution **should be used in** the diagnosis of organic disease. That nothing in fact **was more improbable** *a priori* than the syphilitic character of the lesions **of this patient, and that** specifics saved him from certain death. **He** asks: "**Is not one authorized, in** the presence of one such extraordinary **fact, to** lay down **the absolute rule that** iodide of potash should be employed in all neoplastic **lesions of the rectum?**"

To my own **mind** the case conveys a **very different lesson from the one** intended. **It seems** to me to prove nothing with regard **to the effect of** internal medication in syphilitic stricture, and to be one more example **of** a diagnosis of stricture based upon the fact that a bougie **met with an** obstruction at a point beyond the limit of touch and vision. **It** may be a case of syphilitic stricture cured by treatment, but the history does not prove it.

There are various means by which the comfort **of** these sufferers may be greatly increased without recourse to operative treatment—and since in many cases the surgeon is limited to these means in his efforts to afford relief it is well that they should receive **careful** attention. The most effectual of them will be found to be a careful regulation of the diet, the administration of laxatives on occasion, **and rest.** The diet should consist mostly of **fluids, preferably** milk. **If milk is** complained of, soups may be substituted. **A certain** amount **of** farinaceous food may also be allowed, such as toast, crackers, and mush; but milk is the basis of the diet, and the other things **are only** intended **to make** that diet endurable.

---

[1] Op. cit., p. 306.

Many patients will assert from the outset that they cannot take milk, but nearly all can take it, and considerable quanities of it, daily for an indefinite period if a little care is exercised in its administration.

The bowels should move daily without straining. Should any medication be necessary to secure this daily evacuation a mild laxative will be found all sufficient. The mineral waters or Rochelle or Glauber's salts answer every purpose. Purgatives are always contra-indicated in stricture of any variety, because they cause straining and tenesmus, increase the tendency to congestion and its consequences, and because where obstruction actually exists or is threatened, they may do great harm by exciting violent peristaltic action in an already weakened and ulcerated bowel. The opposite condition of diarrhœa is more difficult to meet and often cannot be controlled by direct medical treatment, depending as it does on the ulceration associated with the stricture. It is best met by diet, rest in the recumbent posture, and bismuth with morphine.

The general strength of these patients is to be supported in every possible way, and in all of them where it can be borne cod-liver oil will be found to answer a good purpose.

When obstruction actually exists, much may be done in the way of general treatment before resorting to operation. Food should be almost absolutely suspended; opium should be given in large doses, to allay the peristaltic action of the intestine, and large poultices covering the abdomen will be found to give great relief to the suffering. Dr. Norman Kerr has derived great benefit from the administration of the extract of belladonna in doses of one or two grains at short intervals, in this condition, but the *rationale* of its operation is not understood. No purgatives should be administered, and the bowel should not be tapped with the aspirator. The dangers of this measure have already been pointed out.

By these means combined, possibly with gentle dilatation, the life of a patient may be prolonged in comfort. I have often been agreeably surprised at the happy results of such measures, where operative interference was either declined or contra-indicated, and they can never be dispensed with, though an operation be performed.

The various surgical procedures at our command for overcoming stricture of the rectum may be considered in the following order: 1. Dilatation. 2. Division. 3. Colotomy.

1. *Dilatation.*—This may be either gradual or sudden, partial or complete. The use of bougies for gradual dilatation is an example of a good practice originating in false ideas. It was first adopted with the idea of destroying the stricture by the effect of medicinal substances applied in this way; experience, however, soon proved that simple bougies were not less efficacious than medicated ones, and the improvement was then supposed to be due merely to the mechanical stretching of the part, and the instruments were introduced as often, and allowed to

remain in, as long as possible, an idea still very popular. But as Syme[1] pointed out, "it is the effusion of organizable matter in the cellular texture of the part which causes the stricture, and it is the absorption of this deposit which removes the disease. The bougie, by its pressure, excites the action of absorption; and if the pressure be too great, too long continued, or too frequently repeated, there will be a great risk of causing more than sufficient irritation for the purpose, and of inducing again the very condition it is desired to counteract, the consequences of which must be a confirmation and increase of the disease."

The rules which should guide the surgeon in this method of treatment are now well understood and generally admitted. The dilatation should be intermittent, and not constant. Attempts at constant dilatation by means of bougies of any sort which shall remain permanently in place generally result either in failure or actual disaster. They are not well borne by the patient, and when their use is persisted in, in spite of the protest which nature is pretty sure to make, the rectum becomes irritable, the suffering is greatly increased, and the patient is exposed to the risk of peritonitis and cellulitis.

The dilatation should never be forced. A bougie should be chosen which will readily pass the obstruction without stretching, and if there be any doubt in the operator's mind as to the proper size of the instrument to be used, let one be selected which is too small rather than too large. The instrument should seldom be passed more than every alternate day, and once a week may be often enough. Little is gained by allowing it to rest for any length of time within the constriction.

Practised in this way, much good may be done by this treatment. The patient may be greatly relieved, and made very comfortable; but it must be continued indefinitely. For this reason, I suppose it is not infrequently used under false pretences in cases of hypothetical stricture in hypochondriacal patients; and most of the reported cases of cure will be found reported by the laity. It has happened to me more than once not to be able to find any stricture after a patient had submitted to a long course of supposed dilatation, and there is but one way of convincing the patient under such circumstances. It consists simply in passing a full-sized instrument its whole length into the bowel.

In cases where the stricture is associated with much ulceration, dilatation by bougies is very apt to make matters worse instead of better, and in such cases I seldom employ it in my own practice and have seen much suffering caused by it in the practice of others.

This treatment by gradual dilatation, perhaps on account of the recent great advances which have been made in the treatment of stricture, has, to a certain extent, been superseded by more radical measures. It is not long since a well-written article on rectotomy in one

---

[1] Op. cit., p. 120.

of our periodicals was begun by the statement that the treatment of stricture by dilatation was acknowledged to be a failure. This is by no means the case. The measure may not be curative, but it is, perhaps, as valuable a palliative as is at the command of the surgeon. It need not always be done with a bougie; for the patient's own finger or that of a careful nurse is often better than any instrument. It is applicable to all strictures, malignant or benign, which are within reach of the anus. When the disease is high up, it is not free from danger, and can scarcely be recommended, on account of the uncertainty and difficulty of its application.

I have said that this treatment by gradual dilatation was not curative, and must be continued indefinitely. I have seen no exceptions to this rule, though many of them are reported. In years gone by, this treatment and that of forcible dilatation or divulsion were about the only means of dealing with this affection. Now we have better ones which will shortly be described.

Divulsion. The dilatation, instead of being gradual, may be sudden and complete. For this purpose, various instruments have been in-

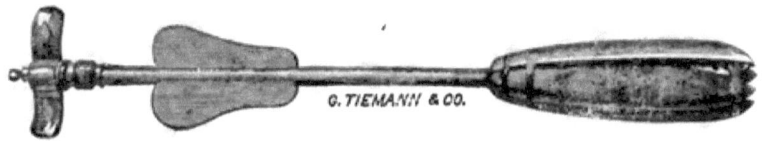

Fig. 49.

vented, all of them with the idea of tearing open the constriction by the use of a considerable amount of force. One of these is shown in Figure 49. More recently, advantage has been taken of fluid pressure, and an instrument has been invented by Wales, which is shown in Figure 50.

Of all the instruments for forcible dilatation, this is perhaps the best. There are now several cases on record where forcible stretching with the fingers, either with or without previous nicking with a knife, has been followed by immediate relief to obstruction and fæcal accumulation.[1]

What may be accomplished by this method is well shown in the following successful case from Smith.[2] "I was called by Dr. Vine to see a military officer, aged 40, who had returned from India in the most miserable plight. He had suffered for several years from chronic diarrhœa, and had not got relief from any measures, and six months previously he had been recommended by a medical board to go by sea to England. On his arrival at Southampton, on his way to Edinburgh, his native town, he was so ill that he determined to stop in London, and

---

[1] Smith, op. cit. Dr. J. M. Matthews, of Louisville, Ky., has recorded one remarkably successful case of this kind.

[2] Surgery of the Rectum.

when he arrived there he sent for Dr. Vine, who, on hearing his history, at once suspected something wrong with his rectum, and making an

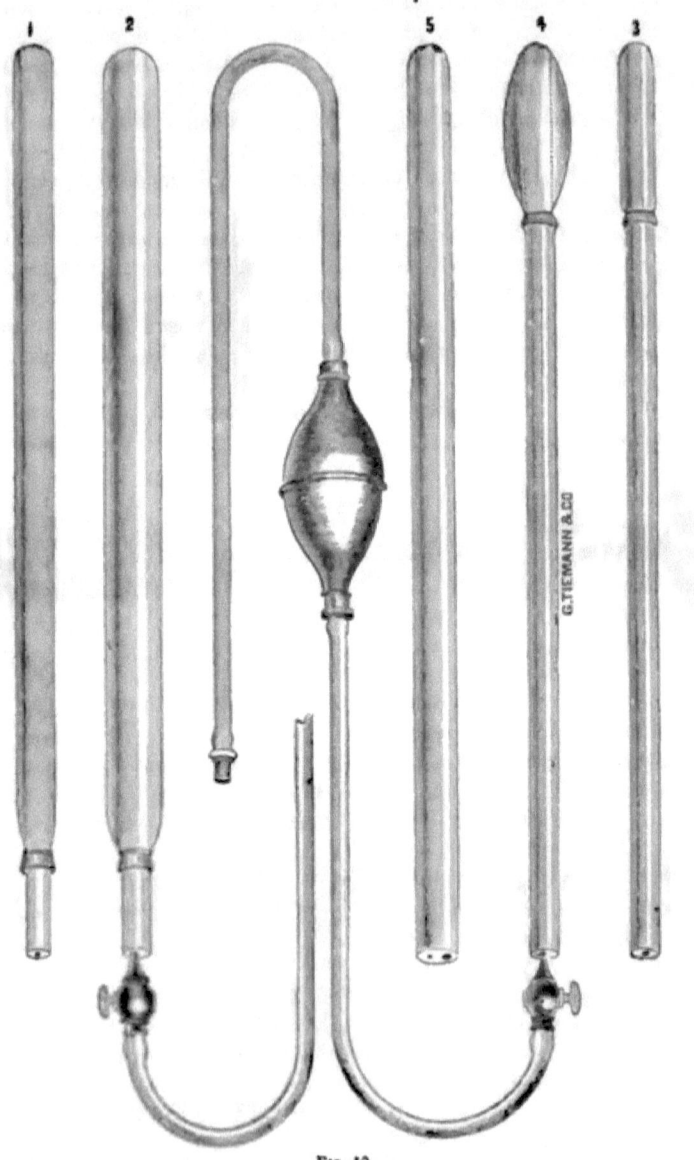

Fig. 50.

examination, found an obstruction. I was requested to see him, and I found the patient exactly in the condition of one suffering from strangulated hernia; he was constantly vomiting, complaining of pain, and the countenance was anxious, and he was much emaciated; the abdomen was immensely distended, and it was clear that, if some relief were not soon given, this gentleman would die.

"In conjunction with Dr. Vine, I made a most careful examination, and I found, on introducing the finger into the bowel as far as possible, that it met with an obstruction, but after some time I discovered what appeared to be the opening of the stricture, more like a dimple than aught else. I was enabled to introduce through this a No. 10 gum-elastic catheter, and through this instrument some fæcal matter and air came. I was thus made to see that I had got beyond the stricture.

"On the following day, the patient was placed under chloroform, and I guided a long, straight, probe-pointed knife very carefully along the side of my left index finger, and fortunately got its point into the orifice of the stricture. I nicked this on either side, and then got the point of my finger into the obstruction, and dilated the orifice as much as I could, whereupon an enormous quantity of fæcal matter was emitted, deluging the bed, and placing myself and my assistants in a most unenviable position. The abdomen became quite flat, and the patient became at once immediately relieved. No bad results followed this operation; in three days we commenced dilatation by bougies, and I was soon enabled to pass a full-sized rectum-bougie through the stricture. In a fortnight I took my leave of the patient, recommending Dr. Vine to pass the bougie daily. I heard a few weeks afterwards that the patient had gone to Edinburgh convalescent, and able to introduce the bougie for himself."

In spite of a few such successful cases as the one above, this method of treatment has but few upholders, because it has been found to possess no advantages over more gradual dilatation, and to be in itself by no means devoid of danger. The dangers are hæmorrhage, laceration and rupture of the bowel, peritonitis, and abscess. The relief obtained is not permanent, and the operation involves the subsequent use of gradual dilatation to preserve the calibre gained. Even when applied to the lower three inches of the bowel, the operation is rough, uncertain, and unsurgical, and above this point it is scarcely admissible. Nevertheless, it has occasionally served a good purpose, and a few happy results are recorded in cases of linear contraction.

Division of the Stricture.—The practice of nicking a linear stricture in two or three places as a first step in the treatment by dilatation is a good one, and generally devoid of danger. It can usually be done entirely by the sense of touch with a straight, blunt-pointed bistoury passed along the left index finger as a guide.

The operation of internal proctotomy consists in dividing the whole of the stricture tissue in the median line, either anteriorly or posteriorly. It is called internal because the incision is confined within the rectum, and does not involve the sphincter; and it is generally performed with the knife in preference to the cautery or écraseur.

Regarding this operation, there is not very much to be said. It involves no new principle **of treatment,** and would seem to rank rather with the older procedures, such as nicking and dilatation, than as a substitute for colotomy. **There have been** many unpublished cases, especially in New York, **and I** should probably express the general feeling of the profession, **were I** to say that it is not looked upon with very great favor. Though at **first** sight it might appear less serious than the external operation, **it is** probably the more dangerous **of the** two—the sphincter preventing the free discharge **from** the wound and increasing in this way the liability to pelvic inflammation. This muscle should at least be stretched as a primary step in the operation, and, when possible, a large drainage-tube should be left in. The danger of hæmorrhage is not **very** great when the incision is confined to the median line, but, should there be trouble from this cause, the advantage of a free external wound in controlling it will at once be manifest. When the cut is anterior **as** well as posterior, the anatomical relations must be borne in mind, lest the peritonæum in the female, or the bladder in the male, **be** wounded. The following case represents my entire experience with the operation, which I abandoned after once trying, being convinced of the advantages of the external incision, next to be described.

CASE XIX.—Mrs. ——, age twenty-six. This patient was a woman with a syphilitic history. The stricture was of eight years' growth, and had previously been treated both by nicking and by gradual dilatation. As a result of this treatment, she describes an attack of "inflammation of the bowels," which made her very dangerously **sick. The stricture** was two and one-half inches from the anus, was of just sufficient calibre to engage the end of the index finger, and did not involve more than one inch of the bowel, though there was the usual amount of ulceration above it.

I divided the stricture by a single, deep, posterior incision, which did not implicate the sphincter, and the operation was followed by an attack of pelvic peritonitis, which very nearly cost the patient her life. This may have been due to the operation, or it may have been due to attempts **at** subsequent dilatation which was begun early and followed with perhaps too great vigor; but it was certainly excited by the patient leaving her bed, going down-stairs, indulging freely in wine, and submitting to the embraces **of her** lover.

Three months **after the** operation, I completely lost track of the case. At **that** time the **calibre** of the stricture was so much increased as to permit **of** easy digital examination of the parts **above.** The increased size seemed due entirely to a deficiency in the old cicatricial tissue at the point of incision; the rest of the circumference of the part having much the same feel as before the operation. The act of defecation was much less painful, and her condition was altogether much better.

I never counted the case as proving anything concerning the value of the operation until a few months ago, and more than four years after its performance. In fact, I had little doubt that the contraction had returned, and supposed the patient had either succumbed to the disease or submitted to colotomy. At that time, however, the woman was in perfect health and spirits, and since then I have thought better of the operation. I would have given much for a rectal examination after so long an interval, but it could not be obtained.

Other cases of similar operations have been reported in this country with equally good results.[1]

External proctotomy involves not only the division of the stricture, but of all the parts below, including the anus. This is the operation usually accredited to Nélaton, and more recently advocated by Verneuil, Panas, and others. It may be performed in several ways, and with the knife, galvano-cautery, or écraseur. The operations with the galvano-cautery and écraseur were invented by Verneuil,[2] and have been practised by him more than by any other surgeon.

The operation as performed by him consists in passing the left index-finger through the stricture as a guide, and then plunging a trocar from a point in the median line, just in front of the tip of the coccyx, into the rectum, on to the tip of the finger above the stricture. After drawing out the trocar, a fine bougie is passed through the canula into the rectum, and brought out at the anus. Removing the canula, the bougie is replaced by the chain of the écraseur, and the operation is completed.

The same section may be accomplished by repeated strokes of the galvano-cautery or thermo-cautery knife. Both these measures are intended simply to prevent hæmorrhage, and have no other advantage over the knife, and by any of the methods all of the stricture tissue and the parts below may be divided.

---

[1] Whitehead—Old fibrous stricture: anterior and posterior incision with bistoury, followed by dilatation. Two months later, much improved; passages large and natural; dilatation continued. Amer. Jour. Med. Sc., Jan., 1871. Lente—Fibrous stricture and fistula; incision followed by dilatation. Three months later, much relieved, with prospect of entire cure by continuing the use of bougies. Amer. Jour. Med. Sc., July, 1873. Beane—Probably syphilitic; incision both anterior and posterior, followed by use of dilators. Seven months after, cure of ulceration and of many bad symptoms, but tendency to recontraction. Amer. Jour. Med. Sc., April, 1878.

[2] Verneuil : Des rétrécissements de la partie inférieure du rectum, et de leur traitement curatif ou palliatif par la rectotomie linéaire, ou section longitudinale de l'intestin à l'aide de l'écraseur. Gaz. des Hôp., October 26th, 29th; November 7th, 9th, 12th, 16th, 19th, 1872. Traitement palliatif du cancer du rectum au moyen de la rectotomie linéaire. Gaz. Hebdom. March 27th, 1874.

Nélaton's method was the simplest of **all**, and was to introduce the left index finger as far as the stricture, and with this as a guide, to **pass** in a blunt bistoury, and **divide all** the soft parts below the stricture **as** nearly as possible in the median line. By pulling open the lips of this incision, the stricture comes **plainly** into view, and may be divided by a second incision.

In performing the operation, I prefer the knife to all other methods of cutting, and **have** had one specially adapted for the purpose, which is shown in Fig. 51.

It is simply the lithotomy knife of **Blizard, made heavier in the back** and at the handle, for with an ordinary bistoury there is **great risk of** breaking the blade in the midst of the stricture tissue, **which is often as hard as cartilage,** and thus having an **awkward** accident. The **blunt point on the end** of the blade is a great convenience in passing the knife **along the** index finger, avoiding as **it does, all risk** of wounding **the operator.**

The best position for the patient is the lithotomy position, and the whole incision may be made at one stroke. The blade should be passed fairly through the **stricture before the** cutting is begun, then the stricture is divided **completely, as near as** possible in the median line

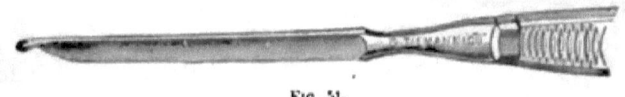

Fig. 51.

posteriorly, and **finally** the incision is **continued downwards and outwards,** growing **deeper** as it approaches **the perineum,** till finally all the soft parts are severed between the anus and **the tip of the coccyx.** In this way, a large triangular wound is made, the **apex** being within the rectum, above the stricture, and the base at the skin, and all the stricture tissue is completely cut through.

There will generally be a free **gush of blood** when the cut is made, but I have never seen so much **as to make me** prefer the écraseur or cautery operation in preference to the **knife.** The rectum should at once be packed in the manner already described, without waiting to try any other method of stopping the bleeding. This is a precaution which should never be omitted.

This **operation may be modified in** various ways to fulfil any special indication. In extensive cancerous disease, I have sometimes made two such cuts, and **taken out a** considerable mass of the growth between them, merely **for the purpose of** opening **the canal.**

It may be asked, Why should **so** large an incision be made, and so much tissue be divided **below** the actual disease? The answer is simple. In the first place, this incision provides for free drainage and discharge in **the most effectual** of all ways, by furnishing **a** dependent gutter-shaped opening which cannot become closed. This is better than any number of drainage tubes, and it is this alone which makes the external operation a safer one than the apparently slighter internal incision.

In the second place, by this incision, the sphincter is completely divided, and another great point is gained. The operation we are now considering, it should be remembered, is nothing less than a substitute for colotomy in the same class of severe cases for which that operation is generally considered the only relief. One point which is exceedingly well brought out by a study of these cases is the important part played by the sphincter muscle in the sufferings accompanying severe cases of stricture and ulceration, and the relief which may be obtained by its simple division without interference with the stricture itself.

In one case of Verneuil's, for example, there was a stricture high up, and yet, under a mistaken diagnosis of spasmodic stricture at the anus, the sphincter was cut through with the galvano-cautery, while the real cause of the trouble was untouched, and yet there was entire relief from suffering. The same experience has been repeated often enough to establish the general principle, that free division of the sphincter is not only a justifiable therapeutic measure for the relief of the pain attendant upon either benign or malignant stricture or ulceration, but is often the best means at the surgeon's command for allaying suffering.

By the external operation, then, the obstruction is divided, and one great cause of suffering is abolished, and both are effected by the same stroke of the knife.

The after-treatment of the incision is very simple. When the rectum has been tightly packed with picked lint, it will usually cause more or less uneasiness on the following day, unless the patient be under the influence of opium. For this reason, I generally remove enough of it on the following day to give the patient ease, and the remainder is allowed to remain until suppuration has commenced. It may usually all be picked out by the third or fourth day without causing any pain. The subsequent treatment of the incision itself consists wholly in cleanliness, which may be obtained by gently syringing the part with warm water and a little carbolic acid. No particular attention need be given to regulating the passages. The first one after the operation will often be the only comfortable one the patient has experienced for years, and unless there is some special reason for interference, they may be left entirely to nature.

The case which follows will give a very fair idea of what may be hoped for from this method of treatment :

CASE No. XX.—Mrs. ———, age 35, mother of one child twelve years old. The patient has always suffered from obstinate constipation, and several years ago was relieved artificially of impaction of fæces. Her husband, a physician, assures me that there is no venereal history, nor is there any reason to suspect any such. The symptoms of rectal trouble began six years after marriage, at which time she was operated upon for large internal hæmorrhoids. Soon after this she began to suffer with the usual symptoms of ulceration of the rectum.

The examination revealed advanced ulceration of the whole circumference of the rectum, with a stricture about an inch and a half up, which just admitted the end of the **index finger.** In connection with the stricture there were two fistulæ. For this condition the **patient** had submitted to the usual **treatment** by dilatation, but without relief. Her general condition **was such** as is usually seen in advanced rectal disease. She had lost flesh and appetite, and the suffering was **extreme.** What she most dreaded was an action of the bowels, so great was the pain attendant upon it.

The operation which I have described was performed. **One of** the fistulæ was also cut, but the other **was** left to the chance of spontaneous closure, since it communicated with both rectum and vagina, and the usual operation for recto-vaginal fistula would have been necessary had any interference been practised. The operation was attended **with** considerable hæmorrhage, which was controlled by stuffing the rectum **with** picked lint, after the ulcerated surfaces both above and below the stricture had been renovated **by** scraping them with the handle of a scalpel.

The subsequent **treatment consisted** merely in absolute rest in bed and milk diet, **with a** dressing **of the** wound by the introduction of picked lint. No attempt was made at passing a bougie, and the stricture was left entirely **to itself.** The immediate **effect of the** operation **was a** most marked and satisfactory relief of **the most painful symptoms.** The passages were involuntary, but were *painless* **and always preceded** by a warning sensation, which gave the patient ample time to prepare herself. At the end of six weeks she had improved greatly in general condition, and was more comfortable than at any time since the trouble began. The passages were of normal shape and occurred painlessly once a day. They were under the control of the will, but there was incontinence of wind. In this condition the patient returned to her home in the West under the care of her husband.

Six months later, she again came to New York for treatment, not from any return of the pain, but because of the discharge from the bowel, and the occasional annoyance which arose from the incontinence of wind. Her general condition was excellent, and, except for the two things mentioned, she would have considered herself in perfect health. An examination showed a very marked decrease and softening down in the stricture tissue; the wound made with the knife had never entirely healed, the patient having exercised freely and constantly while at home, and there were two distinct lines of ulceration within the anus; one on the anterior surface, superficial, about half an inch broad and an inch and a half long; the other, at the site of the cut behind, deeper, and running further up the bowel. Otherwise the old ulceration was entirely healed, and its site marked by a thin, shining bluish-white cicatricial surface.

Attention was at once turned to the treatment of this ulceration. The patient was put upon almost absolute **milk-diet**, and after awhile was also confined absolutely to her bed. The remnant of the old incision was induced to heal by daily dressings of lint and balsam of Peru, and the **ulceration** above was **treated** by applications of bismuth, opium, nitrate **of silver**, balsam of Peru, iodoform, and oxide of zinc, alone and in combination. At the end of a couple of months she was so nearly well that attention was turned to the recto-vesical fistula. The openings into the rectum and vagina were both small, but there was a considerable abscess cavity in the recto-vaginal wall which discharged into each canal. This cavity was freely laid open into the rectum. At the end of three months the ulceration on the anterior wall of the rectum had entirely healed, that on the posterior wall had nearly healed, the incision had cicatrized, and the abscess cavity had closed except an exceedingly fine and tortuous canal leading from the rectum into the vagina. The discharge from the rectum had practically ceased, and in this condition, which certainly warranted a prognosis of complete and speedy recovery, she returned to her home to continue the treatment for a few weeks longer till she should be entirely well. Two months later I again heard from her, and the report was most favorable.

This case is certainly worthy of a careful consideration. When the lady applied to me, all the supposed resources of rectal surgery had been exhausted except colotomy. I do not think I exaggerate when I say that most surgeons would have at once decided in favor of colotomy, and would have been justified, of course, in so deciding, for colotomy is still the recognized mode of treatment in these cases. In my own mind, colotomy was always present as the *dernier ressort*, but having tried proctotomy in several instances, and been more or less satisfied with its results, I determined to make this a test case. The result was most happy, and yet there is nothing exceptional in that result, though the great tractability of the patient, and her determination to do all that was asked of her, alone rendered it possible.

In an analysis of cases made some time since,[1] I found that in eighteen cases of non-malignant stricture treated in this way, all the patients were greatly relieved as to general health, or local condition, or both. In eight, kept under observation for a period of from three months in one case to four years in three cases, the cure was absolute, there being no return of the contraction, and in some a disappearance of all induration. A tendency to recontraction is mentioned in four, due in two to the fact that all of the stricture was not divided.

Brief notes of some of these cases are given below.

---

[1] External Rectotomy as a Substitute for Lumbar Colotomy in the Treatment of Stricture of the Rectum. The N. Y. Med. Journal, March, 1880.

External Rectotomy with the Knife.

1. PANAS.—Female, aged 33. Syphilitic stricture, very dense and painful; eight years' duration. Incontinence for three months after operation. Eighteen months later, described as completely cured.—*Gaz. des Hôp.*, Dec., 1872.

2. WHITTLE.—Hard annular stricture, very close; one fistula. Operation as for ordinary fistule. Hæmorrhage troublesome and controlled by thermo-cautery. Three weeks later, "general health completely restored and local condition greatly relieved."—*Lancet*, June 1st, 1878.

3. PANAS.—Woman, aged 40. Stricture probably syphilitic. Two previous operations by slight internal incision, and two attempts at cure by dilatation. Patient very feeble; suffering from abdominal distention; signs of approaching occlusion; ovarian tumor; diarrhœa and vomiting. Operation followed by relief of pain and by great comfort; no tendency to return; vomiting and diarrhœa continued till death, some time after, from exhaustion. Post-mortem examination showed the complete success of the operation, and the division in the fibrous tissue.—*Gaz. des Hôp.*, Dec., 1872.

External Rectotomy with the Ecraseur, Galvano-Cautery or Thermo-Cautery.

1. TRELAT.—Ano-rectal syphiloma, of several years' duration, with great thickening, ulceration, and fistulæ. Operation (kind not stated) five years before, unsuccessful. Galvano-cautery. Nine days after operation, pneumonia and facial erysipelas. Death in three weeks without local accident.—*Prog. Méd.*, June 22d, 1878.

2. VERNEUIL.—Stricture of several years' duration; great induration and tumefaction, and twenty fistulous tracts. Three operations; first, on one-half the fistulæ; second, on remainder; and third, on the stricture with écraseur. Four months later, "wound healed and functions of the rectum entirely re-established."—*Gaz. des Hôp.*, 1872, p. 1,028.

3. VERNEUIL.—Previous syphilis; great constitutional disturbance; scrotum enlarged to three times its natural size by fistulous tracts, of which there were twelve. Ecraseur through one of the fistulæ—others operated on a month later. Two years later, parts had regained their suppleness, and all traces of disease had disappeared.—*Loc. cit.*

4. VERNEUIL.—Patient in bad general condition. Two operations with écraseur at six weeks interval. First, posterior rectotomy with division of posterior fistulæ; second, anterior rectotomy with division of anterior fistulæ. Incontinence lasted only a few days. There was marked tendency to recontraction, due to the fact that the stricture was so extensive that the chain was not carried to its upper limit, and a distinct zone of cicatricial tissue was left.—*Loc. cit.*

5. VERNEUIL.—Woman, reduced to last degree of marasmus, with

hectic. Stricture complicated with **much ulceration** above and **below**, and three or **four fistulæ**. Operation followed by great relief of all symptoms. After several years, again examined: general condition still good, but a very appreciable recontraction of a year's duration.—*Loc. cit.*

6. VERNEUIL.—Stricture very close and hard; previous dilatation without effect. Phlegmon existing on one side, and old fistula on the other. Abscess laid **open** and **chain** passed through it into gut above stricture. Four years later, died of phthisis, having been entirely free from symptoms in mean time. Before death, stricture admitted two fingers easily.—*Loc. cit.*

7. VERNEUIL.—Constriction very hard **and close**; also fistula. It was found almost impossible to pass trocar beyond the contraction, on account of its great hardness, and this was finally accomplished only by boring **a tract with a pair of curved scissors**. The écraseur required three-quarters of an hour to cut through. Several months later, general state very satisfactory; rectal wall had partly regained its suppleness; no difficulty in defecation, but a still appreciable contraction, due to the fibres which were too high up for the chain.—*Loc. cit.*

8. VERNEUIL.—Previous syphilis. **General condition bad.** Stricture consisted of a limited **contraction of the posterior and** upper fibres of the sphincter, and **disappeared** on prolonged **pressure** with **the finger**. Two previous operations, one by internal incision, the other by nicking and dilatation. Division by trocar and écraseur; incontinence for a few days; after three weeks, passages natural and all symptoms relieved. Three years after, again examined, and found suffering from rectal syphiloma developed since operation, together with tertiary eruptions.—*Loc. cit.* [History completed by Tison in *Thèse de Paris*.]

9. VERNEUIL.—Previous syphilis; stricture annular; much constitutional disturbance, great pain, diarrhœa, colic, and discharge of pus. Operation of *internal* rectotomy with thermo-cautery, followed by phlegmon. Abscess opened and external operation done with thermo-cautery through abscess cavity. One month later, relief of all symptoms; return of suppleness in parts; stricture admitted two fingers easily; tendency to recontraction in posterior part of rectum; anterior part healthy.—Tison, *Thèse de Paris.*

10. VERNEUIL.—Rectal syphiloma; anæmia and loss of **flesh**; great **tenesmus**. Thermo-cautery. Incontinence for three weeks. Reported completely cured after three months.—Tison.

11. VERNEUIL.—Stricture, **probably** inflammatory, **with** several fistulæ. Thermo-cautery. **Incontinence** for three weeks. After five weeks, appetite and **strength returned**; passages easy and painless.—Tison.

12. GOSSELIN.—Syphilitic. Forced dilatation three years before. General condition very bad from excesses of all kinds; passages very frequent and painful. Thermo-cautery, followed by temporary relief. Four

months later, condition same **as** before, with signs of commencing phthisis.—Tison.

13. TILLAUX.—Valvular stricture, posterior, with ulceration; anterior portion healthy; several fistulæ. Galvano-cautery. Three years later, complete cure, and no return.—Tison.

14. TILLAUX.—Old **stricture,** probably syphilitic, with general cachexia—so great **as to** resemble that of cancer. **Ecraseur.** Four years later, remained completely cured.—Tison.

15. TILLAUX.—Probably syphilitic; previous rupture of perinæum; enormous dilatation of anus; incontinence of fluid fæces; general condition exceedingly bad; signs of occlusion; operation undertaken without hope of cure, but to relieve worst symptoms. Galvano-cautery, from without inward, with cautery knife. Life prolonged five **months,** with freedom from suffering.

16. VERNEUIL.—Dysenteric contraction high up, twelve centimetres from anus. Under mistaken diagnosis of spasmodic stricture of the sphincter, that muscle was divided with the cautery. Entire relief from pain, but continued symptoms of retention.—Tison.

17. LABBÉ.—Probably **syphilitic;** much pain; abscesses; fistulæ. Division with galvano-cautery, **followed** by considerable hæmorrhage and tampon. After a **time, slight return of** contraction at margin of anus, the rest of gut remaining **supple. Second** operation by Verneuil with thermo-cautery, followed in the course **of six months by** prolapse of the rectum, which **was cured** by cauterization of **the posterior** edge of **the** anus. Considerable amelioration **of suffering.**—Tison, quoted from Cerou, Thèse de Paris.

18. VERNEUIL.—Syphiloma of long standing; great **anæmia;** intolerable pain; constant purulent discharge; previous dilatation **unsuccessful.** Ecraseur, followed by dilatation. Four years later, absolute cure. No induration; sphincter acting well.—Tison, Thèse de Paris.

19. FOCHIER.—Stricture of many years' standing. Patient feeble and emaciated; great gastro-intestinal derangement; two fistulæ. The constriction was first divided with a bistoury caché to admit the finger, and operation completed with écraseur. Control of sphincter after the first few days. Left hospital ten days after the operation, with appetite and digestion good, and general health much improved, having soft passages of the size of the finger.—Lyon Méd., Feb. 20th, 1876.

Cancers.

1. VERNEUIL.—Cancer. **Ecraseur,** followed by immediate relief; decrease in induration; **recovery of** appetite and strength. Death from subsequent operation of excision.—*Gaz. des Hôp.*, 1872.

2. VERNEUIL.—Cancer reaching beyond point of finger; sphincter continually in contraction, and violent pain caused by slightest touch; attempts at dilatation followed by phlegmon and fistula; constant pain

and tenesmus, with bloody passages; insomnia; rapidly approaching fatal termination. The operation consisted merely in dividing the sphincter with écraseur without touching the cancer, and the relief was so great that the patient left hospital believing himself cured.—*Gaz. des Hôp.*, Nov., 1872.

3. VERNEUIL.—Cancer, with all the usual symptoms, and approaching occlusion. Ecraseur; death on ninth day from peritonitis.—*Gaz. des Hôp.*, Nov., 1872.

4. VERNEUIL.—Cancerous stricture high up, and very close; constant suffering from discharges of gas and pus. Ecraseur passed as high as possible, but not high enough to divide upper portion. Considerable relief; **cessation of pain;** passages easy for several months. Death finally from progress of disease.—*Gaz. Hebdom.*, Mar. 27th, 1874, p. 196.

5. VERNEUIL.—Epithelioma involving right half of rectum, and reaching too high for extirpation; ulceration; loss of flesh and strength; **great** pain on defecation; **retention**. Sphincter divided with chain on **left side in** such a way as **not** to involve the cancer. One year later, **freedom** from pain; general state good; incontinence following operation disappeared; difficulty in passage of solids overcome by seltzer; gradual advancement of cachexia.—*Gaz. Hebdom.*, Mar. 27th, 1874, p. 196.

6. VERNEUIL.—Cancer high up, involving prostate and vesiculæ seminales. Continued diarrhœa and incontinence, and bad general condition. **A double** posterior external operation was done with the chain, **and the** portion included between the **two** incisions cut away, with the **idea of** relieving pain and retention and opening a passage for the subsequent application of escharotics to **the cancer.** Operation followed by immediate relief of **worst** symptoms. — *Gaz. Hebdom.*, March 27th, 1874.

7. NÉLATON.—Operation done with bistoury. Relief continued till death, eighteen months after, from extension of malignant disease to the pelvis.—Panas, *Gaz. des Hôp.*, 1872, p. 1,149.

8. FOCHIER.—Cancer of posterior part of rectum, reaching to height of ten centimetres. **Great pain and tenesmus; fœtid** and bloody discharge; **loss of sleep.** Complete division with écraseur. Left hospital ten days after, believing himself cured. **After two months,** had no more pain and no incontinence, **except when suffering with** diarrhœa. Had two regular passages daily, **and complained only of not** regaining his strength. In this case, the section extended to the unusual height of twelve centimetres from the anus.—*Lyon Méd.*, Feb. 20th, 1876.

I have performed this operation in various other cases, and have every reason to be satisfied with its results. In malignant or non-malignant stricture and ulceration, I have never seen it fail **to** give immediate relief to suffering, and, as a means of relieving the pain of the disease, I believe it to be fully equal to colotomy. It also fulfils the other great indication for colotomy, the overcoming and prevention of obstruction.

Too much must not be expected of the operation, however. I have seen several cases, one in **my own** practice, and several where I have advised the operation in **consultation** with **others**, which have led to disappointment for this **very reason**. An old stricture of the rectum with extensive ulceration **is a well-nigh** incurable disease. Proctotomy may be relied upon with certainty to relieve the pain and prevent fæcal obstruction even in **the worst** cases, and in more favorable ones it may effect a practical cure **by** opening the canal, causing a diminution in the induration, and allowing the ulceration to heal, but it will **not** cure them all. Nothing at present known to surgery will. A rectum which has once been diseased to this extent is never **again** a healthy one, though it **may** be made a very comfortable one.

Another point which must not be overlooked is, that, after proctotomy as after colotomy, there is still a diseased rectum which must be treated by every possible means; and that the incision may be only the first step in the cure. The stricture is easier to overcome than the ulceration which accompanies it. **In the case** given above, I succeeded ultimately by long and patient **effort in** curing **that** also, but it cannot be done in every case. In many **of these cases the** ulceration must be treated as ulceration with **the same results, both good and** bad, as usually attend the treatment of **that** most painful, **obstinate, and** often incurable condition. But the chances of curing it, and **at all** events **of** relieving it, are infinitely better **after** the operation than **before**.

It is understood that I do not advocate the operation **in cases of** disease high up in the bowel, though it may be safely done **at a considerable** distance from the anus, and where **an** incision involving **the** anterior wall would be unjustifiable, for the anatomical reason that the peritoneum extends so much lower in front than behind. For other literature upon **this** subject, the reader is referred to the bibliography given below.

BIBLIOGRAPHY.

Panas: Du traitement des rétrécissements du rectum par la rectotomie externe, Gaz. des Hôp., December, 1872, p. 1,148.

Muron, A.: Des rétrécissements de l'extrémité inférieure du rectum, et de leur guérison par la rectotomie linéaire. Gaz. Méd. de Paris, January 4th, 1873.

Fochier, A.: Sur l'application de la rectotomie linéaire aux rétrécissements tres-étendus du rectum. Lyon Médicale, February 20th, 1876.

Pinguet: Des rétrécissements du rectum; appréciation des diverses méthodes thérapeutiques. Thèse de Paris, 1873, No. 17.

Tison: Nouvelles **considérations** sur la rectotomie linéaire. Thèse de Paris, 1877.

Turgis: Foreign **Body in Rectum. Bull. de la** Soc. de Chir., tome iv., No. 10, 1878, p. 789.

Cerou: Thèse de Paris, 1875, No. 390.

Whitehead, W. R.: Case of Fibrous Stricture of the Rectum Relieved by Incisions and Elastic Pressure, with Remarks. Amer. Jour. Med. Sc., January, 1871.

Whittle, G.: Stricture of the Rectum Divided by the Knife. Lancet, June 1st, 1879, p. 788.

Lente, F. D.: Report of a Case of Non-Malignant Stricture of the Rectum, and Remarks on the Surgical Treatment of this Disease. Amer. Jour. Med. Sc., July, 1873.

Beane, F. D.: Case of Specific Stricture of the Rectum; Antero-Posterior Linear Rectotomy; Recovery; Remarks on the Operation. Amer. Jour. Med. Sc., April, 1878.

Discussion sur les rétrécissements du rectum. Bull. de la Soc. de Chir., Paris, 1873, p. 83.

Verneuil, *et al.*: Rectotomie et colotomie (Soc. de Chir., Paris). Prog. Méd., January 7th, 1882.

Excision.—The operation of excision, which is generally applied only to cancerous strictures and which will be fully described under that head, has also been applied to simple strictures; and, though I have never done it myself, I have seen a few cases which seemed particularly adapted to it. One such case is reported by Dr. Lowson[1] in which the result was comparatively good, though no better than that obtained by proctotomy.

The operation performed by him consisted in dividing the external sphincter posteriorly, so as to arrive at the stricture, pulling it down through this wound when possible, dividing the bowel above and below it, dissecting it out from its attachments, and uniting the two ends of the bowel by sutures. In this case there was considerable difficulty in the subsequent union of the parts, and after healing had occurred, there was considerable contraction, but the condition of the patient was greatly improved.

Colotomy.—This is the last resort of surgery in dealing with ulceration or stricture of the rectum. In ulceration it may be a curative measure; in stricture it is only palliative, and it should therefore not be undertaken till other measures have failed. It is intended to fulfil two important indications, the relief of pain, and preventing or overcoming obstruction, and we have already seen how both of these may be met in many cases by other means which, even when only partially successful, are much preferable.

When none of the methods already pointed out serve to assuage the suffering, and when it is probable that the suffering is not due to an irritable sphincter muscle, or to pressure on neighboring nerves from the mass of the deposit, cancerous or otherwise (in which latter case colotomy cannot be expected to afford relief), and when none of the means already described for preventing or overcoming obstruction can be applied, colot-

---

[1] Case of Stricture of the Rectum, treated by Excision of the Stricture. Lancet, April 12th, 1879.

omy may be resorted to. There is, however, but one class of cases in which obstruction may not be overcome by attacking the stricture itself, instead of the bowel above it, and that is where the stricture is too high to be safely reached by the knife, and where, even then, dilatation is too painful or too dangerous to be admissible.

Judged by these rules, colotomy would be limited to a small proportion of cases. It would be tried after division of the sphincter and of the stricture had each failed to give relief in disease near the anus; and practically would be limited to disease high up in the bowel. Such restrictions as these would greatly limit the number of operations especially in the United States, and I am not sure that this might not be done with advantage. We seldom see in the reports of this operation in current literature any other reason given for its performance than the mere existence of obstructive or painful disease; and yet I doubt if the mere presence of a stricture of the rectum, malignant or benign, is a justifiable reason for the performance of this repulsive and serious operation. It has yet to be proved that colotomy delays cancerous growth, though it certainly prolongs life by diminishing pain and overcoming obstruction. But the relief to the pain may be and often is only partial, for a small amount of fæces which has passed the artificial anus may cause as much suffering and tenesmus as the natural quantity.

In almost direct proportion as the operations of proctotomy and of partial or complete excision of strictures have become popularized and their advantages in suitable cases have become manifest, the operation of colotomy has been limited and the natural objections to it, both by patient and surgeon, have been allowed more weight in influencing the treatment. Especially is this the case in France, the birthplace of the operation, and in Germany, while England, as represented by Allingham, is plainly following in the same course. In this country alone does colotomy still hold its sway—partly for the reason that its substitutes have never been so thoroughly tried here as on the other side of the water.

It would be easy at the present time to collect a much larger table of cases of this operation than was accessible to Mason when he published his paper on this subject, but I do not know that anything would be added to our general knowledge of the subject by such a labor. Allingham had operated at the time of his last edition twenty-seven times. His best result was obtained in a man with a scirrhous growth filling up the pelvis, in whom life was prolonged four and a half years after the operation. Another case, a woman, lived nineteen months, twelve of them in wonderful comfort. Only three of his patients died within a fortnight of the operation, one from phlegmonous erysipelas, another from exhaustion; and the third, in nine days, in whom there was complete obstruction at the time of the operation; and in whom paracentesis abdominis was performed immediately after the colotomy; acute pleurisy being the immediate cause of death. Curling has performed the operation eighteen times with seven **fatal results; two from** chloroform, one from already existing peritonitis, another from **peritonitis** arising independently of the **operation, but** immediately succeeding it, one from pyæmia, and two from exhaustion, one on the **sixth, and** the other on the twelfth **day.** Bryant records fifteen operations of his own, four for vesico-intestinal fistula; **two for** pelvic tumor; and nine **for** stricture, **cancerous** and otherwise. **Of** these latter, one lived eighteen **months in** comfort, dying at last supposably of cancer of the liver; two **lived two and four** months respectively; one lived thirteen days, and two three **days; in these** cases the operation having **been** undertaken too **late to prolong life.** One died of peritonitis due to the operation, and three were alive at periods varying from one to **three years.**

Bultean[1] has collected **one** hundred and forty two cases of lumbar colotomy from the statistics of Doliger, **Mason,** Hawkins, and Heath. Of these ninety-two recovered and fifty died. These figures are about the same as those reached by D'Erckelens.[2]

These figures show as well as would a more elaborate collection of **cases** the general results of the operation itself, the dangers which attend it, and especially the danger of postponing its performance till the patient is at the point of death. These patients sometimes sink with unexpected rapidity at **the** end, **and** when seemingly no worse than for weeks before are often very near death. In my own experience I have had a patient die in the night upon whom I intended to operate in the morning.

Although an artificial anus is justly regarded as being only a substitute for death itself; and although many patients will deliberately choose the latter to the dangers and results of the **former;** it is astonishing how comfortable a patient may be with one **where** the retention of fæces is **good.** Bridge's case,[3] in which the prostitute followed her customary avocation after its performance, is certainly an exceptionally favorable one, but it illustrates what may be **done.** Still we have Allingham's[4] testimony that "**this** operation, though doubtless it may prolong life, **should not be resorted** to without due consideration, because one cannot **fail to see in** many cases the remedy proves a most objectionable one; an opening in the left loin through which the fæces escape is very harassing **and** nothing but a great desire to live or the fear of immediate death would lead me to submit to such a proceeding. I presume after years **the patients get used to** the discomforts and loathsomeness of their condition. My patients who have lived long seem to have had some pleasure in life; indeed, **two women** were married after the operation; but with all that I entertain repugnance to the operation greater than I formerly used,

---

[1] De l'occlusion intestinale **au** point du vue du diagnostic et **du** traitement. Thèse de Paris, 1878.

[2] Arch. für Klin. Chirurg., vol. xxiii., 1 Heft, 1878.

[3] Loc. cit.

[4] Loc. cit., p. 253.

and latterly have mostly performed it **as a last** resource or for total obstruction."

The operation has already been described. A free discharge of **fæces** may follow the opening of the bowel, or there may be only a slight escape of fluid. **It is better** for **the patient** that the evacuation should be postponed till **the edges of the wound** have become agglutinated, as in this way the danger **of extravasation is** diminished. Morphine should **be** given hypodermically **to** keep the bowels as quiet as possible till cicatrization is complete. Only **the** simplest dressings and perfect cleanliness are necessary in the way of local treatment. The sutures **may** be left in till they commence to cause suppuration. **If the bowels are slow to** empty themselves, **an** enema may be administered, or **a scoop used** through the new opening and a purgative may be given by the mouth. No change **is** necessary in the ordinary diet after the second day. The patient should be kept in bed for two or three weeks till cicatrization is complete, and then a pad must be arranged to cover the new anus and prevent leakage of fæces and prolapse of the mucous membrane. Bryant says some of his patients have found great **comfort** from the **use** of an india rubber ball with one of its sides cut away **sufficiently** to cover the new opening. It holds any little fæces which **may come away,** besides preventing the escape of flatus and serving as a **pad.**

Annoying prolapse **is not as apt to occur** with the oblique incision as with the old vertical **one, nevertheless, it** may be expected in some degree, and the patient should be taught to exercise the **greatest** regularity in relieving the bowels **early in** the morning.

Should fæces **pass** the artificial opening, as they **are apt** to do, they must be removed by enemata, for a very small quantity **will cause great** pain and a constant demand for their removal.

It will at once be seen that the treatment of a stricture high up in the rectum or in the sigmoid flexure must be conducted on entirely different principles from one within reach of the finger. In the latter case, the disease itself may be directly attacked with the bougie or the knife; in the former, both are nearly out of the question, and the surgeon is in reality limited to attempts at warding off the natural effects of the malady; in other words to preventing the occurrence of intestinal obstruction, and forming an artificial outlet for the contents of the bowel when obstruction is threatened. The medicinal means of preventing obstruction, and of overcoming it when actually impending, have already been referred to in the chapter on prolapse and invagination. In cases of cancerous disease, attention must be given to cleanliness as well after as before **the operation,** and this is best secured by frequent injections of an unirritating disinfectant, as the permanganate of potash. In cases of non-malignant ulceration, the diseased surface may be treated after the operation as before.

CHAPTER XI.

CANCER.

General Characters of Malignant as Distinguished from Benign Growths.—Malignant, Semi-Malignant, and Benign Adenoma.—Encephaloid.—Colloid.—Melanotic Cancer.—Osteoid Cancer.—Age at which Cancer occurs.—Symptoms.—Diagnosis.—Treatment.—Excision: History and Results of Operation.—**Conclusions Regarding** Excision.—Modes of Performing the Operation.—**Excision of Cancer of the Sigmoid Flexure**.—Palliative Treatment.

In a general way it is undoubtedly true that new growths in the rectum, when benign, increase slowly, tend to grow away from the wall of the bowel, to form pedicles for themselves, and to project into the calibre of the canal, to remain movable, and not to involve surrounding parts; while with cancerous formations the tendency is just the opposite. In this way the diagnosis between a benign polyp and a cancerous nodule in the wall of the rectum is generally easy.

But there is a class of tumors which occupy the border line between the benign and the malignant, in which the dignosis either clinically or with the micrscope may be difficult and even impossible. In fact recent careful study of these rectal tumors goes far to break down the lines between the varieties which are usually drawn, and Cripps,[1] who has done such careful and valuable work in this department, is inclined to group nearly all of them under the single head of adenoma, holding that all are primarily affections of the glandular element. The true nature of the growths may perhaps best be gleaned from a comparison of Fig. 52 with Fig. 45, the latter being a benign polypus, and the former a malignant growth, but both being adenomata.

According to Cripps the names malignant, semi-malignant, and simple adenoid will cover both the benign and cancerous growths of this part of the body, except possibly the form of colloid. Generally, but not always, it is possible to distinguish between them both clinically and microscopically.

After speaking of the innocent growth which is soft, has a fairly

---

[1] Cancer of the Rectum, London, 1880. Also Adenoid Disease of the Rectum. Trans. Path. Soc. of London, 1881.

marked pedicle, and projects into the cavity of the bowel, he says: "In the more malignant varieties, the new growth frequently spreads as a thin layer between the muscular and mucous coats. In this form it often occupies several square inches of the bowel, while its thickness does not exceed a quarter of an inch. At first the mucous membrane lies intact over such a layer, but eventually it gives way by ulceration. This ulceration sometimes begins at more than one point, so that the mucous membrane becomes honeycombed, and portions of the subjacent growth may even sprout through it. The destructive process not only destroys the mucous membrane over the surface of the growth, but after a while the new growth is itself destroyed by ulceration. While destruction is proceeding toward the centre, the growth is advancing towards the circumference. In this way a crater-like mass of disease is produced, the centre of which consists of dense fibrous tissue belonging to the muscular coat of the bowel, which appears for long to resist the ulcerative process. The

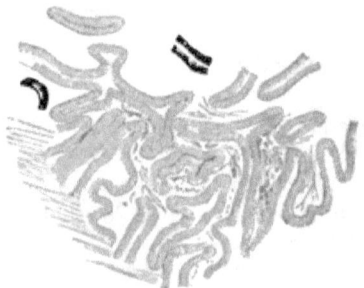

FIG. 52.—Cancer of the rectum—Malignant adenoma (Stimson).

margin of the crater consists of the mucous membrane of the bowel, heaped up by the extending growth beneath it, tucking it over in such a manner as to overlap the healthy membrane. The border is at times so irregular as to represent a series of nodules rather than a continuous line."

Stimson[1] has also made a careful study of these growths. He says: "If it is admitted that cancer of the rectum is essentially a glandular or epithelial affection, one having its origin in the mucous membrane, the borders of the growth, as being the freshest, most recent portions, must be examined, as in carcinoma of other organs, for evidences of primary changes and mode of development. These changes consist of hypertrophy of the mucosa by hypertrophy and hyperplasia of its epithelial elements, together with an abundant development of embryonal connective tissue between the tubules. They are the same as those found in a variety of neoplasm of recognized benign character known as polyp

---

[1] A Contribution to the Study of Cancer of the Rectum. Archives of Medicine, August, 1879.

of the rectum or polypoid adenoma. The formation of a pedunculated growth with a tendency to isolation in the one case, and of a flat growth with a tendency to spread laterally and into the underlying tissue in the other, may be explained partly by mechanical causes and partly by the degree of intensity of the changes in the submucous connective tissue. If the primary change occupies a limited area upon a natural fold of the mucous membrane, and if the muscularis mucosæ remains unbroken until the young embryonal cells produced below it, in consequence of the neighboring irritation, have had time to develop into adult fibrous tissue, the natural retraction of this new tissue narrows the base of the fold, giving it at once a polypoid form and opposing by its greater density a stronger barrier to the extension of the epithelial formation in this direction. The pedicle once formed, the neoplasm increases in the direction open to it, that is, into the lumen of the canal in all its diameters, and the dragging to which it is subjected by the constantly recurring passage of the fæces lengthens its pedicle and tends towards its final separation.

"On the other hand, if a broader area is occupied by the primary change, or if the processes are more intense and rapid, the pedunculation is absent or less perfect, and the epithelial growths of the mucosa break through immediately, or after an interval spent in overcoming the greater resistance offered by the partial pedunculation, into the submucous tissue. Once established in that region the spread of the disease is easy, and its ultimate generalization a question only of time.

"The second and final barrier to generalization is presented by the muscular coat of the intestine, but it is a barrier in which are many gaps, large ones along the lines of the vessels, and innumerable small ones in the fine meshes of connective tissue which separate the muscular bundles and are continuous with the submucous tissue on one side and the para-rectal tissue on the other. Here, too, the intensity of the process materially affects the rapidity of its extension, for if the proliferating connective tissue, which is most easily implicated while it is in the formative stage, is allowed time to reach its full development, to become fibrous, it forms, as it were, a second line of defence capable of offering a certain resistance after the first line has been carried."

With a full appreciation of the importance of the conclusions which Cripps has reached, it may still be well, in a work of this kind, to call attention to some of the clinical characters of some of the different forms of malignant disease as found in this part of the body.

Of all the varieties of true cancer, the one most frequently met with is *epithelioma*, and this presents itself, here as elsewhere in the body, under two forms distinguishable with the microscope and clinically. The first (cancroid, lobulated epithelioma) contains the characteristic onion-like nests of squamous epithelium, and is the same form so commonly seen in the lip, though rarely about the anus. It has its point of origin at the anus, and not within the rectum, and begins as a hard, dry, warty nodule. It is slow in progress, covered at first with firm epidermis, and only begins to ulcerate late in its course. It seldom spreads far up the rectum, but **tends rather** to involve the integument, which it may destroy to **an extent similar to that** sometimes seen in the same variety of disease about **the face.** In the other variety (cylindrical epithelioma), the cells **are columnar,** and the growth resembles in minute structure the **mucous membrane** from which it springs. This variety, on the contrary, chooses the rectum proper for its development, and is found above the internal sphincter. It is easily distinguished from the former, but not so easily from a scirrhus which **has** begun **to ulcerate. It is** softer than **the** other, more vascular, **and** therefore more prone to bleed and undergo extensive degeneration and ulceration; and it rapidly infiltrates surrounding tissues. Early in its course it is movable on the subjacent tissues, but it is seldom seen by the surgeon at this stage. **At** a later period it presents itself as a soft, friable mass seated on a hard, infiltrated base; ulcerated in spots, the edges of the ulcers being hard and raised. At this **stage** the growth will yield on pressure the well-known cancer juice **containing cells and** nuclei, and it may be difficult to distinguish it from **a tumor which began** in the submucous tissue as a hard mass, and **subsequently underwent** degeneration.

Next to epithelioma, *scirrhus,* **or** hard cancer, is the variety most frequently met with **in** the rectum. **It arises, not, like** epithelioma, in the mucous membrane, but in the submucous connective tissue; therefore in the early stages of its growth the membrane is found normal and movable over the hard mass beneath. **When cut into it shows** the characteristic, raw potato-like hardness of scirrhus, and there is **no distinct** line of demarcation between it and the adjacent tissues. From the original tumor are often seen, and sometimes felt, hard fibrous bands spreading out in various directions, generally longitudinally in the bowel —the processes or claws from which cancer takes its name. These tumors may soften down in parts and slough or ulcerate away. When ulceration has begun, a cavity with an irregular outline is formed in the midst of the hard cancer tissue, from which issues a fetid discharge mixed with more or less blood and pus. Although a large part of the growth may die in this way and be discharged, the steady increase in the disease is not checked. Indeed, the growth often seems to be most rapid in the bed of the part which has been destroyed.

This form **of cancer** is said **to be most** apt to show itself first on **the** anterior wall **of** the rectum, **near** the prostate,[1] and "to increase most on the side **of** the **chief arterial** supply, and in that toward which, by lymphatics **and veins, its** constituent fluids most easily filter."[2] It

---

[1] Allingham, Mollière.
[2] Moore, see Bryant's Surgery.

spreads by infiltrating all the adjacent parts, eventually involving all the coats of the bowel, and extending both in surface and in thickness till, instead of appearing as a hard, movable spot under the mucous membrane, it involves a great part or the whole of the circumference of the rectum, inclosing it in a dense, contracting sheath. The hardness and contractility of this form of disease are the chief clinical facts upon which a diagnosis rests; and yet, leaving out of consideration the history of the case, it will often be impossible to distinguish between the gross appearances of scirrhus and those of simple fibrous stricture. I have now under treatment, at the Infirmary for Diseases of the Rectum, a case of stricture which I believe to be dysenteric in origin, in which the extent of the disease is fully as great as in any hard cancer I have ever met with, and yet which has been eighteen years in developing.

Encephaloid has its primary seat in the glandular tissue of the mucous membrane. It is inclosed in a capsule of connective tissue, from the internal surface of which spring trabeculæ which divide the mass into lobules. On section, it may be comparatively firm or nearly fluid, and almost white or stained red with blood. It is often very vascular; large vessels may sometimes be seen on its surface, and large blood extravasations may be found in its interior. The name fungus hæmatodes has been applied to a variety of this disease in which, after the capsule has burst, the mass has protruded. The material composing it may resemble brain tissue (from which it is named), or it may be more spongy and shreddy, like placenta. On squeezing a section of the tumor, a large amount of juice may be obtained, and this, when thrown into a vessel of water, is uniformly diffused through it, giving it a milky hue. This is given by Paget as an exceedingly valuable rough test of the nature of the growth. These cancers are rapid in their increase, and may attain an immense size, fairly filling the pelvis. They quickly affect the neighboring lymphatics, and, when enucleated, speedily recur. The results of removal are, however, particularly favorable for a short time, as shown by the immediate improvement in the general condition of the patient, and the disappearance of the cancerous cachexia. The extreme softness of the tumor, and the deceptive sense of fluctuation imparted to the finger, may cause a mistake in diagnosis, which may be avoided by the use of the aspirator, or even the hypodermic syringe. When the fluid thus obtained is examined under the microscope, it will be found to contain cells and nuclei, with more or less blood.

In *colloid* cancer (alveolar sarcoma), the structure is essentially the same as in the last variety, except that the alveolar meshes are filled with a mucous, glue-like material, which in its most natural state is glistening, translucent, and pale-yellow. This variety of cancer has its origin in the follicles of Lieberkühn, or the crypts which surround the rectum. It is not very rare in this part, and appears in the shape of large, lobulated, fungus-like tumors, which are soft and easily broken down. Under the microscope, the mucous contents of the alveoli will be seen to contain cells of various forms, the most characteristic being large, round, and flat, with a nucleus and concentric laminæ. The growth rapidly infiltrates the surrounding **tissues, and** secondary deposits will often be found in the neighborhood **of the** original mass, the whole tending **to** undergo cystic degeneration. **The** malignancy of these tumors varies in degree, some of them being comparatively benign; **they do** not always recur after removal, **nor** do they readily infect the lymphatics and viscera, being in this respect about on a par with epithelioma. **The term** colloid is used without much exactness, being applied to almost any growth which consists in part of large, cellular spaces filled with glue-like material. The following description of a **case illustrates** very perfectly the general characteristics of colloid:

CASE XXII.—"The patient was an old woman, and the case **was** peculiar, in that the colloid material was contained in cysts of **various** sizes, pressed firmly one against the other, so that the disease **might be** called multiple cystic colloid degeneration. The anus was surrounded with a large number **of tumors of** unequal size, of which several, larger than the rest, were **surmounted by** smaller ones in such a way that the anus occupied **the bottom of an** extremely deep infundibulum. Two superficial ulcerations **were to be seen at** the margin of the anus. The finger recognized **at a short distance** above the anus an ulceration in the form of a zone, **which was deep, had destroyed all the** thickness of the rectum in a part **of its** circumference, **and** communicated with fistulous tracks, which penetrated into the substance of the diseased **skin** adjacent to the anus.

The degeneration, which had given **the rectum an enormous thick**ness, ceased abruptly nine or ten centimetres from the **anus.** Immediately above, the rectum presented considerable hypertrophy in the muscular layer. This affection, which had all the characters of colloid degeneration, presented an arrangement in its upper two-thirds which I had never before met with, and which I will try and describe. Let one imagine a number of acephalocysts of unequal size (some of them as large as pigeons' eggs) squeezed firmly one against the other, and held in a fibrous network, and one will have an exact idea of the change. Only these were **not** acephalocysts. The covering of each cyst was fibrous, very thin, **and yet** very strong; the matter contained in them exactly resembled **currant** jelly, on **the** surface of which had been deposited a cretaceous **matter** exactly similar **to that** which sometimes covers the excrement of birds. This cretaceous matter contained calcareous concretions. In **the centre of the jelly-like substance, two or** three blood-vessels were to be **seen,** similar **to those which form** in a hen's egg— vessels without **walls,** ending in an enlargement of one extremity.

The fibrous network in the midst of which these cysts were inclosed was evidently made up of the transformed coats of the rectum. I could recognize the longitudinal fibres of the rectum. There was also adipose tissue, an evident proof that the degeneration had not only invaded the rectum, but had developed at the expense of the adipose tissue of the pelvis.

The lower third of the rectum presented no sign of a cyst, but an areolar tissue, with fibrous meshes, which occupied all the circumference of the anus; this tissue was filled like a sponge with colloid matter, which could easily be pressed out, and the tissue itself was approaching erosion or ulceration. The areolar and gelatiniform degeneration appeared to me to penetrate into the thickness of the skin of the anal region; while an extremely thin, almost epidermic, pellicle had resisted and covered the swellings on its surface. In the vicinity of the circular ulceration of the rectum, the colloid matter had not undergone degeneration, only it was permeated by an increased number of blood-vessels. Behind the rectum was a colloid alveolar mass, all the areolæ of which contained blood-vessels. This mass had evidently been formed at the expense of the circum-rectal adipose tissue."[1]

Cruveilhier draws this distinction between colloid and encephaloid. The colloid degeneration is not susceptible, as is the encephaloid, of inflammatory action producing gangrene; moreover, if the sanguineous centres are not absolutely foreign to it, it is certain that they are incomparably rarer in colloid than in the cancerous degeneration, properly so called, where effusions of blood are so often met with—apoplectic centres sometimes so large as to conceal the true nature of the morbid tissue.

Colloid alveolar degeneration shows only one mode of destruction—by encroachment in successive layers; this encroachment, sometimes rapid when it occurs in the alimentary canal, permits of the re-establishment of the flow of fæces, temporarily interrupted by the undefined and often very rapid increase in the degenerated parts; so that, to the gravest signs of fecal retention, there sometimes succeeds a more or less rapid separation, with or without diarrhœa.[2]

Melanotic carcinoma, or black cancer, is by some classed among the true cancers, and by others among the sarcomata. It belongs to the class of soft or medullary cancers, and its distinguishing feature is the development of pigment. Whatever may be said of the microscopic characters of melanoma, it is clinically a very malignant growth, running a very rapid course, and very likely to become generalized. Its clinical history, as relates to the rectum, is to be studied from ten cases only, which have been given in full in an exhaustive study by Nepveu, read before the Société de Chirurgie (1880).[3] The cases are reported by the

---

[1] Cruveilhier, Traité d'Anatomie Path. Gén., t. v., p. 67.
[2] Ibid., p. 69.
[3] "Mémoires de Chirurgie," Paris, 1880.

following observers: Schilling,[1] Kopp,[2] Moore,[3] Maier,[4] Virchow,[5] Ashton,[6] Gross,[7] Meunier,[8] Gussenbauer,[9] and Nepveu.[10]

From the six of these cases which are reported with an approach to completeness, several facts of interest are to be gathered. The age of all of the patients was advanced, ranging between forty-five and sixty-four years. Five were in men, one only in a woman. In the microspic examinations which were made in five of the cases, the tumor is in every case described as a sarcoma. There is nothing in the symptomatology to distinguish this form of disease from others, except that in one case the stools were colored black from mixture with the pigment—a point which might aid in diagnosis, were the tumor so high up as to be out of sight. In rectal examinations it was also noticed that the finger was colored in the same way. The location of the disease was once in the sigmoid flexure, twice in the rectum above the sphincter, and four times at the anus. The size of the growth was generally considerable, surrounding the bowel and projecting into its cavity; sometimes it was firm enough to cause tight stricture, at others ulcerated and broken down in parts. The course of the disease is marked by secondary deposits in the adjacent glands or in the viscera, while the original growth my spread in neighboring organs, and by ulceration cause a foul discharge mixed with blood and pigment. To these may be added the usual signs of incontinence and obstruction. The duration of the disease in no case exceeded three years, but it was generally fatal in a much shorter time. The diagnosis is easy if the growth can be seen, and is sometimes assisted by the secondary black deposits. In four cases the tumor was removed, but in none was the return long delayed.

Osteoid Cancer.—Either a sarcoma or a carcinoma in any part of the body may become ossified, and hence pathologists speak of osteo-sarcoma and osteo-carcinoma. It is rare that such a formation is found in any structure except bone or periosteum; and there seems to be but one case on record of bone-cancer of the rectum, which, because of its great rarity, I will quote in part:

CASE XXIII.—The preparation was removed from the body of a lady,

---

[1] Mentioned by Eiselt, obs. v., Prag. Viertelj., Bd. 70 u. 76.

[2] "Denkwürdigkeiten in der Ärztlichen Praxis," Bd. iv., Frankfort, 1838, pp. 305-313.

[3] Medical Times, March, 1857.

[4] Berichte über die Verhandlungen der Naturforschenden Gesellschaft zu Freiburg, 1858, No. 30, p. 516.

[5] "Pathologie des Tumeurs," Paris, 1867, t. ii., p. 281, note.

[6] Ashton, T. J., "Prolapsus, Fistula in Ano," etc., 3d ed., London, 1870, p. 162.

[7] "System of Surgery," Phil., 1872, vol. ii., p. 589.

[8] "Bull. de la Soc. Anat. de Paris," 1875, p. 792.

[9] "Ueber die Pigmentbildung in melanotischen Sarcomen und einfachen Melanomen der Haut." Virchow's Arch. f. path. Anat. u. Phys., lxiii., 1875.

[10] Op. cit.

aged about fifty-four, **who died January 18th, 1869, under the** care of Mr. Collambell, of Lambeth. **The history of the case** pointed to the existence of disease in the rectum **for about twenty years (during** which time she had occasionally complained of pain, irregularity **of the bowels, and a discharge of blood and mucus). . . . The specimen includes the whole pelvic viscera. The rectum is laid** open posteriorly, but **rather on** the right **side, and shows a cancerous** mass projecting into its interior **at a distance of about four or five inches** from the anus. The principal mass, of about the size of a walnut, is situated directly **at the** back, and occupies **nearly the whole calibre of the** rectum, but the disease involves, more **or less, the entire circumference of the** intestine **upon a** level rather above the larger mass. A small opening, large **enough** to admit a goose-quill, **is found in the sigmoid** flexure, about **twelve** inches above the cancerous growth, **and communicates with a** circumscribed abscess cavity within the peritoneum, above **the pelvic viscera, and behind the** pubes, and this **again** communicates **with the** rectum immediately below the obstruction. At the time of the post-mortem this peritoneal abscess contained very little fluid, but what there was was pus **discolored** with fecal matter. **There is also** a large, foul, burrowing abscess, situated **in the submucous tissues,** almost completely surrounding the **rectum at the seat of disease,** communicating **freely with** its cavity and **directly continuous with the** intra-peritoneal **abscess.**

When first laid open, **the** surface **of the** cancer generally presented a nodulated, **red** appearance, but the larger **or** posterior **mass was** roughened in its lower **half by** numerous sharp spicules **of** bone which projected from its **surface. The** cut **surface showed** the growth involving the thickened **muscular coat as a hard, contracting mass** and from its **base firm fibrous** bands ramified into **the neighboring** fat, just as **from the base of an ordinary** scirrhous **tumor. That** portion which projected **into the cavity of** the rectum **was** softer, and **its lower** part was **occupied** throughout by numerous spicules **of true** bone. On the surface, **the** softer structures having sloughed away, the bony constituents were exposed. The growth did not extend **to** the sacrum, which was perfectly healthy, **and the other bones of** the pelvis **were** also free from disease.

The other viscera were examined and appeared **healthy.** The lymphatic glands were not carefully examined, **but in the** parts which were removed **there was** no glandular enlargement **to be found.** The ulceration **in the sigmoid** flexure seemed to be of a simple character; **there was no evidence of malignant deposit** elsewhere than in the obstructed portion of **the rectum.**

On examining the growth in the rectum it was found **to be** firm in **the** deeper parts, where it involved mucous and submucous tissues, but, nearest to the surface, where the spicules of bone were evident, it had **the appearance** and **character, to the naked eye, of a fibro-fatty struc-**

ture. In the deepests parts, however, **where** it was firmest, it had **not** any very great hardness. The parts involved in the ossification lay exposed in the rectum, and seemed, from their shreddy, softened appearance, to have been **recently** sloughing. Upon section, a quantity **of** juice was readily obtained, **and showed** under the microscope an immense number of **free nuclei and cells** of **all** shapes and of variable sizes, though the greater **number** were elongated or oval, and about half **the** size of the **columnar** epithelium of the neighborhood. There was a large quantity **of** molecular matter and oil, and the nuclei were indistinct. The solid **portion of** the growth was composed of **cellular and** muscular structures **imbeded in** a granular matrix. Bands **and fibres,** composed almost altogether of nuclei, ramified in the growth, and **could** be traced as continuous with the osseous portions. It **appeared that** the **nuclei became** darker, granular, and harder **in** outline as **the** examination was **carried** toward the ossified parts; the intervening matrix became more fibrous, and the processes of bone branched out into this. The bony **spicules** contained numerous lacunæ, whose size was about that of the ordinary nuclei of the growth. They were of various forms, generally branching, and **were arranged with** no regularity, but in the **manner** usually found in **adventitious bony** deposits in tumors. The **matrix was** granular.

The interest **of this case lies chiefly in the** fact of bone being found ramifying **through parts of the structure;** and that this bone was the result of ossification **of the scirrhous growth seems** evident from the manner in **which it could be** traced **under the microscope.** That **it** was not an original formation apart from **the scirrhus must be** admitted, for its histological characters show its **definite relation to** the elements of the tumor, the lacunæ replacing the nuclei, **and the** rest **of the** bone occupying **the** place of the intervening matrix. **And a** primary **bone** tumor in this position is difficult to imagine. The occurrence of true **bony** deposit in medullary tumors is not altogether infrequent; but then **it is** found in the deeper parts, and is almost always in connection with some bone. In scirrhous growths, however, I do not find any mention of ossification occurring, except where starting from bone. I have no history of any case of any kind of tumor of the rectum in which bone formed an element of **a** primary growth.[1]

These **are the** rarer forms of cancerous disease in the rectum and their recognitition presents little difficulty. Most malignant growths are included under Cripps's classification of adenoma or under the older terms of epithelioma and **scirrhus.** Hecker[2] found twenty-one cases of epithelioma in **thirty-four cases of cancer.** Cripps says, "I have failed to discover" **(in the rectum)** "any growths or tumors consisting entirely

---

[1] Wagstaffe, "Trans. of the Path. Soc. of London," vol. xx., p. 176.
[2] Schmidt's Jahrbücher, 1870.

of the characteristic structure which pathologists designate as scirrhus or medullary cancers, or as belonging to the various varieties of sarcoma. Considering the eminence of many careful observers who have applied such names to these growths, it would be quite unjustifiable to assume that such distinctive structures never form the entire bulk of the tumor; but I feel bound to state that with, perhaps, a more than average opportunity of examining such growths from the rectum, I have been unable myself to discover tumors composed entirely of the distinctive features appertaining to these diseases."

Cancer of the rectum, like cancer elsewhere in the body, generally occurs in middle life or old age. There are, however, some interesting exceptions to this rule. Allingham[1] reports a case of encephaloid in a boy of seventeen, under his own care; and another (variety of cancer not stated) under the care of Mr. Gowland, in a boy not thirteen; Mayo[2] speaks of one at the age of twelve, and Godin[3] of one at fifteen years; and Quain[4] quotes one, reported by Busk, at sixteen. After the age of twenty the cases increase rapidly in number. With regard to the relative frequency in the sexes, different statements will be found in the works of different writers, according to the experience each has had, and considerable reasoning has been indulged in to explain why the disease should be more common in the one sex than in the other. In a collection of one hundred and seven cases, I have found fifty in males and fifty-seven in females.

The locality in which the disease first appears varies. Quain[5] says: "I have most frequently met with the lower margin of the deposit at the distance of from two to three inches above the orifice of the bowel. The part between that just indicated and the anus is next in order of frequency as the seat of the disease, and to this succeeds the lower end of the colon." This perhaps expresses the facts of the case as well as they could be stated in a few words. The upper limit of the rectum, where it joins the sigmoid flexure, is a common site of the disease, and here it runs a more rapid course then elsewhere, and is more apt to be suddenly fatal on account of the increased liability to obstruction which the anatomical condition favors.

The symptoms of cancer of the rectum may be classified as follows: pain; those due to contraction, to ulceration, to invasion of neighboring parts; and, lastly, the generalization of the disease and the cachexia.

A cancer of the rectum may, and often does, begin so insidiously that its existence is not suspected by the patient till it has made irreparable

---

[1] Diseases of the Rectum, London, 1879, p. 265.
[2] Injuries and Diseases of the Rectum, London, 1833, p. 188.
[3] Mollière: Traité des Maladies du Rectum et de l'Anus, Paris, 1877, p. 580.
[4] Proc. of the Path. Soc. of London, 1846-'7.
[5] Op. cit.

progress. This will be the case particularly when the disease is well **up in the bowel beyond the reach of the sphincters.** The slight sensitiveness of the mucous membrane **of the rectum** proper which permits the existence of extensive ulceration, **the** application of escharotics, and the performance of surgical **operations** without pain has been already referred to. On the other hand, **cancer of the** rectum is usually attended with great pain, and the **suffering in itself may be** made of great assistance in diagnosis.

Attention has been called to the point in **diagnosis that the existence** of pain or **cramp** in the lower extremity **in cancer of the** rectum **is a** bad sign, suggesting a direct encroachment upon some of the neighboring nerves, either by implication and pressure **of the glands,** or by direct extension of the original disease.[1] In **the later stages of** cancer the pain is often the most important symptom **to** be met **by** treatment. **It may then be** due to the irritation of fæces upon an ulcerated surface, to the involvement of the anus in the ulceration, or to direct pressure on adjacent parts, and each **of** these is to be met by a different and appropriate treatment.

The symptoms **directly referable** to contraction of the bowel are often slight, and differ **in no way from those** caused by the simple, fibrous stricture of the **same** part. It is **often** astonishing to the surgeon to meet with an advanced **case of scirrhus in which the** calibre of the bowel is so nearly occluded **as scarcely to** permit **the passage of the** end of the finger, and yet in which **the** patient has **never had sufficient** uneasiness to call for a direct **rectal** examination.

The **hæmorrhage from** an ulcerated rectum in cancerous **disease is** seldom profuse enough to be dangerous, through by **frequent repetition** it may become **an** important factor **in** the ultimately fatal **result.** The odor of the discharge is the same as that from a cancer of the uterus, and **needs** only once to be appreciated to be remembered.

Above the contraction there often develops an ulceration which is not **to** be confounded with the breaking down of the cancer itself. When the cancer itself once begins to break down and ulcerate, its extension is **limited** by no tissue of the body. **The** bladder may be opened and a permanent fistula result, in which **case the** passage is generally from that viscus into the rectum; but the opposite may be the case—and the pain caused by the entrance of fæces into the bladder and their discharge through the urethra is one of the best of all the indications for colotomy. The prostate **and** seminal vesicles in the male and the recto-vaginal septum in the **female may each be** destroyed; in fact, any part near the disease may **be implicated.** Smith[2] has recorded a case in which the

---

[1] Hilton: Rest and Pain, p. 163.
[2] Surgery of the Rectum, London, 1871.

disease opened into the hip-joint, and Mollière[1] another in which it invaded the soft parts in the loin.

There are two sets of lymphatics which may be involved in malignant disease of the rectum, one coming from the anus and going to the glands in the groin; and one coming from the rectum proper and going to the glands in the hollow of the sacrum and lumbar region. The proper place, therefore, to feel for glandular involvement in disease within the sphincter is along the spine, deep in the pelvis—a simple point which may decide the surgeon for or against operative interference. This implication of the lymphatics is sometimes shown by pressure effects at points quite remote from the original disease, as in the following case from my own case-book.

CASE XXIV.—J. B., aged sixty, has always been strong and well until within a few weeks past, when he has been troubled with obstinate constipation. All he desires now is some "pills" to move his bowels. On closer questioning, he refers casually to the fact that he has considerable pain in the right thigh, and some swelling in the right leg and foot, but "nothing to speak of." On examination, nothing was to be detected by rectal touch, but the pelvis at its upper part was partially filled by firm, nodular masses, which extended deeply down into the right iliac fossa. The patient had no conception of any trouble beyond constipation and "rheumatism," though the whole lower extremity on the right side was œdematous. By careful diet and laxatives the threatened obstruction was avoided, and the man gradually sank with all the signs of the cancerous cachexia, and died three months from the first examination. Unfortunately no autopsy could be obtained.

From what has been said, it is evident that there is little in the history which the patient will give of cancer of the rectum to distinguish it from ulceration and stricture of any other variety, and that the diagnosis must chiefly rest upon a physical examination. To make such an examination thoroughly, and yet safely, requires great care and gentleness, and, to properly interpret the conditions which may be found, no little experience and knowledge. It requires many years of practice to reach the point Allingham has reached when he says: "There is something peculiar about the feel of cancer which the practised finger rarely mistakes even for simple indurated ulceration. I think it is many years now since I mistook the one for the other."

In the majority of cases the diagnosis may be made by the history and by physical examination with the finger alone. Cancer in this locality is a disease of rapid growth, and when a patient says that stricture has existed any considerable number of years the idea of malignancy may be abandoned. Something also may be learned from the general appearance of the patient, but most of all from the digital examination. When the

---

[1] Op. cit., p. 565.

disease is seen in its earlier stages, the hard, more or less distinctly circumscribed new growth which has infiltrated the wall of the bowel is diagnostic. The great difficulty is to distinguish between an advanced case where the rectum is partially occluded by hard masses of disease, and an old case of stricture and ulceration which is not malignant. This may sometimes be impossible except by the microscope, and syphilitic disease of the rectum is not infrequently mistaken for cancer. When a soft friable mass of epithelioma is found seated on a hard, infiltrated base, which is ulcerated in spots, the edges of the ulcers being hard and raised, the diagnosis is also easy.

Cancerous stricture of the sigmoid flexure will show itself sooner or later either by examination through the abdominal wall, or by the signs of intestinal obstruction.

In cases where the condition is more complicated and where secondary deposits—in the liver, for example—have begun to do their fatal work before actual obstruction has begun, these symptoms of stricture may all be obscured by the presence of others which shall more readily attract the eye. In a case which I now have under treatment, I had made the diagnosis of cancer of the liver with ascites and great intestinal disturbance some time before my attention was called to the rectum, and it become evident by examination that the affection of the liver was secondary to malignant disease high up in the rectum, which was also gradually involving the pelvic viscera. The greatest caution should be exercised in the examination for cancerous disease above the lower four inches of the rectum.

Treatment.—The treatment of malignant disease of the rectum is designed to be either curative or palliative. In a small number of selected cases a cure is, perhaps, possible, as with cancer of feeble malignancy in other parts of the body—*e. g.*, epithelioma of the lip. At all events, the disease may be removed, and its return delayed for many years. This fact, we believe, may be accepted as proved by a sufficient number of carefully examined cases, from which the chances of error in diagnosis and subsequent history have been eliminated. Cure can, however, only be effected by excision. All other means may be set aside as hopeless failures.

The operation of excision, which, after being fully described and ably advocated by Lisfranc in 1830, was allowed to fall into disuse, has again, within the past few years, become popular. It would probably be a waste of time to inquire to whom the credit of reviving it is due. Cases of its occasional performance are scattered through the surgical literature of the rectum from the early part of the century to the present, and just now it is at the height of its popularity. Like every other surgical procedure at that point of its history, it is perhaps also occasionally done when it were better to be content with less radical measures. As a result of a careful search among the statistics of this operation, Cripps[1] gives the following

---

[1] Op. cit., p. 166.

figures. Out of a total of sixty-four cases, eleven died as a direct result of the operation; six from peritonitis, one from cellulitis, and four from accidents incident upon any surgical interference.

In the fifty-three cases of recovery, the subsequent history is unknown in sixteen, and in three more the diagnosis was so doubtful as to exclude them from the list. No case is worth much in the consideration of a question such as this where the diagnosis has not been verified by the microscope in competent hands; for there are non-malignant growths of this part which, to the naked eye, strongly resemble cancer. We have then a remainder of thirty-four, in whom the disease returned in twenty; but of these twenty, several were operated on a second time for a recurrence of the growth, or possibly for a small nodule which had not been removed at the first operation, and after this second operation remained free. This leaves, however, a total of twenty-three out of sixty-four operations in which the disease had not returned after an interval varying from a few months to over four years—a limit reached in three cases.

This is certainly an encouraging result for this disease, and the fact that undoubted cancer may be removed and not reappear for such a length of time is decisive. Some operators, however, report better results than these, and some have not been so successful. Curling[1] gives one case of removal of an epithelioma in which there had been no return in the rectum after seven years, though for one year there had been "a doubtful tumor of the pelvis." Velpeau and Verneuil each report cases in which the cure has seemed permanent, and Chassaignac gives several in which there had been no return after six years. Dieffenbach's thirty cases in which the patients lived many years without a return are generally looked upon with suspicion. Allingham,[2] on the contrary, considers the *partial* removal of the circumference of the bowel as unsatisfactory. In all of his thirteen cases in which he was able to follow the progress of the case for one year, there was either a return of the growth in the rectum or the glands in the groin became affected, and there ensued disease in the internal organs. In four cases the disease did not return in the bowel, but in the inguinal glands, proving that it was not due to an incomplete operation. With regard also to his ten cases of *total* extirpation, he speaks very cautiously. He believes that a cure is very uncommon, and not generally to be expected; and he does not commit himself even on the question of the prolongation of life. The mortality, as a direct result of the operation, is generally about twenty-five per cent.[3]

Billroth[4] reports thirty-three cases. Thirteen died of the operation, and the remainder all died within two years, most of them of recurrence.

---

[1] Diseases of the Rectum, ed. of 1876, p. 164.

[2] Loc. cit., p. 277.

[3] Mollière, Traité des Maladies du Rectum et de l'Anus. Paris, 1877, p. 627.

[4] Clinical Surgery. Extracts from the Reports of Surgical Practice Between the Years 1860–1876. By Th. Billroth. New Sydenham Society, 1881.

The deaths immediately following the operation were invariably due to retro-peritoneal suppuration, characterized by acutely septic symptoms. Most of them died within from four to eight days.

Since then, in certain cases, we are justified in expecting recovery from the operation itself, and such a length of life as would not result were the disease left to its natural course, we may ask: 1. What are the dangers, and what is the mortality of the operation? 2. In what class of cases is it applicable? 3. What are its results as a curative and as a palliative measure, and how do these results compare with those of lumbar colotomy? 4. What are the results as regards the subsequent condition of the bowel, and the control of the fæcal evacuations? 5. What is the best method of its performance?

For the purpose of arriving at a knowledge of what experience has already taught in this matter, I collected, a couple of years ago,[1] the re-

---

[1] For the full literature of the cases upon which these conclusions are based, the reader is referred to the following bibliography:

Agnew.—Phil. Med. Times, June 23d, 1877.
Allingham.—Diseases of the Rectum, 3d ed., London, 1879.
Briddon.—Med. Record, January 6th, 1877.
Bushe.—Treatise on Diseases of the Rectum, New York, 1837, p. 294.
Byrne.—Annals of the Anat. and Surg. Soc., May, 1880.
Baumès.—Bull. de l'Acad. Roy. de Méd., t. x., p. 938.
Chassaignac.—Traité de l'écrasement linéaire, Paris, 1856.
Cripps.—Cancer of the Rectum.
Crosse (quoted by Mayo).—Observations on Diseases and Injuries of the Rectum, London, 1833, p. 210.
Curling.—Observations on Diseases of the Rectum, London, 1851. Med. Times and Gaz., March 14th, 1857.
Dennonvilliers.—Gaz. des Hôp., 1844.
Desgranges (quoted by Mollière).—Maladies du Rectum, etc., Paris, 1877, p. 627.
Dieffenbach.—Die operative Chirurgie, Leipzig, 1845.
Dolbeau.—Thése de Fumouze.
Duplay.—Gaz. Méd. de Paris, 1872, p. 486.
Dupuy.—Bull. de la Soc. Anat., Paris, 1872. 2me s., xvii., p. 242.
Emmet.—Principles and Practice of Gynæcology, 1st ed., Philadelphia, 1879, p. 511.
Ewart.—Lancet, June 21st, 1879.
Fenwick.—Montreal Gen. Hosp. Reports, vol. i.
Gay.—Lancet, June 28th, 1879.
Gosselin.—Gaz. des Hôp., 1879, p. 921.
Holmer.—Hospitals-Tidende, March 31st, April 7th, 14th, 1880.
Holmes.—Trans. of the Clin. Soc. of London, 1878, p. 113.
Holt (quoted by Curling), op. cit.
Keyes.—Arch. of Med., August, 1879.
King.—Brit. Med. Jour., June 21st, 1879.
Kumar.—Wiener med. Woch., 1878, p. 1,070.
Labbé.—Gaz. des Hôp., June 4th, 18th, 1880.
Levis.—Arch. of Clin. Surg., February, 1877.

ports of operations up to that time as far as they were then attainable. The list at that time included one hundred and forty cases. At that time I arrived at the following general conclusions concerning the operation, and subsequent study of the question has led me in no way to alter them.

1. *Although there have been a few cases of excision in which the cancer has not returned in a number of years, such a result is so rare as not to justify the exposure of the patient to the risk of immediate death which attends the attempt to remove extensive disease.*

Regarding the question of radical cure, we find difficulty in establishing exact dates, and have to take into consideration the reputation of the reporter. We find, however, that in one hundred cases (deducting those immediately fatal, and seventeen which passed out of observation immediately after operation) we have five cases of reported permanent cure, in which there had been no return for at least ten years. Three of these are reported by Volkmann, and two by Velpeau. March, of Albany, has been credited with another case of radical cure, but the author is much indebted to the present Dr. March for a letter stating that the case of

---

Lisfranc.—Thèse de Pinault, 1829.
Maisonneuve.—Union Méd., 1865. Also Thèse de Cortes, 1860.
Mandt.—Revue Méd., 1836, p. 264.
March.—Trans. of the N. Y. State Med. Soc., 1868; also Med. and Surg. Reporter, June 9th, 1877.
Mayo.—Observations on Diseases and Injuries of the Rectum, London, 1833, p. 212.
Moore.—Med. Times and Gaz., March, 1857.
Mollière.—Thèse de Carcopino, 1879.
Nussbaum.—Aerztlich. Intelligenzblatt, 1863.
O'Hara.—Phila. Med. Times, vol. viii.
Paget (quoted by Cripps), op. cit.
Peters.—Arch of Med., August, 1879.
Pital du Cateau.—L'Expérience, t. vi., p. 27.
Polaillon.—Gaz. des Hôp., 1879.
Post.—Med. Record, July 31st, 1880.
Récamier.—Thèse de Massé, 1842.
Roddick.—Montreal Gen. Hosp. Reports, vol. i.
Schuh.—Abhandlung der Chir. und Operationslehre, Wien, 1867.
Siebold (quoted by Curling), op. cit.
Simon.—Lancet, 1851, ii., 1882.
Simon, of Rostock.—Deutsche Klinik, 1866.
Stimson.—Arch. of Med., August, 1879.
Terrillon.—Thèse de Carcopino, 1879.
Van Buren.—Arch. of Med., August, 1879.
Van Derveer.—Med. Record, September 20th, 1879.
Velpeau.—Nouveaux Elémens de Méd. Opératoire, Paris, 1839, vol. iv., p. 814.
Verneuil (quoted by Marchand).—Etude sur l'Extirpation de l'Extrémite Inférieure du Rectum.
Volkmann.—Klin. Vortrāge, March 13th, 1880.

supposed radical cure reported by his father passed out of observation at the end of one year. There are some other cases which have been included in the category of permanent cures—cases in which the disease had not returned in four or five years—but the great majority recur within the first year and are fatal within two.

2. *The operation* **is chiefly valuable as a** *palliative measure and as* **such** *it compares favorably* **with** *colotomy both in prolonging life and relieving pain.*

The treatment of cancer of the rectum by excision has not yet been accepted by the surgical world as a substitute for other measures even in cases best adapted for the operation, although it cannot be denied that a radical cure has sometimes been obtained, and that in many other cases life has been prolonged beyond what could have been hoped for by any other means of treatment. It is no less true that the operation is one of great danger, and that there are not lacking those whose experience has led them to believe that life was rather shortened than lengthened by it. By these it is claimed that in lumbar colotomy we have a safer method of relieving pain, and delaying the progress of the growth, and in both these ways prolonging life. American and British surgeons hold rather to this latter idea, while the French and the Germans favor excision.

Excision can scarcely be judged in comparison with colotomy, being applicable properly only to an entirely different class of cases. In cancer above four inches from the anus, colotomy or colectotomy are about the only means of relief. In cancer within four inches of the anus almost any other plan of treatment is preferable.

This leads me to call attention to another point—the operation of excision as a palliative measure. In cases properly chosen, where the disease is not so extensive as to render its removal one of the capital surgical operations, we know of nothing better, and this fact cannot fail to be deeply impressed upon the reader of these cases. The statement that all suffering was relieved is almost invariable. In almost every case attention is called to the great improvement in general health, the loss of pain, and the increase in strength. Patients go away believing themselves radically cured, return to their employments, and are reported by the French surgeons as *"parfaitement guéries,"* a few weeks after the operation.

It has been claimed[1] against this operation that even when a good immediate result is obtained, it may shorten life by hastening the return and final progress of the disease. Unfortunately, it is difficult to tell in any particular case how long a patient would have lived had the disease been left to its course; but, accepting as a basis for comparison Allingham's estimate of the average duration of life in cancer of the rectum as two years or less, we are justified in concluding that in all cases where

---

[1] Labbé, Gaz. Hebdom., June 4th, 18th, 1880.

life was prolonged more than one year and a half after the time of operation (the operation generally being done late in the disease), this length of life may fairly be attributed to the surgical interference. This estimate is manifestly a small one, for a study of the cases makes it evident that many who did not live eighteen months after the operation yet gained a considerable length of comfortable existence; and there is nothing to prove that in any case the operation hastened the natural course of the disease.

I have carefully searched the record of cases in which a return of the disease within six months of the time of operation is reported, to discover whether, here also, there was any marked relation between this result and the nature or extent of the disease at the time of operation; but it is especially at this point that the table fails us. A proper answer to this question involves not only a careful report of the extent of the disease, but a microscopic study of its character, and such data are given only in a relatively small proportion of cases. I believe, however, that the cases show a marked relation between the rapidity of the growth before operation and the speedy return after removal.

We can trace no connection between the time of the return and the extent of the disease removed when the removal has been complete; and the microscopic reports are too few for general conclusions to be drawn from them. I know of no writers, except Stimson and Holmer, who have made a careful study of the specimens excised, and have given the results; and, so far as the clinical reports of the German operators go, they would seem to give support to their practice of removing everything involved, no matter how extensive, in the hope that the local return may be long delayed.

3. *When the disease reaches above three inches from the anus, or involves neighboring parts so as to render its entire removal without injury to the peritoneum questionable, the operation is contra-indicated.*

The Germans have apparently no limits to the applicability of this operation. They perform it in cases of the most extensive disease, opening the peritoneum, exsecting the sacrum when necessary to reach its upper limit, and removing the prostate and base of the bladder when they are implicated, balancing the risk of immediate death from the operation against the chance of radical cure, or prolonged immunity from return. Conservative surgeons will hesitate long before accepting this view, for, although very satisfactory results have been obtained in such cases, they can hardly be considered other than exceptional, and a study of cases shows that the frequency of the fatal result is in direct proportion to the extent of the operation attempted. The rules for the selection of cases laid down by Lisfranc were these: when the bowel is movable, in other words, when the disease has not involved surrounding parts, the operation should be undertaken. When, on the other hand, the disease is more extensive, and reaches higher, he leaves the question to be decided by future experience. I believe that experience has now decided against it. In deciding for or against the operation, an examination of the glands in the hollow of the sacrum and in the loins is of great value, for these receive their lymph directly from the rectum, and may be enlarged, while those in the groin, which are supplied from the skin around the anus, may still be uninvolved.

I shall not stop at this time to again discuss the question as to how much of the anterior wall of the rectum is uncovered by peritoneum, but must refer the reader to the chapter on anatomy. The height to which it is safe to go cannot be definitely stated for all cases, the reflection of the serous coat upon the rectum being at a variable point. Fochier[1] reports a case in which he used the écraseur at twelve centimetres without harm, and Allingham,[2] who is always a safe guide, has seen all but the lower two inches of the bowel covered by peritoneum in a female, has opened into it in a male when not more than three and one-half inches were removed, and has taken away fully five inches in a male without bringing it into view.

There is an old rule for applying the trephine, that in every instance the operator should remember that some skulls are very much thinner than others, and he should act on the supposition that the particular point upon which he is operating is the thinnest part of the thinnest skull ever seen. Something of the same kind might be said of the peritoneum over the rectum; and everybody who has studied the anatomy of the part knows how various are the opinions of different authorities on this point. Nevertheless, a line of danger can be marked out, and that line is about three inches from the anus. It is true that more than this amount of the rectum has been removed without encountering the peritoneum, and it has been opened below this point; but I should not, for my own part, hesitate to try to remove three inches of the bowel for a a cancer, and I have refused to attempt to extirpate in an otherwise suitable case because the disease passed this line. The index finger is a good guide. What is well within its reach in a hand of good length, it is safe to try to remove, provided it does not involve surrounding tissues to an extent which renders its complete removal impossible. Whatever may be said of the impunity with which the peritoneum may be opened in other parts of the body does not seem to apply here; for I have been able to find but three cases in which that accident was not followed by a fatal result.

Unfortunately, the disease is but rarely seen at a stage when extirpation is justifiable, that is when it is limited to a circumscribed spot within three or three and a half inches of the anus, when it is movable on the muscular coat, has not invaded the deeper tissues, and before there has been any glandular enlargement.

---

[1] Lyon Méd., February 20th, 1876.
[2] Op. cit., p. 275.

Although there is a very evident relation, which is shown by a study of the statistics of the operation, between the **extent of** the operation attempted and the favorable or unfavorable results obtained; a fatal result will often follow the extirpation of disease which is comparatively slight in amount. The three great dangers of the operation are peritonitis, pelvic cellulitis, and septicæmia. Hæmorrhage may fairly be dropped out of consideration, for the operation may, if desired, be rendered almost bloodless by the use of the écraseur or galvano-cautery.

4. *The operation is **not** followed by any annoying after-consequences which are of sufficient gravity to contra-indicate its performance.*

In a small proportion of cases, there will be complete incontinence, in a greater number there will be partial control over the evacuations, and in a majority the control will be sufficiently complete to prevent the occurrence of any annoying accident.

Stricture to a troublesome extent is also rare, and when it exists, it may generally be overcome by the introduction of bougies. In one case reported by Verneuil, a special plastic opération was performed to relieve this condition, an account of which may be found in the work of Marchand.[1]

Regarding the best way of performing the operation, the surgeon has his choice of several. The first case of extirpation of the rectum of which we have any record was by Faget, in 1739, and was not for cancer, but simply a removal of the lower portion of the bowel, which had been completely surrounded and denuded by an abscess beginning in one ischio-rectal fossa, and subsequently extending into the other. From that time until 1826, the operation, as a means of treatment of cancer, will occasionally be found mentioned in surgical literature; generally, however, only in condemnation. In 1826, Lisfranc performed the first successful operation for cancer; and three years later, his student, Pinault, in a *thèse* reported nine cases, and gave to the procedure a permanent place in literature and practice. In 1833, Lisfranc himself embodied the same ideas in a paper read before the Acad. Royale de Médecine,[2] and from that time the operation became widely known. Since then, it has had its advocates and opponents, and has been subject to many modifications in its performance. For a long time, it was coolly received by British surgeons, but within the past decade it has received a new stimulus from the Germans, and at the time of writing, it seems about to be fairly tried by the surgical world, and judged on its merits.

Almost every surgeon whose name is prominently associated with the operation has had his own favorite way of performing it; and we shall, therefore, speak in detail only of those which have proved most acceptable,

---

[1] **Etude sur** l'extirpation de l'extrémité inférieur du rectum. Marchand, **Paris, 1873.**

[2] **Mém. de l'Acad.** Roy. de Méd., 1833, iii., p. 296.

and first of those described by Volkmann in his *Klinische Vorträge* for March 13th, 1880. He describes **three** different operations, depending on the location of the disease. **The first** is for the removal of a circumscribed spot only. This **is** accomplished by dilating the anus, dragging down the disease, and **excising** it in such a way that the wound shall not cause subsequent **stricture. When the** growth involves the anus, the edges of the wound **are** carefully brought together, stitched with catgut, and a drainage-tube inserted between them. When the growth is entirely within **the** sphincter, the edges are brought together with equal care, but the **tube is** inserted through a track made **for it,** which communicates with the wound above, and perforates the healthy skin at a point outside **of** the border of the sphincter. When dilatation does not suffice, the anus is freely divided down to the coccyx, and this wound is subsequently carefully closed under the antiseptic precautions.

In the second class of cases where the growth involves the whole circumference of the bowel, but not the anus, the latter is divided forward into the perineum, and backward to the tip of the coccyx, when necessary, to give room for manipulation. The latter of these two incisions is carried as far into the bowel as the lower border of the disease, which is then removed. The mucous membrane above is stitched to that below, the preliminary incisions carefully closed, and a drainage-tube left in the posterior one.

In the third class, where the disease involves all, **or nearly all,** of the anus and of the circumference of the rectum, the entire tube is separated and removed in a cylinder. The same preliminary **incisions** may be made as in the second class, and the anus is surrounded **by** a circular cut, which runs outside the sphincter. **From** this as **a** starting point, the dissection is carried parallel with the bowel till the upper portion of the disease is passed. By the use of knife, scissors, and fingers the bowel is completely freed, then drawn down to the anus, and cut off above the disease, the healthy upper end being stitched to the margin of the skin. In case the peritoneum is opened, the wound is at once stuffed with carbolized sponge, and afterward carefully closed with catgut. The coccyx and part or nearly all of the sacrum are removed when necessary to make room, as a preliminary step.

The risk of hæmorrhage is one of the great objections to this operation, and later **on we** shall describe another procedure, which is preferred by many, in **which** the knife is supplanted by other and bloodless instruments. It **is** no **doubt true** that the deep dorsal incision is the key to the operation, and greatly facilitates the securing of bleeding vessels; yet the hæmorrhage may **be** so great as to impede the operator and endanger **the life** of the patient. It will be seen that, at every step in this operation, union by first intention is aimed at, and Lister's methods are carefully followed. If the elements of success in Listerism are, as I believe, cleanliness and drainage, these are certainly better met by a deep posterior wound, which is left open and syringed out frequently, than by carefully closing that safety-valve with catgut sutures and inserting a drainage-tube. It will also be observed that the bowel is always brought down and stitched to the free edge below. To do this much, dissecting is necessary, and but little permanent good is gained, as the stitches soon tear out.

Maisonneuve described, in *L'Union médicale* of 1860, an operation which he named the *procédé de la ligature extemporanée*, and which differs from the preceding in being almost entirely bloodless, although it differs little from the operation previously described by Chassaignac, under the name *l'écrasement linéaire*. In the latter, the rectum is divided into two lateral halves by the chain écraseur, and each half of the disease is then attacked in the same way and removed. In the operation as done by Maisonneuve, a strong cord is substituted for the chain, and the disease is removed in the following manner. The skin and subcutaneous tissue are divided by a circular incision which completely surrounds the anus. The operator is provided with several strong curved needles, each of which is to be threaded through the point as often as used, with a strong silk ligature about a foot in length. One of the needles with the ligature in its point is then passed from the external incision into the bowel above the growth, going wide of the gut to clear the tumor. The loop of string in the eye of the needle is seized within the rectum and drawn out of the anus, while the needle is drawn back out of its own tract. The result of this is a double uncut ligature, passing from the point where the needle entered the external incision, outside of the tumor, into the rectum above it, and then out of the anus; and this manœuvre is repeated eight or nine times at points around the circumference of the anus equidistant from each other. A strong whip-cord or bow-string is the next requisite—about two yards long—and to this all the loops hanging from the anus are attached at points nine inches distant from each other. Each of the original ligatures is then withdrawn by the same course it entered, carrying a loop of the whip-cord with it. When all are drawn out, the rectum above the disease is surrounded by a series of loops of strong cord, and the ends of each loop hang out from the original incision. The ends are then attached to an écraseur, and each loop made to cut its way out in turn. After all have been cut out, the lower end of the bowel and the diseased mass are of necessity completely separated from their attachments.

The operation performed by Cripps is a modification of the two preceding ones, and would seem to possess several advantages in facility of performance. The preliminary dorsal incision is made from within outward, by passing a strong curved bistoury into the rectum, bringing its point through the skin at the tip of the coccyx, and cutting all the intervening tissue. The buttock is then drawn away from the anus to put the tissues on the stretch, and a lateral incision made from the preliminary cut behind, around the rectum to the median line in front. The site of this incision, whether inside or outside the anus, will depend upon the location of the disease, and whether or not the anus is implicated. The cut itself should be made boldly, and deep enough to reach well into the fat **of the ischio-rectal** fossa. The forefinger in this incision will readily separate **the bowel from the** surrounding tissue, except at the attachment of the **levator ani muscle,** which should be divided with the knife or scissors. A piece of sponge is pressed into this cut to restrain the bleeding, while **the** opposite side is treated in the same way. The anterior connections give more difficulty, and the dissection in the male is aided by having a sound in the urethra. The knife and scissors replace the finger in this part of the operation. When the dissection has **been** carried to **a** point above the disease, the bowel is drawn down and **held** while the wire écraseur is passed over it, and the section made at the required level. After this there may be free but seldom serious hæmorrhage. The vessels divided in the first steps of the operation all come from the wall of the bowel, and if ligatured when first cut, are again opened with the écraseur.

When the disease **is located to** one side of the bowel, the operation is modified accordingly. **The** preliminary dorsal cut is the same, and the lateral incision is **made on the affected side.** At the farther end of this lateral incision, away from the **dorsal one, a needle** carrying a cord in its point is passed **around** the **disease and into the rectum** above it. The loop of cord **is** brought out of **the anus, attached to the** chain of the écraseur, and withdrawn as it entered. The chain is **then** made to cut its way out, and a rectangular piece of the rectum **is thus included** between two longitudinal incisions, one posterior with the **knife** and one lateral with **the** chain. In this rectangle is the cancer, **and it** is dissected upward from below, and separated above by **again** using the **écraseur.**

Instead of the chain or wire écraseur, the wire of the galvanic cautery may be used, heated to a dull red, and not a white heat, if the desire is to avoid hæmorrhage. Or again, instead of the wire the galvanic cautery knife may be used, and the operation performed with bloodless incisions. This is the operation favored by Verneuil. The rectum is first divided into lateral halves with the écraseur, as in the method of Chassaignac, the cut dividing both the anterior and posterior walls. Then with the galvanic-cautery **blade the** lateral halves are separated from their attachments **stroke by stroke,** until a point **is** reached above the level of the disease. **The chain is again** slipped over the **end of each,** and the final section **made.**

An ingenious and simple method applicable to certain cases has been recorded by Emmet.[1] The growth in the case in which it was used was

---

[1] Principles and Practice of Gynæcology, **ed.** 1879.

an epithelioma the size of a hen's egg, situated on the posterior wall of the rectum an inch above the sphincter, with considerable surrounding infiltration. The sphincter was stretched, and the mass seized with a double tenaculum, and drawn well down by an assistant. "A steel grooved director, as the most convenient instrument for the purpose, was pushed through the skin in front of the coccyx and just behind the outer edge of the sphincter, into the cellular tissue of the pelvis, and then made to puncture the rectum, in healthy tissue, just beyond the upper edge of the tumor. The end was turned out of the gut, and pushed far enough forward to rest on the perineum while the other end was over the coccyx. Then a second director was pushed around from the outer side of the muscle on one side, through the cellular tissue into the rectum, across to the other side, through the cellular tissue and skin again to the opposite side of the muscle. So that the mass, with a portion of the rectum above, was now brought through the anus and fixed by the two directors, which had been passed behind the mass at right angles to each other, with their ends resting outside on the soft parts. The chain of an écraseur was placed behind these two instruments and slowly tightened till the whole mass, as transfixed, was cut through along the course of the directors. By this means, I removed the entire sphincter muscle, about three inches of the posterior wall of the rectum, and about an inch and a half of the rectal surface of the recto-vaginal septum. The immediate result was a most formidable opening in the connective tissue of the pelvis, about three inches in diameter and cone-shaped from below."

Dr. Rouse[1] has recently called attention to a simple method of avoiding a wound of the sphincter, which is applicable to some of the slighter cases. A curved incision is made parallel with the outer border of the sphincter, and on a line with its outer limit. By introducing the finger through the rectum, the growth may be everted through this incision, and removed with the part of the rectal wall to which it is adherent.

Perhaps the best of all the operations we have spoken of is the combination of the écraseur and galvano-cautery knife, as used by Verneuil. But the operator is at liberty to choose from among them all the one he considers easiest of performance, and most free from the risk of hæmorrhage or of wounding surrounding parts.

A wound into the vagina, though always to be avoided when possible, may often be necessary in order fully to remove the disease. When the fistula thus made is not too extensive, it may be closed immediately after the operation. If large, it must be left. A wound of the urethra in the male, when slight, is to be treated as though the patient had submitted to an external urethrotomy, by the frequent passage of the sound, to prevent contraction. When a large piece has been taken from the urethral

---

[1] Lancet, Oct. 2d, 1880.

wall, a permanent recto-urethral fistula is the necessary result, and the danger of fatal inflammatory action is greatly increased from the presence of the urine in the rectal **wound.** As for the cases reported by Nussbaum and others, **in which the** whole neck of the bladder, the greater part of **the prostate, and the** seminal vesicles have been removed, and the patients have lived **for years in** comfort, **they** are merely curiosities of literature. **That such a thing may** happen has been proved, but that the operation **should ever be** undertaken in **any case** where such a result is necessary for the entire removal **of** the **disease,** has yet to be proved.

It is with **this** operation much **the** same as with proctotomy—by trying to save too much, discharge is impeded and life may be lost. Cases where the whole of the sphincter is removed, together with the skin of the anus, do better than those in which an attempt is made to save the sphincter and drain the wound with drainage-tubes.

The operation of excision has, with the recent advances in abdominal surgery, also been applied to cancer of the sigmoid flexure and descending colon. This operation **to** which allusion has already been made and to which Mr. Marshall[1] **has very** properly applied the name of "colectomy" **has now assumed a definite place** in surgery and marks another of the great advances **of the present century.**

It dates from the **time of Reybard of Lyons,**[2] **who** in 1833 removed a tumor the size **of an** orange from **the** sigmoid **flexure of a** man aged twenty-eight years. In this case the tumor could be **felt** through the abdominal wall in the **left** iliac fossa, **and** the incision **was made** parallel with Poupart's ligament and the **crest** of the ilium. **The tumor was** drawn out through this wound and excised with three inches of the adjoining intestine. The two ends of the bowel were stitched together and replaced within the abdomen and the abdominal wound was completely closed. There was considerable local trouble for a few days, but on the thirty-eighth day the wound had entirely healed and the natural passages were restored. Death occurred ten months after from recurrence of the disease. This case was subject to considerable discussion in the academy, but was finally admitted as authentic.

The operation thus inaugurated **in 1833** has been modified in two essential particulars by subsequent operators, one in the choice of location of the incision, the other in the subsequent disposal of the ends of the divided intestine. Since the first case by Reybard, the operation has been performed at least seven **times.**

Gussenbauer, of Liège, **has** done it twice. The first time in 1877[3] was upon a male patient aged **forty-two** years. The tumor which was asso-

---

[1] Clinical Lecture on Colectomy, Lancet, May 6th, 13th, 1882.
[2] Bull. de l'Acad. de Méd., vol. ix., 1843-4.
[3] Arch. für klin. Chirurg., Bd. xxiii., 1879.

ciated with the usual symptoms of obstruction could be felt in the left side, but an attempt was made to remove it through an incision in the median line of the abdomen. This incision proving insufficient, was enlarged by cutting laterally as far as the lumbar fascia. Another complication arose from the attachment of the growth to the small intestine which was opened, and fæces were allowed to escape into the peritoneal cavity. All the intestinal wounds were closed with sutures, the bowel was replaced within the abdomen, and the abdominal incision sewed up. In this case death followed in fifteen hours. Gussenbauer's second case was performed in 1879,[1] and there had been no return of the disease two years later.

Baum of Dantzic[2] operated between these two dates (1878) upon a male patient, aged thirty-four years, in a case of doubtful nature. He first opened the small intestine to relieve the symptoms of obstruction, and seven days later he discovered the seat of the obstruction in the right hypochondrium. A second operation was then performed. The abdomen was again opened, this time by a longitudinal incision over the tumor, two and a half inches to the right of the median line, and this incision was afterwards enlarged by another running toward the right. The growth was situated at the junction of the transverse with the ascending colon, and was removed together with a piece of the mesentery which contained an enlarged gland. The divided ends of the bowel were invaginated and united, the intestine replaced, and the abdominal wound closed. There was considerable discharge of fæces from this opening, however, up to the time of death on the ninth day.

The next case was by Martini, of Hamburg,[3] in 1879, and was performed with the deliberation and consequent success which arise from a certainty in diagnosis of the character and location of the tumor. The growth was situated in the sigmoid flexure and could be felt both through the abdominal wall and the rectum. The incision was made over the tumor, the intestine below was cut between double ligatures, the mesocolon was divided and the affected glands excised, and finally four inches of the bowel were excised together with the diseased mass and two inches breadth of mesocolon. After the removal of such a section it was impossible to approximate the divided ends of intestine. The rectal end was, therefore, invaginated upon itself, closed with sutures and allowed to drop into the pelvis. The upper extremity was attached to the incision in the abdomen to form an artificial anus. There were no bad symptoms and in a few weeks the man was able to return to his business.

Czerny, of Heidelberg, reported the next successful case in 1880,[4] in a

---

[1] Ztschr. für Heilk., Prag, 1880.
[2] Centralblatt für Chir., 1879, Bd. ii., p. 169.
[3] Vierteljahrschrift für Heilk., Bd. i., 1880.
[4] Berliner klin. Woch., 1880, No. 45.

female patient, aged forty-seven years. In this case also the growth could be felt through the abdominal wall on the left side and the diagnosis was therefore positive. After opening the abdomen over the tumor, the bowel was found to be implicated at two points, one at the transverse colon, and the other at the sigmoid flexure which curved upward to an abnormal degree and was involved in the same disease through a fold of the great omentum. Two and three-quarters inches of the sigmoid flexure, and four inches and a half of the transverse **colon** were excised and the cut ends of each portion were united. The peritoneum was washed out, a drainage tube inserted, the abdominal incision closed except for the drainage tube, and the whole dressed antiseptically. For a time there was a discharge of fæces through the abdominal wound, but this finally closed and the patient was well in four months. **The** return **of the** disease was, however, very rapid, and death was **caused** by it in **about** seven months after the operation.

Billroth operated next in order, in 1881,[1] on a male patient twenty-eight years of age. The operation was done antiseptically, and the incision was the usual one for left inguinal colotomy. The tumor involved the lower half of the **sigmoid flexure**, and there was considerable involvement of the adjacent **mesentery and** of the tissue behind the bowel. The upper section **of the bowel was used** for the formation of an artificial anus. The patient died **in about** thirty-six hours from incipient diffuse peritonitis.

Bryant's case[2] **is next in** order, **and is peculiar in** the fact that the incision was the usual one for left lumbar **colotomy. This, in** fact, was the operation attempted, but after the bowel had **been opened, the** obstruction was found to be above the opening made. **It** was then determined to excise the disease, and this was successfully done through the original incision. The two ends of the bowel were attached to the wound, the upper in the usual manner for forming an artificial anus. The patient recovered, and was well at the time of the publication of the case. The disease constituted a cylindrical stricture of limited extent.

Finally, Mr. Marshall's[3] case has just been published at the time of writing. The patient was a woman, aged forty-nine years, and no positive diagnosis as to the seat of the obstruction could be made. The difficulties attending the diagnosis may best be gathered from his own description.

"The wasting and rapid ageing of the patient, although she took food tolerably well, suggested the presence of a malignant stricture, probably epitheliomatous; but it was difficult to say how far the symptoms were referable merely to **the pain and vomiting** which she had suffered; but,

---

[1] Wien. Med. Woch., March 5th, 1881.
[2] Lancet, Vol. i., 1882.
[3] Lancet, May 6th, 12th, 1882.

whatever the nature of the obstruction, its seat was obscure. The chronicity of the case pointed strongly to the large intestine, but the abdomen was not broad in shape; no tumor or scybala could be felt in either iliac fossa, or elsewhere along the course of the large gut, though both fossæ could be well examined under chloroform. There was no dulness in either loin to indicate a full colon, and no "colonic" note to show that the bowel contained gas. Rectal examination revealed nothing. The long tube passed one foot, and an enema of three pints was easily given, and seemed, from an accompanying diminution of resonance in the left flank, to have entered the descending colon. But as the patient was lying on the left side, it was possible that fluid contents had gravitated into the small intestines lying over the descending colon—a source of movable dulness which, as remarked by Mr. Boyd, is often overlooked. The amount and uniformity of the abdominal distention were sufficient to prove that the obstruction, if in the small intestine, was near the lower end. If, however, the suspicion were correct that the cause of the obstruction was an epithelioma, the probability of its seat being in the large intestine, somewhere beyond the cæcum, was greatly increased."

On account of the uncertainty in diagnosis, the incision in this case was an exploratory one in the median line, and the growth was found in the descending colon, between the lower end of the kidney and the iliac crest. As it was impossible to bring this part of the bowel to the median line, the first incision was abandoned, and a second one made over the tumor, parallel with the last rib, and one inch and a half above the posterior part of the iliac crest. The growth was cut out with the scissors, together with an inch of the bowel above and below, between double ligatures. The open end of the upper section of the bowel was attached to the abdominal wound to form an artificial anus, and the lower end was left projecting from the lower and hinder part of the wound with the strong catgut ligature drawn tight upon it. The patient died of peritonitis on the third day.

Of these eight cases, one-half may fairly be said to have prolonged life, and the others have been fatal within a short time from peritonitis. As pointed out by Marshall in his instructive résumé of the operation, the result undoubtedly depends in a great degree upon the certainty with which the diagnosis is made, or, in other words, upon the exact adaptation of the operation to the end to be attained. In most of the successful cases, the diagnosis as to the seat of the obstruction was made before the operation was begun, and in all of them only a single incision was necessary to reach the tumor. In three of the four fatal cases, two incisions were made—one in the median line, and, subsequently, another to reach the disease. In this way the severity of the procedure was greatly increased.

There seems to be little difference in the mortality whether the ends of the divided intestine be united and the abdominal wound closed; or one end be brought to the surface for the formation of an artificial **anus**. The latter is the simpler procedure; the former, when successful, gives the better result. A great difference in the size of the two ends will sometimes render their union difficult; the upper one being frequently hypertrophied and dilated, and **the** lower contracted.

The study of these **cases leads** plainly to the following conclusions:—

1. In cancer of the descending colon, sigmoid flexure, and upper part of the rectum, when the disease is still movable, an attempt at its removal through the abdominal wall is justifiable.

2. In cases of obstruction where the symptoms point toward this part of the bowel as the affected part, even when the diagnosis is not certain, it may be well to make the exploratory incision in the left groin instead of in the median line, having in mind the possible extirpation of the disease and the formation of an artificial anus.

3. In cases of intended colotomy also, it may be found possible, after the incision has been made, to substitute colectomy, and this constitutes another reason for choosing the inguinal to the lumbar incision in that operation, though, as in Bryant's case, colectomy may be done through the **loin**.

4. The operation of colectomy compares very favorably with colotomy in malignant disease, **and while the** latter may be the more suitable in an advanced case, the former may **give better** results when the disease is in its incipiency.

The palliative **treatment of malignant stricture of the** rectum is in many points the same as of non-malignant. **The relief of pain is** perhaps a more marked indication in most cases. The pain **depends** on two classes of causes—those which make cancer a painful **disease wherever** met with in the body, and those which are due solely to its **situation at** the outlet of the bowel. Among the first, we have pressure upon adjacent parts and involvement of neighboring organs and nerves; and among the second, the passage of fæces over an ulcerated surface and spasm of the sphincter muscle from irritation caused by its direct implication in the cancerous growth, or by the passage over it of irritating sanious discharges from the sore. From this it is easy to understand why cancer is in one person attended by excruciating suffering, while another may hardly be conscious of its presence; and why the pain is in some paroxysmal and particularly aggravated by a movement of the bowels, and in others dull and constant, radiating through the loins and down the thighs. For the relief of this **symptom we** have at our command: *a.* Regulation of the passage, **diet,** and the recumbent posture; *b.* Anodynes locally **and** by the mouth; *c.* Partial destruction of the growth by means of the curette, cauterization, or partial extirpation; *d.* Division of the sphincter; *e.* Lumbar colotomy.

The passages should be kept soft but not fluid, as any approach to diarrhœa always aggravates the suffering. This may be done partly by the choice of food, which needs to be regulated with great care on account of the tendency to gastric disturbance, more or less of which is always present; and by the administration of the mineral waters, which are generally sufficiently laxative for the purpose. Rest in the recumbent posture is a means of palliation of great value, sometimes giving more relief than anodynes. These latter may be given both by the mouth and in enemata, and if possible should be pushed to the point of relieving suffering. This seems so plain a duty which the surgeon owes to his patient, that we need not stop to discuss any possible moral bearing it may have. If the agony of this incurable malady could always be relieved by the administration of opium, the question of operative interference would arise much less frequently than it now does. But, unfortunately, the constant administration of this or any other narcotic will sometimes cause gastric and mental disturbance, harder to bear than the disease. By using the finger-nail, a curette similar to the one used in the uterus, or a scoop such as is used for submucous uterine tumors, the pain may in some cases be greatly relieved by a removal of a part of the growth when of the soft variety. The same may be done by the application of chemically destructive agents or the actual cautery, and even by the partial excision of the mass, merely as a means of relief and where there is no question of cure. I have already called attention to division of the sphincter muscle as a palliative measure in the treatment of rectal disease, and all that was said regarding the treatment of benign stricture applies equally well to cancer.

The *dernier ressort* of surgery for the relief of pain is lumbar colotomy. We have already attempted to limit the scope of this operation. In any case in which the suffering is due to the direct contact of fæces with the diseased surface, and is not due to a spasmodic action of the sphincter muscle, and cannot therefore be relieved by the permanent division and paralysis of that muscle, and is not due to the extension into and pressure of the disease upon neighboring parts, the operation may be tried. There may be such cases, but they are not common—not nearly as common as is lumbar colotomy for cancer. Let it be remembered, however, that after colotomy fæces will still find their way to the tender point, and that the amount of suffering from a small mass of fæces may be as great as from the entire quantity.

With regard to husbanding the sufferer's powers and prolonging life, much may be done by careful nursing and medication. Milk is by far the best diet, and cod-liver oil in small doses the best medicine where it can be borne, for it has a laxative as well as a tonic action. Cleanliness is best obtained by frequent washing out of the rectum with disinfecting fluids, as permanganate of potash and carbolic acid.

The means of overcoming obstruction in malignant disease are also much the same as in benign stricture, and to what has already been said on that subject we must again refer the reader. Before commencing to treat the obstruction as such, it is well to remember that an exceedingly small outlet to the alimentary canal may, with proper care, be made to answer all the calls of nature. We see this constantly in cases of stricture both simple and malignant, where the finger cannot be forced through the obstruction, and yet there is no retention; and in such cases, by the judicious administration of laxatives, life may be made so comfortable that the question of surgical interference shall be postponed indefinitely. When, however, obstruction is actually threatened, **much may be done** by the medical means already pointed out.

When dilatation becomes necessary, it should be of **the** gentlest kind. The cases of fatal accident from perforation of the bowel where the coats have been weakened by ulceration are already numerous enough to serve as warnings for all future time. The best of all dilators in cancerous disease is the finger, either that **of** the patient or the nurse, passed daily; and none of the mechanical means with which we are acquainted **equals** this for safety and comfort.

When the disease is beyond the reach of the finger, a bougie must be **used**, but the dangers are greatly increased, and it may be better at once to make an artificial **anus than to** incur the risk of fatal accident which the use of a bougie **high up the bowel** certainly entails. The frequency with which the **bougie may be used will** depend upon the result of its trial. Should **much irritation, tenesmus,** or hæmorrhage follow its employment, the **patient will soon refuse to submit** to its continuance; while, on **the other hand,** should **the result be favorable, it may be em**employed daily. The softest bougie **is the best, and a candle** often answers admirably.

If dilatation be found too painful or ineffectual, as it sometimes will, recourse may be had to division or partial destruction of the cancerous **mass.** A double proctotomy may be done in case of malignant disease, **and** the section of the growth between the two incisions be removed, in this way opening once more the calibre of the bowel and overcoming the obstruction. I have performed this modified operation with great relief, and I have also found that, after making a single free division of the cancerous mass, large pieces adjacent to the cut could be excised with great facility and without danger. The latter operation is rather the preferable one.

Relief both to pain and obstruction may sometimes be gained in this way by a partial destruction and extirpation of a cancerous growth, where its entire removal is out of the question, and its local return may be expected with certainty. By such measures, the evacuations may be made less painful, the spasmodic action of the sphincter and the rectal tenesmus may be allayed, the cancerous look may for a time disappear, and the patient recover sufficient strength to resume the ordinary occupations of life.

A growth may be attacked in this way, either with the knife, cautery, finger or curette. Caustic applications are of no use, except in cases where a fungous mass has protruded from the anus. This may, at times, be removed with great advantage to the sufferer, by the application of a paste of arsenite of copper, mixed with mucilage. The operations for removing a part of the growth with the finger, scoop, or curette may give great relief in the soft varieties of the disease. The sphincter should first be thoroughly dilated, the anus held open with a speculum, and as much of the diseased tissue as possible torn and scraped away. Hæmorrhage, of course, is to be expected, but this is less where the growth is boldly attacked in its deeper parts than when the surgeon is timid and attacks merely the superficial portions; and may be controlled either by plugging the wound with lint and styptics, or by the actual cautery. Allingham relates a case in which he entirely enucleated an immense encephaloid with his hand, with the happiest results.

As a substitute for partial destruction of the growth in this way, the operation of crushing with an instrument similar to the enterotome of Dupuytren has been proposed. The proceeding is only applicable to a certain class of cases, in which the stricture is annular and not too extensive to be grasped by the instrument, and has no advantages over the other methods.

There is no obstruction within four inches of the anus which may not be overcome by some one or other of these means. What, then, remains for lumbar colotomy? Simply those above the reflection of the peritoneum.

It will often be difficult for the surgeon to decide for or against colotomy in these cases. Two factors enter into the question: 1st, whether or not the patient is likely to survive the operation itself; and, 2d, if this is decided in the affirmative, whether sufficient is to be gained to pay for the risk. The general condition of the patient, the extent of disease as regards secondary deposits, and the amount of pain due to defecation, all have to be taken into consideration. The operation may be indicated to relieve this pain when there is not much chance of actually prolonging life, and it may be indicated to prevent or overcome obstruction where there is no great amount of pain. I am inclined, for myself, to limit the operation to those cases where the pain of defecation is great, and where the disease is still circumscribed, and should not for the choice between death from obstruction and death a few weeks later from exhaustion always have recourse to this extreme measure, but should rather trust to securing a comparatively easy passing away of the patient under the influence of opium. Indeed, many patients will decide the question in this way for themselves when it is explained to them in all its bearings.

It is a curious fact that, by relieving the over-distention of the bowels by colotomy, the obstruction also will sometimes cease, and passages will again pursue their natural course. Such a case is reported by

Goodhart, where three successive operations for opening the colon above the stricture were resorted to to relieve obstruction, and after each one the passages were again restored to the natural outlet.

CHAPTER XII.

IMPACTED FÆCES AND FOREIGN BODIES.

Impacted Fæces.—Intestinal Concretions.—Diagnosis and Treatment of Impaction.—**Foreign** Bodies Swallowed.—Results which may Follow the Swallowing of a Foreign Body.—Ulceration and Abscess.—Foreign Bodies Introduced *per Anum*.—Cases.—Prognosis.—Treatment.—**Dangers** of Attempts at Removal.—**Laparotomy for Removal.—Cases Successful.**

Impaction of Fæces.—The impaction of fæces may be due to several causes, but is most generally a symptom either of intestinal atony in old people or of some paralytic affection such as locomotor ataxia. It not infrequently occurs in women as a result of the entire neglect of the function of defecation for which they are perhaps unjustly celebrated; and they may follow a partial paralysis of the rectum from the long-continued use of large enemata, or the pressure of the fœtal head in childbirth. They may also be formed as a consequence of a painful affection such as a fissure which renders each act of defecation an agony to be avoided by every possible means. The disease is generally one of old people, of hysterical girls, and of careless women; but it has been seen in children, and as a result of improper diet may occasionally be encountered in young and healthy men.

Intestinal concretions may be composed entirely of hardened and stratified or clayey masses of fæces, or they may contain within them as a nucleus a biliary calculus, or indigestible substances which have been hastily swallowed, such as peach-pits, cherry stones, etc. Mollière calls attention to the presence of magnesia which favors the aggregation of fæcal matters, and which also may act as the nucleus of a scybalus; and the frequency of impaction during the famine in Ireland in 1846, when potatoes, and those of a very poor quality, were the only article of diet, is a well known historical fact.[1] In Scotland, where oat-meal is a favorite article of diet, **fæcal accumulations** are said to be of frequent occurrence. Certain other drugs besides magnesia, such as chalk, sulphur, and powdered **cubebs** have been blamed as the cause of intestinal concretions. Intestinal calculi have been seen which were composed of pure cholesterin or of a biliary calculus coated with cholesterin.

---

[1] For description of these cases see article by Dr. Papham in the **Lancet**. 1850.

The usual location of a mass of impacted fæces is the rectal pouch, but it may be situated anywhere between the cæcum and this point. The symptoms to which it gives rise are generally sufficiently well marked to enable the practitioner to reach a correct diagnosis if he be on his guard. The pains which it causes will generally be obscure and may be located anywhere in the abdomen or in the lower extremities; and the signs of disturbance in digestion are not in themselves sufficiently marked for diagnosis, but the one symptom which is characteristic is diarrhœa.

Just as the practitioner has to learn that incontinence of urine may be a sign of a distended and not an empty bladder, so he may have to learn by a disagreeable error in diagnosis that a diarrhœa is sometimes a result of an overfilled and obstructed rectum. This diarrhœa is peculiarly fœtid in character, and the matters discharged may be entirely free from fæces and consist entirely of mucus. In some cases there may be an approach to a daily natural evacuation. The act of defecation is always attended by straining and pain as the fæcal ball is pressed down against the perineum and rises again when the muscular effort ceases. To these symptoms Allingham adds a peculiar ringing, barking cough, morning vomiting (particularly in women), and night-sweats.

Of course errors in diagnosis are easy in such a condition as this, and a mass of fæces in the colon may be mistaken for any and every sort of tumor in the pelvis or abdomen. Liver, spleen, stomach, uterus, and ovaries have again and again been supposed diseased in these cases when a simple digital examination of the rectum, or in women even of the vagina, could not fail to make the diagnosis clear. Unfortunately for diagnosis, the general practitioner is not fond of making rectal examinations, and these cases are not infrequently treated with bismuth and opium as a consequence.

The following instructive case was reported by Dr. Griffith.[1]

In the autumn of 1876, I was hurriedly summoned to an old lady, who had within a few days of my seeing her met with a severe accident in the city, having been knocked down by a hansom as she was crossing the street. All her friends had given her up to die. She was so powerless to move, so prostrated, and so large a tumor, they stated to me, had made its appearance since her injuries. Her age (80) seemed to exclude all hope of recovery; and I was asked to see her—more that it should not be said she had died incapable of making her will and to witness her signature to it, than with any idea that I could benefit her.

I examined the abdomen, and while doing so learned from her that she thought she had been larger on the left side for some time before the accident. I found considerable enlargement of the entire abdomen from flatulent distention, and on the right side a tumor, hard and apparently

---

[1] Fæcal Accumulations Stimulating Utero-Ovarian Tumors, Edinburgh Medical Journal, May, 1877.

irregular, extending from the left hypochondriac into the left iliac fossa, and passing a little way to the right of the median line. At first, I thought it might be enlarged spleen, or a left ovarian dropsy, or an extrauterine fibroid, which had been unnoticed, and was now observed, solely because attention was directed to the left side, where the patient had been struck by the vehicle. I could not at this, my first visit, make a very minute examination, owing to the extreme prostration and depression; but at my second visit, having in the interval built her up and cheered her all I could, I examined very carefully per vaginam, and with equal care explored by the rectum. I then came to the conclusion that there was neither ovarian nor uterine tumors, and that I had to deal with an accumulation of fæces—even though the bowels were moved every day, as the attendant informed me, and that the accumulation had commenced previous to her accident; forming, no doubt, the enlargement which she told me she had noticed before her injury, and which, as the accumulation increased, culminated in the enlargement I found. I swept out the bowels by free purgation, kept up for some days, while I sustained her with light and easily digested nutrients, allowing as stimulant only good tea and coffee.

The next case is also from the same author:

Mrs. G., aged twenty-five, mother of three children; the last being about four months old when I was first in attendance. I was called up to her on the night of **18th June, 1876**. "as she was suffering acute pain in the left side, which she could endure no longer." On examining the abdomen, I found a hard, irregular, exceedingly tender tumor, from which she was enduring great agony, and which was almost as large as an infant's head. I made no further examination that night, contenting myself with ordering her one-half grain morphia suppositories, to relieve not only the pain, but likewise the tenesmus and the passing of mucus. The discharge from the bowels was quite fluid, but distinctly fæcal, occasionally a scybalous mass making its appearance.

Next day, the morphia having taken good effect, I examined with the finger by the vagina, but could make out neither ovarian nor uterine tumor; the sound *in utero* enabled me to make certain that there was no intrauterine growth; but movement of the uterus with the sound in the interior of it was attended with the movement of the mass, which I found lay outside the womb, yet connected to the left and upper portion of it— in fact, attached to it. I gave it as my opinion that, whatever the mass was, it was outside the uterus, and was adherent to it, and that it was not ovarian. I did not, however, express the opinion at which I arrived after the above examinations and after thoroughly exploring by the rectum, viz., that it was a case of impacted and accumulated fæces, which, having set up great irritation, had occasioned inflammation, effusion of lymph, and matting or gluing of the bowel to the left and upper portion or cornu of the uterus, that organ being still enlarged, its involution **after** delivery being not yet completed, probably owing to the irritation, inflammation, **and** subsequent adhesion to which I **have** referred. Taking this view **of the case,** I purged freely and continuously for some days, till **at length, after the** lapse of six weeks, I had **the** satisfaction of hearing **from my** patient—for I did not attend her continuously during **this period—that the** tumor was **all** gone, **and** she **was** quite well; facts **I verified by careful** manipulation when she last visited me. The iodide **of** potassium had been combined with the aperients, as had also anodynes—the former in hope of dissolving **adhesions,** the latter with a view to ease pain. I would **add, to** show the difficulties which sometimes behedge the diagnosis in these **cases,** that this patient had previously had pronounced to her by **three medical men that** operation alone (gastrotomy) could do her any good; **and of this she** had a mortal dread, so that all through I buoyed her up with the hope that the knife might never be required.

The swelling had commenced to be noticed about twelve or **fourteen days** after the birth of her child, was chiefly confined to the left side, **though** sometimes it seemed to enlarge, and to extend higher up and **across** the middle line **towards the** right, and was so large that it was as though she was **at her full time, and** when walking, even across her room, she required **a towel to support** the abdomen; at other times it would subside, **preserving, however, the same shape;** these alterations in size were **synchronous with** the **action of the** bowels, and gave me a valuable clue. The agony had been very great, and she told me nothing had relieved her **for** any length of **time** till she **had used the** morphia suppositories. **At** no period was there a discharge **of matter indicative** of any internal abscess; nor any flux of water either into **the abdominal** cavity or into the bladder, or any way externally, which would demonstrate the existence and rupture of an ovarian or other cystic growth; therefore, the only diagnosis at which I could arrive was that the bowels had become blocked during the confinement period, had not emptied themselves fully, that an accumulation occurred and became greater and greater, being, however, occasionally partially lessened by the aperient action of the bowels themselves, which accounted for the diminution of and subsidence that **had** been noticed in the swelling.

The treatment of impaction is simple, and consists first of all in the entire removal of the mass. In cases of paralysis, where the accumulation has not been allowed to reach any very great amount, and the scybala are small and not very hard, this may sometimes be accomplished by the **use of** injections with a long **tube and** the assistance of the finger of the operator. In women very effectual **aid** may be rendered under similar conditions by pressure from the vagina, by which small masses may be extruded one after another, **each** with a certain amount of pain, but without laceration of the mucous membrane at the anus. This plan of treatment will often constitute one of the regular duties of the attendant upon a **case of** paralysis—a disagreeable duty which must be attended to at certain regular intervals.

In cases of longer standing, however, these means may be entirely inadequate and all injections, no matter what their supposed solvent virtues, will be of no avail even if they are not at once ejected. In such cases the operation of breaking up and removing the mass must be begun by the administration of ether and dilatation of the sphincter. This accomplished, the mass may be attacked with the fingers, an iron spoon, or a pair of lithotomy forceps, and removed piece by piece. When this has been done, an injection may be administered through the long tube and more matter will generally come down from the sigmoid flexure. The impacted mass is often as large as the fist, and sometimes as a fœtal head, and the amount in the sigmoid flexure and colon may be much greater though not as hard; so that at a single sitting an enormous amount may be removed.

After such an operation as this, the patient must be treated by injections and a daily laxative, as will be described in speaking of constipation, till the over-distended rectum has recovered its tone. This may require a considerable time.

Foreign bodies which have been swallowed.—Medical literature is full of curious cases in which foreign bodies have been swallowed, either accidentally or by design, and have in some cases passed the full length of the alimentary canal, and been safely voided with the fæces, or in others have become entangled in the mucous membrane, and given rise to much trouble. Every practitioner is familiar with cases of peach-stones and coins which have been accidentally swallowed, and knows how generally such substances take care of themselves, and cause no symptoms after once passing the œsophagus. Much larger substances, such as whole or partial sets of false teeth, and the various things with which performers in travelling shows entertain an audience, may also be passed in safety.

To show what nature is capable of in this line, it may be well to enumerate the substances which were swallowed and safely voided by a certain lunatic now become famous. The patient stated that she had been swallowing nails, etc., and a dose of castor oil brought away two pieces of faïence, one or two centimetres long and about the same breadth, two nails, and a pebble. During the following six weeks she passed nineteen large pointed nails, a screw seven centimetres long, numerous fragments of glass and china, a piece of a needle, two knitting needles, fragments of whalebone, etc., amounting in all to three hundred grammes. During all this time the patient ate and drank as usual, and seemed in ordinary health.[1]

Prof. Agnew " saw in the dissecting room of the Philadelphia School of Anatomy, a female subject, afterwards learned to have been insane, in

---

[1] Lancet, 1866, Vol. i, p. 23.

whose intestinal canal from jejunum to rectum were found three spools of cotton partially unwound; two roller bandages, one of them 2½ inches wide and one inch thick, the other was partially unrolled, one end being in the ileum, the other in the rectum; a number of skeins of thread, a quantity being packed tightly in the cæcum; and finally a pair of suspenders."

Prof. Gross records the "case of a man who swallowed a bar of lead, ten inches long, upwards of six lines in diameter and one pound in weight, whilst performing some tricks of legerdemain," which was removed by gastrotomy and the patient recovered in two weeks. He also mentioned another case in which a teaspoon was swallowed, whilst the patient was in a paroxysm of delirium, which was removed from the ilium by enterotomy, recovery taking place in a few weeks.[1]

"Henrion, called Cassandra, born in Metz, in 1761. Not satisfied with the various trades which he followed in his youth, he began to force himself, at the age of twenty-two years, to swallow pebbles. Sometimes he swallowed them whole and without any preparation, and sometimes he broke them between his teeth, after having first heated them red-hot and then suddenly plunged them into cold water. In this manner he palmed himself off as an American savage. For several years he had fixed his residence at Nancy, and there continued the same habits which he had not interrupted, swallowing daily a large number of pebbles, sometimes as many as thirty or forty. The largest pebbles equalled in volume a large nut, but they were usually smaller, and Henrion demonstrated their presence in the stomach by the collision which he obtained by percussing the epigastric region. With the aid of salts, he passed them in twenty-four hours, and often made them do duty for the next day. He also swallowed live mice, though only one in the course of a day, as well as crabs of moderate size, after their claws had been cut. When the mice were introduced into the mouth, they threw themselves into the pharynx, in which they were soon suffocated, and their deglutition was then facilitated by that of a nail. Upon the following day it was passed from the rectum, flayed, and covered with a mucous substance. At another time three large pennies were successively put to the same use, and Henrion found them later, scraped clean and mixed with fæcal matters.

He continued this calling until 1820. At this time he swallowed some nails, and then a plated iron spoon measuring five and a half inches in length and one in breadth, for a moderate sum. He died seven days later."[2]

Napoleon relates a case of considerable historic interest where the alimentary canal was used for the purpose of secreting dispatches.

---

[1] Randolph Winslow, Maryland Medical Journal, March, 1880.
[2] Arch. Gén. de Méd., 3e Série, 1839, p. 353 (Poulet).

"When I commanded at the siege of Mantone, shortly before the surrender of this fortress, a German was arrested while endeavoring to enter the city. The soldiers, who suspected him of being a spy, searched him without success; they then threatened him in their own language, which he did not understand. Finally a Frenchman was called who spoke German slightly, and who threatened him, in bad German, with instant death if he did not at once disclose all he knew. He accompanied this threat with furious gestures, drew his sword, placed the point of it upon his belly, and said he was going to slit him open. The poor German, frightened and not understanding the jargon of the French soldier, imagined, when he saw him threatening his belly, that his secret was disclosed, and cried out that it was unnecessary to slit him open, and that if he waited a few hours it could be obtained in the natural manner. This gave rise to fresh questions; he stated that he was the bearer of dispatches for Wurmser, and that he had swallowed them as soon as he found himself in danger of being captured. He was carried to my headquarters, whither several physicians were summoned. It was proposed to administer a purgative, but they stated that it was best to await the operation of nature. He was then confined to a room under the surveillance of two staff officers, one of whom was constantly near him. After several hours the expected object was found. It was inclosed in wax, and was as large as a nut. When opened it was found to be a dispatch written in the hand of the Emperor Francis, and which requested him not to be discouraged and to hold out a few days longer, when he would aid him with a strong column." Napoleon, upon these indications, left with his troops and completely defeated Alvinzi at the passage of the Pô.

It would be beyond the scope of a work such as this to attempt to deal with the whole question of foreign bodies in the alimentary canal, and the accidents which may attend them. In a general way, the prognosis is good unless the foreign body be a very ragged one or a large sharp one like a fork; and the treatment consists in giving a diet like bread and fruit, which will cause copious stools, with little drink, and the avoidance of exercise such as walking. If complications arise, they must be treated on general surgical principles; and at the present day no patient would be allowed to die from the effects of a foreign substance in the stomach or intestines without a surgical operation for its removal, provided only the diagnosis were clear.

The complications which may attend the detention of such substances in the rectal pouch just above the internal sphincter are ulceration with perforation, hæmorrhage, and abscess. Ulceration may be caused by the pressure of a large body, and may cover a considerable space, or it may be caused by the pressure of the sharp ends of a smaller body, in which case the spots of ulceration will be smaller, and may be located at two opposite points in the rectum. As a result of ulceration, there will be

---

[1] Memorial de Sainte Hélène, t. ii., p. 468 (Poulet).

more or less pain, purulent discharge, and perhaps also a sharp hæmorrhage from the erosion of a vessel. When perforation of the wall of the bowel has occurred, inflammatory action is almost sure to be excited in the surrounding parts, and this may vary greatly in its extent and gravity. If the injury be above the point of reflection of the peritoneum, it may cause either a localized or a general peritonitis. A general peritonitis caused in this way will be fatal, as it is also generally accompanied by more or less extravasation of fæces. A circumscribed peritonitis with formation of an abscess is a less fatal complication. Under these circumstances the usual signs of pelvic abscess will be present—fever, pain on pressure, tympanites, painful defecation and urination—and by careful examination a tumor may be discovered, either through the rectum or at the bottom of the iliac fossa. Such cases, when the tumor is on the right side, are often mistaken for cases of perityphlitis, but the tumor is not in the same location. It is deeper and nearer the median line.

Such an inflammation may terminate in resolution, provided the cause be discovered and removed; but the usual termination is in suppuration, and the pus, if not removed by the surgeon, may find its way into the general peritoneal cavity or into the bladder or rectum. Abscesses of the superior pelvi-rectal space have already been described, and those which are due to foreign bodies in the bowel do not differ from them in general characters.

Spontaneous cure may follow the rupture of such an abscess into the rectum or bladder, but an incurable fistula is more apt to result even after the foreign body has been discharged. In one such case I was able to withdraw the pus through the abdominal wall with the aspirator, and subsequently, when the abscess cavity again filled up, I incised it through the rectum just behind the prostate. This opening was kept from closing by the daily introduction of the end of the index-finger, and the abscess finally healed very kindly—a result which was in great measure due to the fact that the patient was a child of twelve years, and not an adult.

When the focus of inflammation is located below the reflection of the peritoneum, the prognosis is less grave. Phlegmonous abscess may form in the ischio-rectal fossa, and must be treated according to the rules already laid down; but here the difficulty is well within the reach of the surgeon, and a cure may confidently be looked for by proper care.

Foreign bodies introduced per anum.—A classification of these cases is useless. The foreign bodies may be introduced through traumatism: by the patient in an honest endeavor to relieve himself of piles or prolapse; by the surgeon for the purpose of relieving rectal disease. They are often introduced in a spirit of revenge or of trickery; and most often of all they are lost in the practice of an unnatural vice. Edward II. is said to have met his death by having a red-hot iron thrust into the rectum. ["We seized the king," said one of the murderers, "and threw him forcibly upon the couch, and, whilst I kept him there by the assistance of a table, with a pillow on his face, Gurney inserted through a horn-tube a red-hot iron into his bowels." Gross, Vol. ii., p. 627.]

The case of the prostitute into whose rectum the students of the University of Göttingen introduced a pig's tail, butt end first, is as follows:

"Some students had formed the plan of playing a practical joke on a prostitute; they determined to push into her anus a frozen pig's tail. They cut the hairs very short in order to make them sharper and rougher, then dipped it in oil, and forcibly introduced it into the woman's anus, with the exception of a portion three fingers' breadth in length, which remained outside. Several attempts were made to extract it, but, as it could only be withdrawn against the hairs, the bristles entered against the mucous membrane, and gave rise to excruciating pain. In order to relieve it, various oily remedies were given by the mouth, and the attempt was made to dilate the anus with a speculum in order to extract the tail without violence, but it was unsuccessful. Severe symptoms developed, violent vomiting, obstinate constipation, very high fever, and intense pains in the abdomen. Marchettis was summoned on the sixth day. This physician, having been informed of what had happened, invented a very simple and ingenious device. He took a hollow reed, one end of which he prepared so that he could easily introduce it into the anus, and completely inclosed the pig's tail in this reed, in order to withdraw it without pain. For this purpose he attached to the tail, by the end which projected from the anus, a stout wax thread which he passed into the reed. With one hand he pushed this form of canula into the rectum, and held the cord in the other, to prevent the tail being pushed in still further. He succeeded in completely inclosing the tail, and promptly relieved the patient."[1]

A punishment for adultery among the Greeks is said to have been the introduction into the rectum of a peeled radish, covered with hot ashes; and cases in which patients have fallen upon sharp and fragile objects, such as the wooden pickets of a fence, which have broken off and remained in the rectum, are on record.

The list of foreign bodies which have been lost in the rectum by ignorant persons, in attempts to check a diarrhœa or to prevent the descent of piles or prolapse, is a very long one, and includes such substances as bottles, sticks of wood, and round stones, some of them of a size relatively enormous; and the use of the rectal pouch by criminals for the purposes of concealment is well known to the police.

In the Museum of Anatomy and Pathology, at Copenhagen, is a longish oval flat stone, about $6\frac{3}{4}$ inches long, $2\frac{1}{2}$ inches wide, $1\frac{1}{2}$ inches thick, and weighing nearly two pounds, which a patient in Bornholm introduced into his rectum to prevent prolapse, from which he had for a long

---

[1] Hevin, p. 339.

time suffered. The stone was extracted by a surgeon, Frantz Dyhr, in 1756.[1]

Reali operated in 1849, in the hospital at Orvieto, on a peasant who nine days previously had introduced a piece of wood into the rectum for the purpose, as he said, of economizing his food, and preventing it from passing out too quickly. He had violent pain. On exploration, the finger could feel the base of the piece of wood lying in the hollow on the sacrum, and surrounded by the broken mucous membrane. As repeated attempts at extraction led to no result, Reali made an incision in the right iliac region, and found that the foreign body lay in the sigmoid flexure, which it had dilated and pushed to the middle line nearly as far as the umbilicus; he incised the intestine, removed the foreign body, and closed the intestinal wound by Jobert's method. The patient was treated by purgatives (!) and had entero-peritonitis and abscess in the iliac fossa, but recovered, and two years afterwards was in perfect health. The foreign body was a piece of chestnut wood of the shape of a truncated cone, 10 inches long, and about $3\frac{1}{2}$ or 4 inches in diameter.

A little case with very ingenious housebreaking and other thieves' instruments was found by Dr. Closmadeuc at the necropsy of a man if the prison at Vennes. The man had died of acute peritonitis, from which he had suffered seven days. During his illness, a hard, rather large body was felt in the left side of the hypogastrium; he said that it was a piece of wood containing money, which he had introduced into the rectum; this, on exploration in the mean time, was found empty. On section, the case, which was cylindro-conical in form, lay in the transverse colon, with its apex directed towards the cæcum; it was of iron, and was wrapped in a piece of lamb's mesentery; it weighed about 23 ounces, was about $6\frac{1}{2}$ inches long, and $5\frac{1}{2}$ in circumference, and contained 13 tools and some coins.[2]

"A monk, desiring relief from a severe colic from which he was suffering, was advised to introduce into the rectum a bottle of Hungary water, in the cork of which there was a small opening, through which the water gradually distilled into the intestine (these bottles are usually long). He pushed it so far that it entered the rectum altogether, whereat he was greatly astonished. He could neither have an evacuation nor receive an enema, inflammation and death were apprehended. A midwife was consulted in order to see whether she could introduce her finger and extract the bottle, but she was unable to do it. Forceps, a ripping-iron, and anal speculæ were useless. It could not be broken; this would have been more disastrous as the pieces of glass would have wounded him. Finally, a little boy, eight or nine years old, was found,

---

[1] Bull. de la Soc. de Chir., 1878, p. 660.

[2] London Med. Record, Dec. 15th, 1878. Abstract of Studsgaard's paper read before Soc. de Chir., Paris, Oct. 9th, 1878.

who introduced his hand, **and had sufficient address to cure the** good monk."[1]

A depraved sexual appetite has been mentioned as accounting for the presence of many foreign **bodies.** It is known that sexual orgasm **may** be excited by stimulating the reflex power of the rectum, and it is probable that at the moment when the orgasm is at its height, the body used to produce it is allowed to escape from the hand and is lost within the bowel. **This is a habit** which will never be acknowledged by its victims, **but which** may often be assumed **to exist** by the surgeon in depraved patients. The bodies used for this purpose are generally smooth, long, and round, such as glass bottles, and pieces of wood. The following case is one in point, and the age of the patient is suggestive, for this vice is said to be more common in old men than in others—men whose physical powers have not kept pace with their desires.

"On the afternoon of March 1st, 1848, a young man consulted Parker with regard to his father, whom he had brought into the hospital. After beating around the bush and manifesting considerable shame and embarrassment, he stated that his father, named Loo, who was sixty years old, had passed the previous night in a house of prostitution. Overcome by drink and opium, the old debauchee conceived the strange notion of pushing a goblet, two and a half inches in diameter and three and a half inches long, into the vagina of his partner. During the night, while Loo was completely intoxicated, the woman attempted to revenge herself. She carefully introduced the bottom of the goblet into the rectum, placed the end of the opium pipe, which was a foot and a half long, into the goblet, and pushed it into the rectum. The goblet disappeared and had been retained twenty-four hours. A piece of the edge, about half an inch long, had been broken off by the friends in attempts at extraction. The glass was firmly fixed, and it was very difficult to pass the finger between it and the rectum. Parker, determining to break it, employed a cephalotribe and removed it **in pieces, taking care to protect the parts with cotton.** The most **difficult part was the extraction of** the glass, which was very irritating. It **was done,** but not without difficulty, by making it see-saw from side to side. Considerable hæmorrhage occurred, which was arrested with **sulphate** of copper and alum. The man recovered **in two weeks.**"[2]

It would be interesting to enumerate the foreign bodies which have been removed from this part of the body and the list would be startling from the strangeness of the different articles; but enough has been said to indicate that almost anything from a conical stone to a club or a coffee cup may **be encountered by the surgeon, and to indicate the** size of the **body which the sphincter will allow to** pass. Among them may be

---

[1] Mém. de l'Acad. de Chirurgie.
[2] Amer. Journ. of the Medical Sciences, 1849, p. 409.

mentioned beer glasses, mushroom bottles, wooden pepper boxes, wine bottles of all kinds, lamp chimneys, and a part of the wooden handle of a baker's shovel twenty-two centimetres in length.

A foreign substance may remain in the rectum for a considerable time and finally be expelled spontaneously as in the following case reported by Weigand.[1]

"A farmer, aged sixty-eight years, of a robust constitution, but somewhat stupid, introduced into the anus a cylindrical piece of wood for the purpose of relieving his obstinate constipation. However, he performed the manipulation so unskilfully that the piece of wood broke and remained partly within the rectum. All attempts made to remove the foreign body failed; two days later, he suffered from abdominal and lumbar pains, dysuria, and constipation. Weigand being consulted by the physician, recognized the symptoms of enteritis. As the introduction of a finger into the rectum did not demonstrate the presence of a foreign body, he restricted himself to combating the inflammatory symptoms and pain (calomel, enemata, narcotics, leeches). On the eleventh day a purulent, sanguinolent, fetid fluid was evacuated, after which the patient felt markedly relieved; but it was impossible to discover any trace of the piece of wood. Weigand then expressed serious doubts as to whether a foreign body was really contained in the rectum; but as the patient resolutely maintained that he continued to feel the piece of wood, renewed search was made, until the finger being introduced far in, encountered a rough, hard object which it was impossible to seize for want of proper instruments. As circumstances did not indicate a necessity for more active treatment, Weigand contented himself with giving the patient from time to time two or three spoonfuls of castor-oil which always produced the discharge of a small amount of muco-sanguinolent fæces. At this time the lumbar and abdominal pains again appeared more frequently, and, on the other hand, the patient's former appetite being gradually restored, he walked about and attended to light domestic duties. On the 31st day after the accident, after having taken three spoonfuls of castor-oil, he stated that he had an intense desire to go to stool, when in addition to blood and pus, the piece of wood made its appearance, 0.1357 m. long, 0.027 thick, cylindrical, serrated at the broken end, and roughened on the cylindrical surface; in fact it was the end of a pole with which bean-vines are propped. The patient recovered entirely without having been subjected to any further treatment" (Poulet).

Prognosis.—The prognosis in cases of foreign bodies will depend greatly upon their size and nature. A long body like a piece of wood may go so far up the bowel as to do fatal damage before its removal; and a fragile body like glass may cause fatal injury in the attempt to remove it. Again the prognosis depends in great measure upon the surgical

---

[1] Schmidt's Annalen, 113, 'iv., p. 95, 1862.

ability of the one in charge of the case. A little bungling in the treatment may at any moment change a case which promises well into a fatal one. **Finally, much** will depend upon the length of time during which the body has remained in the rectum; and it is not very uncommon for patients who have met with an accident in the practice of this secret vice to conceal the real nature of the trouble which they well understand till they are forced by suffering to confess. In this way a **week's** valuable time may be lost and a fatal amount of injury be done.

Treatment.—Each case of foreign body must be treated by itself, and besides a few general principles which apply equally to all cases, the surgeon will be left entirely to his own ingenuity. The one guiding principle should be to avoid doing fresh injury in the attempt at removal. Only the smaller and least friable of bodies can be removed without a previous dilatation of the sphincter under ether, and in most cases it will be advisable to incise the anus in the median line down to the tip of the coccyx as a preparatory measure to all treatment. This step will sometimes render a body movable which before was absolutely immovable and thus open the way for its extraction.

Having opened the way to the body, it may sometimes be removed by passing the whole hand into the rectum and seizing it. At other times forceps may be used with advantage and these may be of any shape which seems best to answer the purpose intended, including the obstetric forceps which have been found useful in many cases. If a bottle has been introduced with the mouth downward a string may be secured around the neck for the purpose of traction, but, unfortunately in almost all cases the position will be reversed. In cases of long bodies the lower end is not infrequently firmly wedged in the hollow of the sacrum—so firmly as to resist all efforts at dislodgment. Under such circumstances fatal injury may easily be done by the operator by persistence in the attempt.

Above all things the surgeon must avoid breaking such a substance as a cup, for experience has proved that after this has happened, removal without causing great injury is almost impossible.

Certain complications may at any time arise in the treatment of these cases, one of which is recorded by Desault.[1] A man, aged forty-seven years, entered the Hôtel Dieu, on April 17th, 1762, in order to have a crockery vessel extracted from his rectum, which he had introduced a week previously in order to overcome, as he said, his obstinate constipation. This vessel was a preserve jar, the handle of which was broken and the bottom detached. It was conical in shape, and three inches long; it had been introduced by the smaller end, which was two inches in diameter.

When the patient presented himself at the hospital, he had already made efforts to extract the foreign body, but an escape of blood and the

---

[1] Journal de Chir., T. iii., p. 177 (Poulet).

excessive pains had compelled him to suspend his efforts. The **upper** part of the rectum was infolded and invaginated in the vessel, and formed a very hard tumor, which filled it completely. The surrounding parts **were** inflamed, and this fact rendered the extraction more difficult. Desault made the patient lie upon the side, and then, separating the **intestine** from the **walls** of the vessel, he succeeded in seizing the latter with a **strong extractor, which he** pushed up as far as possible and which was held by **an assistant.** By means of this point of support, and with another extractor introduced in the same manner, he succeeded in breaking the vessel and in extracting it in small pieces without wounding the rectum. The operation was neither long nor painful, though it was necessary to introduce the extractors a large number of times. After all the pieces had been removed, Desault pushed back the inverted portion **of the** rectum by means of a charpie tampon six inches long and two and a half in diameter, which he pushed in altogether after having covered it with cerate. Below this were placed a large amount of charpie, several compresses, and a triangular bandage which supported the whole dressing. The dressing was renewed twice a day, on account of the relaxation **which** did not cease **till the** sixth day. Then the intestine no longer protruded when the **patient went to stool,** and such large tampons were not required. They **were discontinued** entirely after the tenth day, when the ruptures had **cicatrized, and the man left the** hospital entirely cured two weeks after the **operation.**

In cases where **a long body has become firmly wedged** into the lower end in the hollow **of the** sacrum, the proper **treatment consists** in opening the abdomen **and** this should be done after an attempt to **remove it** *per anum* has been continued a reasonable time, and before injury has **been** done in such an attempt. It is not necessary to describe the operation of laparo-enterotomy in this connection. The incision may be made either in the median line or in the groin. In the Surgical History of the War of the Rebellion, T. II., p. 322, there is a history of one such operation performed upon a sailor who had introduced a stone five and a quarter inches long by three wide. The colon had been perforated and the stone was removed from the peritoneal cavity by an incision near the umbilicus. The man recovered. The oldest known case[1] was reported by Réalli in the Bull. dei Soc. Médich., and Gaz. Méd., July, 1851, and is as follows:

CASE XXV.—"On the 18th of December, 1848, a peasant **was** brought in the hospital of Orviéto in a condition of extreme weakness. Nine days previously, having hit upon the ingenious idea that, if he prevented the discharge of food he could limit the quantity to be swallowed, he introduced a piece of wood into the rectum; all his attempts at re

---

[1] For this and many other interesting facts in connection with this subject the reader is referred to Poulet's work on "Foreign Bodies in Surgery." Wood's Library of Standard Medical Authors, 1880.

moval only served to push it in still further. The finger could only touch the end of the object and it was firmly fixed in such a manner as not to yield to any tractions which could be made upon it with such a slight purchase.

After the failure of all **attempts at removal,** the foreign body completely obliterating the intestinal cavity, and the patient being threatened with death from his atrocious sufferings, Réalli decided **to operate.** After having cut the abdominal walls on the left side, he could distinctly feel the stake in the descending colon. He desired to push it down to the anus, but the attempts proved unsuccessful, and he was compelled to incise the intestine. Only after this was done could he remove the body, which was ten centimetres long and more than three centimetres in diameter at the base. The point was rounded and very soft. No fæces were retained above the plug, but the mucous membrane was blackish, the peritoneal coat strongly injected, and the thickness of the intestinal wall markedly increased.

The wound in the intestine was united by a suture, which was applied according **to Jobert's plan.** The lips of the wound in the abdomen were united by means of an interrupted suture. Cold, and then iced applications were made over the operated region. Two doses of castor-oil were administered. There was a purulent discharge from the anus. During the first **few** days, the tumefaction of the walls of the intestines prevented the advance of fæces, and caused meteorism and vomiting. Three bleedings, two applications of leeches, and a few doses of castor-oil put an end to these symptoms, which had acquired an alarming character. The evacuations from the bowels were again passed on the fifth day. Towards the fourteenth day, the wounds had cicatrized. Two years later, the health remained perfect."

In a paper read before the Soc. de Chirurgie,[1] Studsguard, of Copenhagen, reports the following similar case:

CASE XXVI.—"J. F., footman, aged thirty-five years, was admitted on January 10th, 1878, to the Copenhagen hospital, and left cured on April 16th, 1878. The night before entering, he had introduced an empty mushroom bottle into the rectum, the neck of the bottle being uppermost, in order, as he stated, to relieve a rebellious diarrhœa, and on the morning of January 10th, he was obliged to call a physician, acute pains being experienced in the abdomen.

He was anæsthetized with chloroform, but the bottle, which, previous **to the** narcosis, had been felt in the rectum, slipped further up. He **was** exhausted by the passage and the increasing pains; vomiting of mucus. The bottle could be felt through the somewhat tense abdominal wall along the median line on **the left** side, the bottom being near the

---

[1] Bull. de la Soc. de Chir., 1878, p. 662.

horizontal ramus of the pubis. In the evening, profound narcosis and posterior linear rectotomy; the hand was introduced as far as the third sphincter, which was not forced, on account of its resistance. The bottle was then pressed from the outside down into the pelvis, but it descended in a loop of the intestine in front of the rectum. Immediately afterward, **antiseptic** laparo-enterotomy, through the median line, by an incision **ten** centimetres long, commencing at the umbilicus. A loop, which was thought to be the sigmoid flexure, was extracted, and the bottle was **then** slowly removed through an incision four centimetres long, which was made upon the orifice and upper part of the neck. The entire circumference was protected by sponges and compresses between the fæces, and the intestinal incision was closed by twelve to fourteen catgut sutures, according to Lambert's **method**, the peritoneal surfaces having been freely washed. In order to be on the safe side, the sutures were tied with three knots; the intestines were then introduced, and the abdominal wound united with eight silk sutures, tied alternately **with** knots and the figure of eight.

The operation lasted an hour.

The bottle **was seventeen** centimetres long, the diameter of the bottom was five **centimetres, that of** the neck three centimetres; the opening contained a **notch, which was** evidently of old date, about half a centimetre long, **and presenting cutting edges.** The recovery occupied a long time, and **the prognosis was uncertain for a** very protracted period, on **account of** a local **peritonitis with abscess** formation, which I incised both upon the median line and through **the rectum, upon** the posterior wall of which it projected. Gas began to pass two days after the operation; from the ninth day on, he had spontaneous **evacuations,** which were well-formed, and contained no traces of pus."

One other case of this kind has been placed on record[1] by Verneuil, **and** those four, I believe, make up all the literature of the subject.

CASE XXVII.—A man, aged forty-five, had been in the habit **of** stopping up his rectum to overcome an incontinence of fæces which had resulted from two previous attacks of dysentery. For this purpose, he used various large bodies, taking the precaution to tie to them a piece of cord, the ends of which were left hanging outside. But one day, he had no cord, and a cylindrical piece of wood, ten centimetres long and about eight in diameter, escaped into the upper part of the rectum, and **could** neither be forced down **nor** reached with the finger. All the **efforts** which were immediately **made by** a physician of the place only forced the body further from **the anus.**

In this condition, the patient entered the service of M. Verneuil. There were few **signs of retention, but the finger** could not be made to reach the foreign body; only with the hand on the abdomen could it be felt in

---

[1] Prog. Méd., May 15th, 1880.

the left iliac fossa. It was so high that linear proctotomy could give no assistance, and therefore laparotomy was decided upon. The plan of operation was the following: Through a small abdominal incision to search for the sigmoid flexure, in which the body was probably lodged; to draw the sigmoid flexure outward, and, if healthy, to incise it, remove the body, sew up the gut, and replace it in the abdomen. If, on the contrary, it was diseased, to stitch it to the abdominal wall, and make an artificial anus. But the foreign body was so fixed in the upper part of he rectum, with its long axis from behind forward, as to be immovable, and by reason of this immobility of the rectum, the former plan of operation had to be abandoned.

Fortunately, it was possible to dislodge the body from this fixed position, and M. Lucas Championnière, who, at that moment, practised the rectal touch, received it upon the end of his finger. While an assistant fixed the body by pressing on the abdomen, M. Verneuil endeavored to seize it with the forceps of Muzeux, or to fix it with a gimlet, but without success. Linear proctotomy was then resorted to, and M. Verneuil succeeded in moving the body with one of the blades of a lithotomy forceps, bringing it down, and seizing it with another pair of strong forceps. The instrument slipped many times on the bark of the wild cherry wood, and it was only after many long and painful attempts' practised with a very defective stock of tools, that the foreign body was finally withdrawn. It was followed by a discharge of very fœtid fæcal matter and a little blood. The result of the operation, thanks to the precautions taken during the manœuvres and the treatment subsequently employed, surpassed all expectations. The abdominal wound healed by first intention under Lister's dressing, and a soft-rubber catheter, kept permanently in the rectum, through which chloral was injected every two hours, prevented any complications in that part.

These four cases indicate with sufficient clearness the general rules which should guide the practitioner. The operation is applicable only to bodies high up in the rectum. The point of incision may be in the median line, over the sigmoid flexure in the left loin, or over what seems to be the most prominent point of the foreign body, wherever that may be. If the intestine is healthy, it may be closed and returned into the body. If not, an artificial anus should be made at the point of incision.

It is worthy of note that all of the cases thus far recorded have ended in recovery.

CHAPTER XIII.

PRURITUS ANI.

Pruritus generally a Symptom of some **other** Disease.—Description.—Causes.—Relation of Internal Hæmorrhoids, Fistula, Worms, Parasites, and Eczema to Pruritus.—Treatment of Eczema.—Herpes and Erythema.—Constitutional Conditions causing Pruritus.—Dependence upon Constipation.—Treatment of Constipation.—General Treatment of Pruritus.

PRURITUS ani—itching at the anus—is generally a symptom of **some** other disease such **as hæmorrhoids** or eczema, but it is often present in **a** marked degree **when no cause for** its existence can be discovered. It is an exceedingly painful **and annoying** affection, and one which will often tax the powers **of the surgeon to the utmost** for its cure. It is met with in both men and **women and** seems **to** be dependent upon no particular general state, being found in rich **and** poor, the overfed and underfed, the professional man of nervous constitution and the laborer, **alike.**

The disease is marked by an itching at the **anus** which is more or less constant, but is generally worse after the sufferer has **become warm in** bed at night. The itching causes an attempt at relief by scratching, and **the** scratching, though it may be controlled during the day, is generally practised unconsciously during sleep to an extent which causes laceration of the skin. The itching in bad cases, even when constant, is marked by exacerbations and remissions, and may cause an amount of suffering which is simply unbearable.

The disease is attended by certain changes in the appearance of the parts. The skin becomes thickened and parchment-like or else eczematous and moist from exudation. It may be red from the scratching, or there may be quite a characteristic loss of the natural pigment of the anus. In the latter case the skin becomes of **a** dull-whitish color, and this will oftened be noticed where the disease is of long standing and severe. The exudation may be **very marked where the** itching is slight, and may be attributed **by the patient to** trouble **within the** rectum instead of to its real **source.**

Causes.—The cause of pruritus may sometimes be easily discoverable and in such cases a cure rapidly follows its removal. For example, pruritus is often a symptom of internal hæmorrhoids and is easily and effectually cured by their removal. **Again it is often a** symptom or complication of a fistula with a small external opening such **as** may easily be overlooked in a cursory examination; and is cured by the ordinary operation and the consequent cessation of the discharge upon which it depends. It is often dependent upon the presence of the *oxyuris vermicularis* in the rectum and in every case these should be carefully looked for. If they are present they may generally be seen like small pieces of white thread between the radiating folds at the margin of the anus, especially at night when the itching begins. They may generally be eradicated by certain simple measures, the best known of which is an enema of limewater, or of carbolic acid, ℨ i.; glycerin, ℥ i.; and water, ℥ vij., injected after each passage. Turpentine and tincture of iron may be used for the same purpose and are both very effectual; but the parasites are much more easily removed in children than in adults, and I have had one case which was exceedingly intractable, and in which I have never been able to keep the worms from returning for any great length of time. A single examination should never be considered as proof of the absence of this parasite in an obstinate case of pruritus.

Instead of a parasite located within the rectum, pruritus is occasionally easily accounted **for** by the presence of pediculi. In such **a case the diagnosis and cure are alike easy.**

Again the parasite may be vegetable instead of animal, and **the itch**ing may be due to the disease known as *eczema marginatum*. In this case the diagnosis will rest upon **the finding of** the spores under the microscope in the epidermis scraped **from the** edge of the affected spot and moistened with glycerin. The most effectual remedy for this condition is a wash of equal parts of sulphurous acid and water frequently applied with a soft cloth, and gradually increased in strength, if necessary, up to the pure acid, which latter is, however, generally a painful application. Strong tincture of iodine applied with a brush is also an effectual remedy in eradicating the plant.

Pruritus may also be dependent upon other skin diseases, among which chronic eczema is perhaps the most common, and this is to be treated **exactly** here as elsewhere in the body, first by general measures directed to the constitutional state, and second, by local applications. The congestion and the thickening of the skin **must first be** remedied, and for this purpose very hot **water, compound** tincture of green soap, and if necesary a solution of caustic potash may be applied. The water, to be of any use, must be as hot **as the fingers can bear,** and should be applied to the part with a **soft cloth and held there till it** begins to **cool.** This may be repeated **half a dozen** times, but all rubbing should be carefully avoided **both during the** application and **in drying** the parts after it. This is **a favorite remedy** with most dermatologists; it should be used just before **going** to bed, and is often in itself sufficient to insure a good night's **sleep.**

If there be thickening of the skin from effusion, a stronger application than hot water will be **necessary; or** the solution of potash (gr. v.–℥i.) or liquor potassæ may be resorted to with caution. The formula **for the** compound tincture of **green soap is** the following:

℞ Saponis **viridis,**
    Olei cadini,
    Alcohol.. āā ℥i.
M.

It is a much stronger preparation than **the** simple green soap and also a much more disagreeable one, but it is very effectual and should be well rubbed into the part once a day. These remedies should **be followed at** once by soothing ointments, or lotions. A good ointment is **the ordinary** oxide of zinc made soft and applied gently, and one which is pretty certain to allay itching is that made of chloroform (ʒi.–℥i.). This soon loses its power by the evaporation of the chloroform **and** should on this account be kept in a wide-mouthed glass bottle, tightly corked, and should be frequently **renewed.** Another favorite application, and one which is very generally **effectual,** consists in a lotion of carbolic acid. The formula is:

℞ Acid. **carbolici** .... ℥ss.
    Glycerinæ .... ℥i.
    Aquæ .... ℥iij.
M.

This may **be** applied at night, and if found **to be too strong may be** diluted by the patient. In a more dilute form it may also **be continued** for a considerable time after all symptoms have ceased.

For the sake of those who have never encountered an obstinate case of this disease, but who are pretty sure at some time to have both knowledge and ingenuity taxed to the utmost, I will give one or two more formulæ which have been found reliable. The following comes from Allingham, and by it alone he has "**seen** a bad case cured in forty-eight hours."

℞ Liquoris carbonis detergens (Wright's),
    Glycerinæ .... āā ℥i.
    Pulv. zinci oxidi,
    Calamin. prep .... āā ʒss.
    Pulv. sulph. precip............................... ʒss.
    Aquæ puræ .... ad ℥vi.
M.

The part affected is to be thickly painted over with this once or twice a day and allowed to dry. **The** white precipitate ointment made soft with vaseline or glycerin is also a good application, and the following lotion, also from Allingham, will often work well in allaying irritation:

℞ Sodæ biboratis................................... ʒij.
Morph. hydrochlor ....gr. xvi.
Acidi hydrocyanic. dil.......................... ℥ss.
Glycerinæ .... ℥ij.
Aquæ ....ad ℥viij.
M.

This should be applied to the part four or five times in the twenty-four hours. Dr. Bulkley[1] has also recommended the following as being useful, and I have often found it so.

℞ Ungt. picis .... ʒiij.
"   bellad................................... ʒij.
Tr. aconit. rad................................. ʒss.
Zinci oxidi .... ʒi.
Ungt. aquæ ros................................. ʒiij.
M.

An ointment of chloral and camphor, a drachm of each to the ounce, is also at times effectual in allaying itching.

There are two other skin diseases either of which may be the cause of pruritus—herpes and erythema. Herpes at the margin of the anus is the same as when seen on the lips. In the latter case it heals spontaneously, in the former a dressing may be necessary. This may consist simply of a dry powder such as zinc or bismuth, or of one of the lotions already mentioned. Erythema will be found chiefly in fat people where it is due to contact of the opposing cutaneous surfaces. It also is best treated by the application of dry powders, and by separating the opposed surfaces by a layer of dry sheet lint or old muslin.

These are the most palpable and perhaps also the most common causes of pruritus, but there are many cases in which the cause is not so easily discoverable, because it is a constitutional and not a local one. Where no local cause can be detected, a careful inquiry must be instituted with regard to the patient's general health and habits. If chronic constipation be present, this must first of all be overcome, for this is in itself an efficient cause for the disease. The treatment of chronic constipation is by no means a simple matter. It may be begun with a purgative such as three compound cathartic pills, for the sake of opening the way for future treatment, but here the administration of purgatives should end, for their repeated administration is calculated to do harm rather than good, by substituting an occasional over-action for the daily one which indicates a healthy state of the intestinal tract. The following suggestions may be found of use in the treatment of this condition which is one that must be overcome at the commencement of the treatment of any rectal affections with which it may be associated.

Constipation may be due to deficient action of either the small or the large intestine, and this deficient action in either case may be the result either of deficient secretion or deficient nerve power.

---

[1] The Med. Record, December 18th, 1880.

Deficient secretion is very apt to be associated with hepatic disturbance, and is marked by dull headache, bad taste in the mouth, viscid secretion from the buccal **glands, etc.** This is a condition pretty sure to be aggravated by cathartics, **for the** reason that the temporary increase in secretion which they cause **is** followed by a corresponding decrease, **which** serves only to make **the patient worse** than before. For **the** purpose of increasing the natural **secretion of** the small intestine, the fruits containing citric acid, **such as** oranges; and other fruits, such as figs and apples, when the **patient can** digest them, all serve a **good** purpose. Water is also an excellent remedy, and two tumblerfuls **of it** taken in the morning will often be very beneficial. To it may **be added a** slight saline, which decreases its capability for absorption ($\mathfrak{z}$ ss.–O.i.), **and,** therefore, increases the peristalsis; and the addition of a single grain **of** quinine is said **to** greatly increase the effect.[1] This treatment, if patiently persisted **in for** a few weeks, will generally be followed by a good result.

Deficient innervation will be found in most cases of constipation in old people, people of sedentary habits, and those who have little exercise. **It is** generally attended by deficient action of the skin and a sallow complexion. In such **cases water will** be found only to weaken the digestive power, unless it **can be combined** with a different mode of life and abundance of **out-door exercise. Cold** bathing, however, cold against the spine and **abdomen, plenty of exercise in** the open air, and nux vomica will generally be found **to give relief.**

In constipation dependent upon the **large intestine, the** trouble will generally be found **to** be due to deficient innervation **rather than to** any lack in the secretion. It is best treated by keeping the **rectum** empty, by nux vomica, **or** belladonna in doses sufficient to cause **dryness of the** throat, and by electricity. The latter **should be in the form of the** Faradic current, one pole being placed over the spine and the other **passed** up and down along the track of the colon.

Infantile constipation may be due, as pointed out by Jacobi, to the disproportionate length of the sigmoid flexure. In children it is not unusual to find two, or even three, flexures in the lower part of the colon, in which the fæces may remain until they become hard and friable, and when such an anatomical formation is associated with a deficiency of the intestinal secretion, a very obstinate constipation, and even impaction, may result. In such a case oat-meal is to be given in preference to tapioca, rice, or barley, and with it an abundance of water. Purgatives should never be administered except in extreme cases, enemata being preferable.[2] Fæcal accumulation is not very uncommon in young children.

In chronic constipation, the patient should first of all be instructed to have a regular time for the daily evacuation, and the best time for this purpose is immediately after breakfast. The time being fixed, the patient

---

[1] Thompson, New York Med. Record, May 5th, 1877.
[2] N. Y. Med. Record, Sept. 25th, 1880.

is to go to the closet whether the **desire for a passage** be present or not, and pass a certain time upon the commode. I generally recommend the time immediately after the morning meal for this purpose, because the breakfast itself often **acts a stimulant** to this function, especially in those **in the habit** of taking **a morning cup** of coffee. If the patient be a man **in the habit of** smoking, the first few whiffs of smoke often act in the **same way; and** there are many men to whom the morning cigar or cigarette **is an essential to the daily evacuation.** In such a case it must be a **very decided opponent of the weed** who would object to its continuance **in moderation.**

If the plain cold water taken in the **morning has** no effect, the mineral **waters** may be tried in its place with great advantage; and the patient may select the one most agreeable to **the taste** and which most effectually **accomplishes the** desired end. The morning meal **may** consist of whatever **the** patient most desires, **but** a dish of oat-meal or coarse cracked wheat and milk should always be an essential **part** of it.

A laxative bread may be made of equal parts **of coarse** Scotch oatmeal, whole wheaten flour, and coarse ordinary **flour,** with yeast or baking power. This may be eaten once or twice daily.[1]

I have almost always found that where perfect regularity **in the daily** life with regard to eating **and** exercise can be established, **the function** of defecation will also be performed regularly, provided the **diet be plain and rather** coarse in quality. **To** have a copious, well-formed evacuation **it is** necessary first of all that the diet should be composed of substances which leave a considerable quantity of waste, and chief among these are **the coarser grains and the** vegetables. **In** women **a certain** regulated amount of daily out-door exercise should **be** insisted upon, in spite of all excuses and professions of disability. If necessary, this may be small at first, and gradually increased; and in **a woman** who has lost the habit, and, **perhaps,** almost **the** power of walking, considerable tact and **firmness on** the part of the physician may **be required to** carry out this part of the treatment, but it will be found to be **care** well spent.

In addition to **these** dietetic and **hygienic** rules, certain medication may and often will be found necessary. **This** should be of the mildest possible kind which will accomplish **the** object. A pill which I have found to act very effectually and pleasantly **under** these circumstances is made after the following formula:

℞ Pulv. aloes soc...............................gr. iss.
   Ext. nucis vom ....gr. ss.
   Ext. belladonnæ..........................gr. ¼.
M.

One of these should be taken at **bed-time, and will** generally be followed **by an** easy passage on the following morning. If this does not **work satisfactorily,** various other remedies may be substituted, amongst

---

[1] W. H. Taylor, Lancet, May 31st, 1879.

the best of which is the compound licorice powder, the rhubarb and soda mixture, or the dinner pill; the object being to find one among the many laxative preparations which, without causing pain or diarrhœa, will give an easy and natural evacuation of the bowels once every day.

The use of enemata for chronic constipation should not be commenced till all other means have failed, for the reason that when once the bowel has become accustomed to this form of stimulus it will be found very difficult to discontinue its use. In some cases, however, their employment may be a necessity and they are always much less harmful than purgatives. Instead of the ordinary enema of soap and water, the introduction of a harmless foreign body into the rectum will sometimes excite peristalsis. Small fragments of soap or of candles are preferred by many for this purpose to fluid injections.

In cases where enemata have lost their power from prolonged use my own practice is to resort to the use of a long rectal tube two or three times a week; but this should not be trusted to the patient for fear of accidents. Most patients will find it impossible to introduce them easily and will not care to make the attempt. With a long flexible tube of small calibre a pint or more of water may easily be thrown into the sigmoid flexure and colon and the bowel be thoroughly emptied.

Another not infrequent cause of pruritus is derangement in the function of the liver. This may or may not be associated with the constipation which we have just considered. It must be treated by general dietetic measures, the dilute mineral acids, occasionally by doses of podophyllin, active out-of-door exercise, and cold and friction applied to the hepatic region. In women uterine disorders must be looked for and cured before very much will be accomplished in the treatment of pruritus; and in women also the urine must be examined for sugar in obstinate cases, for diabetes will sometimes give rise to incurable pruritus.

In case none of these causes can be found to account for the itching, errors of diet must be searched for, and corrected when found. Anything like excess in smoking or in alcoholic drinks will keep up the disease, and in men these habits must be carefully regulated if indulged in at all. The disease will sometimes be encountered in stout full-blooded persons who live well and perhaps incline to the gout, and who show no other signs of disorder. In such, active exercise and plainer living with cold bathing of the part at night and morning and the use of a lotion of carbolic acid will often effect a speedy cure. On the other hand, the disease may be present in exactly the opposite class of persons, the overworked and worried professional or business man, and it is in this class of cases alone where the itching seems to be purely a nervous symptom that arsenic is indicated. It may be combined with quinine and cod-liver oil and carried up to its full physiological effect. As a relief for the intolerable itchings at night, Allingham recommends the introduction of "a bone plug shaped like the nipple of an infant's feeding bottle, and with a circular shield to prevent its slipping into the bowel." Its benefit is explained by the pressure it exerts upon the terminal filaments of the blood-vessels and nerves of the anus.

In this way then the physician must undertake the cure of a case of pruritus ani; and not by the administration of any single lotion or ointment to allay the itching which is but the symptom of some local or general condition. In every case the cause must be found and removed if success in the treatment is to be gained. I know of no disease of the rectum or anus in which there is a better chance for the practitioner to show his general knowledge and skill. If a case be undertaken in this way, and the treatment be intelligently followed by both doctor and patient, a cure may generally be effected; sometimes in a very few days, but at others only after prolonged effort and many discouragements. The prognosis should, therefore, be guarded at the outset lest the patient be led to expect a too speedy relief, and in some cases, in spite of the best of care, the disease will frequently return and the patient can scarcely at any time consider himself as perfectly cured.

CHAPTER XIV.

SPASM OF THE SPHINCTER.—NEURALGIA.—WOUNDS.—RECTAL ALIMENTATION.

Spasm Without Other Disease.—Cases.—Authorities.—Symptoms.—Treatment.—Neuralgia.—Cases.—Diagnosis.—Treatment.—Wounds.—Complications.—Spontaneous Rupture.—Treatment of Wounds.—Alimentation.—Physiology of Absorption.—Nutritive Enemata.—Nutritive Suppositories.

SPASM of the sphincter without the presence of any other rectal affection is undoubtedly rare. Its general character may perhaps best be shown by the citation of the following cases.

CASE XXVIII.—Physician, aged twenty-eight. The patient was a man decidedly given to thinking about his own health, and though generally well, not at all robust. He came to me complaining of a sense of discomfort about the rectum, accompanied by difficulty in defecation. The discomfort seldom amounted to actual pain, and he had noticed that when he was away on his summer vacations he was always better and in fact perfectly well. Nevertheless, the trouble in defecation had increased so markedly during the past few months that he was fully convinced that he was suffering from actual stricture.

An attempt at digital examination caused the most exquisite suffering, forcing the patient to cry out in agony, and yet there was entire absence of any lesion.

The treatment was based upon the fact which he had himself noted, that when his general condition was improved the local trouble ceased; and the patient was cured by purely general measures looking toward the building up of the system.

CASE XXIX.—Professional man. Age, thirty.

In this case also the only symptom complained of was pain on defecation, sometimes severe, sometimes slight. The history given pointed so strongly toward the existence of a fissure that I etherized the patient, fully expecting to cure him by stretching the sphincter. He was entirely cured by stretching the muscle, but, to my surprise, a most careful examination revealed no disease; and, being dubious myself about the existence of spasm without fissure, the examination was a very thorough one. This patient was also a man of sedentary habits and of rather a nervous character.

The following case is taken from Syme, and is characterized by him as a remarkable instance of the affection.[1] "I was asked to see a gentleman, about sixty years of age, who stated that, a few weeks before, after sitting out a long debate in the House of Commons, he had felt extreme difficulty in evacuating the bowels, having previously for several years experienced more or less uneasiness from this source; that he had consulted a physician and surgeon in London, who prescribed laxatives without affording relief; and that his complaint had continued so as at length to confine *him to bed*. I proposed an enema, which was at once objected to on the ground that the anus would not admit the smallest-sized tube. Suspicion being thus excited, the anus was examined and found to present the characteristic features of spasmodic stricture. Having explained my views of the case, I gently insinuated the narrow sheath of a *bistoury caché*, which I happened to have with me, and then expanding the blade, withdrew it, so as to make an incision on one side of the orifice. A copious stool immediately followed, and the patient was at once completely relieved from his complaint."

With regard to this much disputed affection, a citation of authorities may be useful. Syme[2] believed that spasm existed as an independent condition without morbid change; that, though there could be no doubt that spasm and fissure frequently existed together, it was not reconcilable with the facts met with in practice that spasmodic stricture was always of secondary origin and dependent upon the fissure. He says: "In a considerable number of cases, I have found the sphincter firmly contracted without any perceptible fissure or abrasion of the surface."

Mayo describes spasm of the sphincter as a kind of cramp which often comes on suddenly, sometimes at night during sleep. The paroxysms may occur daily or two or three times a year; and the attack may come gradually and cause uneasiness for two or three days, and then pass away, or its coming and going may be sudden. He says: "There are cases in which the disease produces long-continued and permanent suffering; in which the anus becomes permanently contracted and hardened, constituting, therefore, a permanent stricture, and generally combining both permanent and spasmodic contraction. The motions are passed with an effort and with pain, and all the common symptoms of stricture of the rectum are present.

Allingham[3] says: "Spasm of the sphincter has been said to be the cause of impaction, but I have more often thought the reverse was the case; and the impaction the cause of the spasm. I must, however,

---

[1] Diseases of the Rectum. Edinburgh, 1838, p. 138.
[2] Loc. cit., p. 134.
[3] Op. cit., p. 210.

acknowledge that spasm is often the cause of the constipation which is the **forerunner of** impaction. In impaction, spasm of the sphincter always exists; in some instances to such a degree that, when the patient strained, I have observed the anus protruded like a nipple, and an injection returned in a fine **stream as** if **coming** out of a squirt. I have certainly met with **cases of idiopathic** spasm of the sphincter usually in elderly, nervous, single **women, and** though no impaction was present, costiveness was."

Quain' concludes that "where pain, brought **on by fæcal** evacuations and continuing after them, happens to be present, the **fault—the** morbid condition—is not in the sphincter, but in the skin **or** mucous membrane covering it, and that the division of the muscle **is** not required in order to remove the patient's suffering." In **other words, that spasm is always** dependent upon fissure. Boyer² treats **of** "constriction **with fissure**" and "constriction without fissure."

Dupuytren³ says: "The gravity of this affection (fissure) depends chiefly on the painful spasm of the sphincters; the fissure is only an accident, as is proved by the existence of painful spasm without fissure, **which, according to** well-known surgical authorities, is found in proportion to the other **of one to four."** And, "the spasmodic constriction is the true lesion, and **the fissure only an** epiphenomenon." Sir B. Brodie⁴ held the same **views.**

The **symptoms of spasm of the sphincter are** pain on defecation and for a time after; **more or less uneasiness about the anus,** especially when sitting; fulness in the perineum; **often more or less trouble** with the bladder, as **shown** by frequent micturition, **sometimes attended by** smarting in the urethra and constipation. **The disease is** generally attended by exacerbations and remissions. A digital examination of the anus is always painful, and the contraction may be so great as to leave hardly a **trace** of **the** anal orifice. Any anxiety or distress of mind, a generally irritable nervous condition, and everything which has a tendency to irritate the rectum, or the parts around, will aggravate the complaint. It may easily be confounded with the affection next to be described, neuralgia, but is generally distinguishable from it by the marked dependence of the pain upon the act of defecation, which is not seen in neuralgia without spasm.

The treatment consists in attention to the general health of the patient, in allaying any nervous extitement, in the administration of a cathartic to empty the **bowel** when **the** spasm **is** present; and in ano**dyne** injections, such as, **for** example, of twenty drops of laudanum in

---

¹ The Diseases **of** the Rectum. London, 1854, p. 167.
² Traité des Maladies Chirurg., etc., fourth edition, t. x., p. 139.
³ Leçons orales de Clinique Chirurg., t. iii., p. 284.
⁴ Lectures on Diseases **of** the Rectum. London Med. Gaz., vol. xvi., p. 26.

an ounce of water. Suppositories may cause renewed irritation. Even in the more aggravated form, the disease will often yield to such measures as this, but, if it does not, a cure may always be effected by forcible dilatation of the sphincter under ether. If the patient will not submit to this, the next best thing will be found to be the introduction and retention of a bougie.

Neuralgia.—Neuralgia of the rectum is generally met with in nervous people, especially females, such as are subject to neuralgia in other parts of the body. The following cases show its general character.

CASE XXX.—Professional man, age 49. The patient was slight and pale from sedentary habits, but was generally well. Thirteen months before consulting me he was operated upon for fissure, and after the operation he had for some time been entirely well, but he now has what he describes as a dull, wearing pain in the rectum, coming on while at his daily work, lasting a longer or shorter time, sometimes all day, but generally passing away after he has reached his home and become quiet and rested. He has noticed that the pain has a direct connection with the state of his general health, and that, when he is away from his work and rusticating, he is entirely free from it. The pain is no greater at the time of defecation than at any other, and is never so severe as to be unbearable. A careful examination of the part failed entirely to show any lesion.

CASE XXXI.—Woman, aged 65, married. This patient had been treated for fissure, for ulceration, and for coccygodynia, and had refused to submit to excision of the coccyx. Her general health was fair, but there was decided gastro-intestinal disturbance. The pain of which she complains has been present for about eighteen months. She suffers chiefly when sitting, sometimes finds it impossible to lie upon her back, and is apt to have a sharp twinge when she starts suddenly from her chair. The pain is no worse at defecation, is not increased by pressure upon or movement of the coccyx, and is entirely unconnected with any lesion of the rectum or anus. The greatest sensitiveness to touch seemed to be located well within the sphincter, upon the posterior wall of the bowel. There was enlargement of the womb and misplacement.

From these cases, which are both good examples of mild forms of the affection, it is evident that the disease may vary greatly in its severity. In some persons, it will cause the same suffering as the most intense neuralgia elsewhere. The pain is apt to be paroxysmal, but may be continuous, and is independent of the act of defecation. In cases of well-marked periodicity, a malarial element should be looked for, and the disease may be a manifestation of the gouty diathesis. In the former case, quinine, and in the latter, colchicum may be of the greatest service. In all other cases, the treatment will often be found unsatisfactory, and is to be conducted on general principles. The first care should be for the general health, the second for the regularity of the bowels, and after this, local applications of cold **water**, ointment of belladonna (ℨ i.-ℨ i.), and blistering over the sacrum **may be** tried. Besides this local treatment, the case must be managed exactly as would be a case of neuralgia in any other part.

The diagnosis from coccygodynia and from spasm must both be made with care.

Wounds of the Rectum.—**Wounds of** the rectum may be either **contused** and **lacerated or incised.** The latter most frequently result **from** surgical operations, and may be intentionally inflicted as in the operations for fistula, or for the removal of tumors, or the result of accident, as in the operation for stone. Contused and lacerated wounds are generally the result of accident, and perhaps the most frequent cause of such **an** injury is the perforation of the bowel with an enema tube, a **bougie, or** a urethral sound. The gravity of this accident will depend upon **two** factors—whether the perforation of the bowel is above the peritoneum, and whether the enema has been deposited in the perirectal tissues. **The** latter complication will be followed by abscess and peritonitis, and **will result** either in death or in stricture and fistula. If the wound be uncomplicated by **the injection,** the mere puncture may heal spontaneously. It is oblique from **below** upwards, and this greatly favors spontaneous healing without fæcal extravasation.

Esmarch **has met with four cases of** this injury, none of which were fatal though attended **by much local trouble.** Velpeau describes eight cases, six **of** which ended **fatally. Passavant** observed five cases, one fatal. Chomel has had two fatal results. **There are** two preparations in St. Bartholomew's Hospital showing the results of this accident, one in a man, the other in a child ten years of age (Esmarch).

Besides these most common injuries, many others **may be enumerated.** The person may fall upon a sharp body, as **the** point **of an** umbrella (Bushe[1]), may be caught upon the horn of an animal (Gundrum;[2] Ashton), or may be impaled upon a spike (Esmarch[3]).

In such cases, the accident may be immediately fatal from collapse, and the wound in the rectum may be complicated by a wound **of the** peritoneum, or of any of the adjacent organs. The body which **has** done the injury may also be so firmly implanted as to require great force and an anæsthetic for its removal.

The rectum is not infrequently lacerated in child-birth, and although such wounds are generally of slight extent, Bushe[4] relates a case in which the child's head was passed through the anus. It has also happened that, in **a** violent effort to expel a mass of hard fæces, **the** rectal wall has given

---

[1] Op. cit., p. 80.
[2] Detroit Lancet, Oct., 1879.
[3] Op. cit., p. 43.
[4] Op. cit., p. 80.

way. Mayo[1] relates one such case in a woman of forty, in whom the rupture was in the recto-vaginal septum, about two inches within the bowel. Ashton[2] reports a similar case and Bushe[3] another. Such a rupture may be either vertical or transerse, will be marked by sharp pain at the moment of the accident, and will be followed by a discharge of blood. It is doubtful whether it ever occurs without previous disease of the wall of the bowel.

The consideration of gun-shot wounds comes more properly within the scope of military surgery. They are always complicated with injuries of other parts, and are generally fatal from extravasation of urine or fæces.

The complications which may attend a wound of the rectum have already been hinted at. They are hæmorrhage, either primary or secondary; fæcal infiltration; purulent infiltration; peritonitis; emphysæma; hernia; invagination; and later, stricture and fistula. When fæces are forced out of the rectum into the adjacent tissue, diffuse inflammation and gangrene will probably result, and the condition must at once be met by free incisions and free drainage, as has been described in the chapter on abscess. The danger of fæcal infiltration may be lessened by a diet which shall prevent fluid passages, and by the free use of opium. A dilatation or a free division of the sphincter is also to be recommended, so that a free outlet may be accorded to the contents of the bowel.

Emphysæma, as a result of a perforation, is generally confined to the perineum, but may be diffuse.[4] It is very apt to be fatal from diffuse inflammation and septicæmia, due to the putrid nature of the gas, and is to be met by free incisions.

Wounds of the bladder or urethra communicating with the rectum are to be met by providing for the free issue of the urine. This may be done by catheterism, by aspiration, or by free division of the sphincter.

Where none of these complications exist, a fresh wound of the rectum may close by first intention, and an effort should always be made to secure this by rest in bed, by emptying the bowel, and keeping it empty by frequent washings with water, and by the use of opium. Healing by granulation will, however, be the rule. In some cases, such, for example, as laceration in child-birth, sutures may be at once applied.

Alimentation by the Rectum.—The fact that certain substances may be absorbed into the general circulation through the mucous membrane of the rectum has been abundantly proved by physiological experiment and clinical experience. The close anatomical resemblance between the inverted follicles of the rectum and the intestinal villi render an analogy

---

[1] Op. cit., p. 13.
[2] Op. cit., p. 152.
[3] Op. cit., p. 69.
[4] Lancet, Jan., 1860, p. 89.

in function extremely probable without experimental proof; but such proof is easily obtainable. A solution of salt, in the proportion of one part to eighty of water, injected into the rectum will disappear completely in the course of an hour—so completely, that an evacuation at the end of that time will be found to contain no more than the usual quantity.[1] The fluid extract of rhubarb may be detected in the urine in about an hour after being injected into the rectum by the characteristic red color caused by the addition of caustic potash.[2]

Bouisson,[3] after injecting beef-tea into the rectum, found the lacteals charged with fluid. Savory,[4] in his experiments on the relative rapidity of this absorption by the stomach and rectum, **found that** strychnia in solution acts more quickly by the rectum, **but that in powder** the relation was reversed. Quinine should be given in **larger doses** by the **rectum than** by the mouth, while chloral and belladonna are readily absorbed **by** the former. Curare, on the contrary, acts more quickly by the **rectum** (Cl. Bernard). Cubebs and copaiba both act equally well **by the rectum;** and water charged with sulphuretted hydrogen gas is rapidly eliminated in **the dog** by respiration, as may easily be proved by the usual test with **a salt of lead.**

The fact of **absorption being** admitted, the next question is as to the power of digestion **before absorption,** and upon this point there has been considerable discussion **of late, and** much difference of opinion.

The theory **that the follicles of Lieberkühn** may take on a vicarious action, and **secrete a** digestive fluid **under the** stimulus of albuminous food placed **in** contact with the **epithelium has its** upholders, but has never been absolutely proved.[5]

Another theory is that food introduced into **the rectum excites secretion by** the gastric and intestinal follicles, and that, **in** the absence **of** food **in the** stomach the digestive fluids thus secreted pass down into the rectum and there act upon the injected materials.[6]

Still another theory is that, instead of digestive fluids descending **to act** upon the food, the latter ascends to be acted upon by the fluids **in the** small intestine, and is there fitted for absorption.[7] This theory has

---

[1] Liebig: Animal Chemistry.

[2] Smith: Supplementary Rectal Alimentation, and Especially by Defibrinated Blood, as Applicable to a Large Range of Cases in which Nutritive Enemata have not Heretofore been Employed. Read before the N. Y. Acad. of Med., February 20th, 1879.

[3] Dict. Encyc., Art. Rectum.

[4] Gaz. Méd., 1864.

[5] C. H. Stowell: **Is Food Digested in the** Rectum? The Medical Advance, January, 1879.

[6] A. Flint, Trans. N. Y. Acad. of Med., Feb. 20th, 1879, **and** " Cases Illustrative of Rectal Alimentation, **with** Remarks," Amer. Practitioner, Jan., 1878.

[7] H. F. Campbell: Rectal Alimentation in the Nausea and Inanition of Pregnancy—Intestinal Inhaustion an Important Factor and the true Solution of its **Efficiency.** Trans. Gynæcological Soc., 1879.

grown out of certain facts which have recently come to light regarding the reversed peristaltic power of the bowel. Injected matters such as blood and milk colored with madder may be found on *post-mortem* examination evenly distributed over the coats of the intestine for a considerable distance above the rectum, and this is in itself a simple argument in proof of a reversed action of the bowel. But there are many stronger ones. Dr. Battey, in an article on the "Permeability of the entire alimentary canal by enema, with some of its surgical applications," details some experiments of his own by which he succeeded, in the cadaver, in passing an injection from the rectum through the whole length of the digestive canal, and out of the mouth. He also gives certain cases in which what he has accomplished on the dead subject has been done by nature in the living patient. In this way he accounts for the undoubted fact that patients will often complain of tasting in the mouth a substance like castor-oil which has been administered by the rectum; and for the fact that the ingredients of an enema, or a suppository, have occasionally been actually vomited. Dr. Harris, of Milledgeville, Ga.,[2] has recently reported a case in which clear beef-tea enemata were vomited after an operation for ovariotomy.

Jaccoud records a case of fæcal vomiting which occurred in his wards at the Lariboisière, in 1867, in a young woman who was admitted with hysterical convulsions. For eight days this person, at least once, and sometimes twice, in the twenty-four hours, vomited veritable fæces, dense, solid, cylindrical, of a brown color, and with the normal fæcal odor, coming evidently from the large intestine. Jaccoud witnessed the act himself, and so also did Dieulafoy, and he characterizes it as actual defecation by the mouth. Apart from the passing disgust which followed the act, the patient ate as usual, and continued in her ordinary health, except in the absence of normal action of the bowels. All possibility of deception seems to have been rigorously excluded. Within a fortnight the woman was seized with grave typhoid fever and died. Careful exam-

---

[1] Virg. Med. Monthly, vol. v., 1878.

NOTE.—Dr. Battey makes a claim to priority in having established the "entire permeability of the canal to enema," which though no doubt perfectly just as far as his own experiments go, is refuted in the Med. and Surg. Hist. of the War, Med., vol. ii., p. 836, foot-note, by the following references.

A. Guaynerius: Tractatus de fluxibus. Cap. 2, Lyons Ed., 1534. History of a man who vomited suppository placed in the rectum.

J. Matthias de Gradibus, Practicia de Ægritudinibus stomaci. Cap. 5, de vomitu, fol. 213. Venice E1, 1502; History of girl who constantly vomited her suppositories even after they had been tied with a string to keep them in the rectum.

Morgagni, references to numerous similar cases.

[2] Quoted by Campbell, loc. cit.

ination of the body disclosed **no mechanical** obstruction whatever in the intestinal canal. The ileo-cæcal valve **was** normal.[1]

By one of these three explanations it is attempted to overcome the obvious physiological objections to rectal alimentation which arise from the facts that albumen is not diffusible, or if so at all, only very slowly and in very small quantity; and that to be absorbed it must first be changed by digestion into albuminose. Another and very practical way of overcoming the obstacle has been suggested by Dr. Chadwick[2] which consists in **placing** the enema directly into the small intestine by means of an **aspirator**—a procedure which might be considered as not unattended with danger. Michel[3] has found the obstacle insurmountable and has, therefore, come to a conclusion unfavorable to the absorption of the nutritive matter of the substances injected.

The theoretical difficulty of the digestion of albuminoid **substances** has been practically overcome in a very simple manner which **is** nothing more or less than artificiallly digesting such substances, either before or after their administration, by mixing with them a certain quantity of pepsin or freshly prepared pancreas. Catillon[4] has performed the following instructive **experiments in** this connection. He fed two dogs **for** two months with injections **of eggs**. The first had eggs only and lived with difficulty **and with considerable loss** of weight; the second had glycerin and pepsin **mixed with the eggs and** lived in an apparently normal manner, the weight and **temperature remaining** constant. After thirty-seven days the pepsin was stopped, when the animal began to lose weight and the temperature fell 3° Fahr. The conclusion is **plain** that for nutrition the digestive ferments must **be** associated with **the food, or in** other words that they must be transformed into peptones. In **another** series of experiments the same author has demonstrated that **the same** result is obtained by peptones prepared artificially.

There would seem to be no doubt, in the light of the abundant clinical evidence which has now been accumulated, that life may be supported indefinitely, without loss of weight, by the proper administration **of** properly prepared enemata. Flint[5] refers to one case in which life was so sustained for fifteen months, and in which the feeding had **been** mainly of this kind for five years.

For the convenience of the practitioner, the following formulæ **for** nutritive enemata have been collected. The first is the one used by

---

[1] Van Buren: On Phantom Stricture, etc. Amer. Journal Med. Sci., October, 1879.

[2] Amer. Journ. of Obstet., **viii., Nov., 1875.**

[3] Gaz. Hebdom., 1879.

[4] Meeting of French Ass. for Advancement of Science **at** Rheims, 1880. Abstract in Brit. Med. Jour., Sept. 18th, 1880, p. 485.

[5] New York Med. Record, 1878, p. 56.

Mayet[1] and approved by Brown-Séquard.[2] Take of fresh pancreas of the ox from one hundred and fifty to two hundred grammes, and of lean meat from four hundred to five hundred grammes. Bruise the pancreas in a mortar with tepid water at a temperature of 37° C. and strain through a cloth. Chop the meat and mix it thoroughly with the fluid which has thus been strained after separating all the fat and tendinous portions. Add the yolk of one egg. Let stand for two hours and administer at the same temperature after having cleansed the rectum with an injection of oil. This quantity is estimated by Brown-Séquard to be sufficient for twenty-four hours' nourishment and should be administered in two doses.

Where the pancreas cannot be readily obtained, the following formula may be found useful.[3] To a basin of good beef-tea add one-half a pound of lean, raw beefsteak pulled into shreds. At about the temperature of the body add one drachm of fresh pepsin and half a drachm of dilute hydrochloric acid. Place the mixture before the fire and let it remain for four hours, stirring frequently. The heat must not be too great or the artificial digestive process will be stopped altogether. It is better to have the mixture too cold than too hot. Sometimes a little more pepsin may be needed which may be ascertained by stirring with a spoon. If alcohol is to be given, it should be added at the last moment. Eggs may also be added, but should be previously well beaten. This preparation is said to be well borne for a long time.

The formula of the late Dr. Peaslee was as follows: Crush one pound of beef muscle fine, and add to it one pint of cold water. Allow it to macerate three quarters of an hour and then raise gradually to the boiling point. Allow it to boil two minutes and no more. The favorite injection of Dr. Flint is milk ℥ ij., whiskey ℥ ss., and the half of an egg. This he administers every three hours, day and night. But these simple enemata, no matter what their merits may be or may have been in the past (and we are inclined to wonder whether all attempts at alimentation before the admixture of pancreas was thought of, have been as useless as Catillon's experiments would indicate) are now generally replaced by those of artificially digested meat.

In the year 1878, many experiments were made in New York with defibrinated blood as an enema, and the conclusions reached were embodied by Dr. A. H. Smith in the paper already referred to and were as follows:

"1. That defibrinated blood is admirably adapted for use in rectal alimentation.

---

[1] Gaz. Hebdom., Nov. 21st, 1879.
[2] Gaz. Hebdom., Nov. 14th, 1879.
[3] Rennie: Case of severe cut throat; with some remarks on the administration of nutritive enemata. Lancet, Oct. 22d, 1881.

2. That in doses of sixty to one hundred and eighty grammes (two to six ounces) it is usually retained without any inconvenience, and is frequently so completely absorbed that very little trace of it can be discovered in the dejections.

3. That, administered in this way once or twice a day, it produces, in about one-third of the cases, for the first few days, more or less constipation of the bowels.

4. That, in a small proportion of cases, the constipation persists, and even becomes more decided the longer the enemata are continued.

5. That in a very small percentage of cases irritability of the bowels attends its protracted use.

6. That it is a valuable aid to the stomach whenever the latter is inadequate to a complete nutrition of the system.

7. That its use is indicated in all cases not involving the large intestines, and requiring a tonic influence which cannot readily be obtained by remedies employed in the usual way.

8. That in favorable cases it is capable of giving an impulse to nutrition, which is rarely, if ever, obtained from the employment of other remedies.

9. That its use is wholly unattended by danger."

However useful and nutritious these enemata may be, there is one practical objection to them which I have occasionally met with and have been unable to overcome. The sight of the blood, its administration, and its subsequent voiding are not calculated to impress the mind of a nervous and delicate lady pleasantly—on the contrary, they sometimes excite the most profound disgust.

No one form of enema should be continued for too long a time, and, as a rule, patients will be found to thrive the best upon an alternating diet of milk and egg, with preparations of beef and pancreas; alcohol being given as it is indicated. The rectum proper will seldom accommodate more than six ounces of fluid, and this is the usual quantity for an enema; but the sigmoid flexure will hold much more than this; and for myself, I much prefer what may be called the colonic to the rectal method, because the injections are better retained, cause less irritation, may be given in larger quantity, and hence need not be so often repeated. The best apparatus for this purpose is a small-sized, soft-rubber, flexible rectal bougie, the end of which will accommodate the smallest end-piece of the ordinary Davidson's syringe. This should be well oiled, and the fluid to be injected should be forced through it once or twice till it is well warmed, and the air is entirely forced out. The tube is introduced into the sigmoid flexure after the syringe has been connected. In this way, all over-distention of the rectum and consequent desire of the patient to immediately evacuate what has been administered is avoided. The enema should be administered slowly, and by the physician himself rather than the nurse or relative of the patient; for the operation is one requiring judgment and skill, and on the success of the method depends the life of the patient in most cases. It is always well to empty the bowel by a simple enema before administering nutriment at least once a day. With proper care in using the syringe, the rectum and sigmoid flexure will generally be found to submit kindly to this method of treatment, but when once they become irritable, unless the injections can be intermitted for a day or so, and suppositories of opium be substituted, the treatment is practically at an end. In a few cases I have succeeded in re-establishing a tolerance by rest and careful treatment, but it is much better so to manage the case from the first that no irritation be excited. An enema, for this reason, should never be administered at a lower temperature than that of the body.

Dr. Spencer[1] has described a suppository which he recommends in the place of enemata. It consists of the extracted product of artificially digested meat, from which the insoluble matter has been removed, mixed with a little wax and starch. Twenty ounces of meat thus prepared may be made into five suppositories, one of which should be given every four hours.

---

[1] Practitioner, Feb., 1882.

INDEX.

Abscess, boundaries of ischio-rectal, 73
    causes of deep rectal, 73
    classification of, 71
    clinical history of superficial, 72
    course of pus in deep, 74
    diagnosis of, 76
    distinction between treatment of, and of fistula, 77
    due to diffuse inflammation of subcutaneous tissue, 72
    due to disease of urinary organs, 74
    due to disease of neighboring parts, 74
    due to foreign bodies, 73
    due to perforating ulcer, 74
    due to rupture of the rectum, 74
    due to stricture of rectum, 74
    due to submucous inflammation, 74
    due to suppuration of hæmorrhoid, 72
    early incision in, 76
    following surgical operations, 73
    horseshoe, 75
    how to avoid the formation of fistula in, 76
    involving skin of anus alone, 71
    not to be cut into the rectum, 77
    of superior pelvi-rectal space, 73
    of ischio-rectal fossa, 73
    originating in cutaneous glands, 71
    prognosis of, 76

Abscess, reasons for not healing spontaneously, 75
    results of deep, 76
    rupture of, into neighboring organs, 75
    symptoms of ischio-rectal, 74
    symptoms of pelvi-rectal, 74
    treatment of deep, 76
    treatment of superficial, 72
Absorption by the rectum, 283
Adenoma, malignant, 218
Adenomatous polypus, 138
Albumen, digestion in rectum, 285
Alimentation by the rectum, 282
Allingham, case of spasmodic stricture, 182
    ligature holder, 84
    on spontaneous cure of fistula, 81
    on treatment of deep rectal abscess, 76
    operation for hæmorrhoids, 106
    results of colotomy, 216
    spring-scissors for fistula, 87
    symptoms of ulceration, 174
Alveolar sarcoma, 222
Amussat, operation for imperforate anus, 39
Ano-rectal syphiloma, 149, 172
    syphiloma, cause of stricture, 140
    syphiloma, definition of Fournier, 149
    syphiloma, primary seat of, 149
    syphiloma, use of anti-syphilitic treatment for, 149

Anus, abnormal, 34
    absence of, 32
    congenital malformation of, 30
    description of, 5
    double, 34
    erectile tissue of, 5
    imperforate, 31
    imperforate, child living thirty days, 37
Arteries of rectum, 13

Baum, case of colectomy, 244
Benign fungus, 148
    fungus, composition of, 148
    fungus, hæmorrhage from, 148
    fungus, treatment of, 148
Billroth, case of colectomy, 245
    report of thirty-three cases of excision of cancer, 232
Bivalve speculum, 59
Blood for rectal alimentation, 286
Bougies, how to pass, 57, 192
    varieties of, 56
Boyer on fissure, 160
Bridge, case of colotomy, 216
Broadbent, on puncture for relief of obstruction, 132
Broca, fatal case of excision of polypus, 142
Bryant, case of colectomy, 245
    villous polypus, 137
Bulteau on invagination, 125
    statistics of colotomy, 216
Bursa mucosa coccygea, 10
Byrd, case of formation in anus in natural position after colotomy, 45

Calculus projecting into rectum, 4
Callisen's operation for imperforate anus, 41
Cancer, 218
    age of patients, 228
    causes of suffering in, 247
    causes of mortality after excision, 232
    causing œdema of lower extremities, 230
    caustic applications in, 249
    chances of radical cure by excision, 234
    comparative frequency of, in the sexes, 228
    cure of, 231

Cancer, diagnosis of, 229
    dilatation of, 249
    distinguishing marks from benign polypus, 218
    division of sphincter for, 248
    examination for, 230
    excision of, 231
    excision of, as a palliative measure, 235
    excision of, bibliography, 233
    excision of, contraindications, 236
    excision of, compared with colotomy as a palliative measure, 235
    excision of, dangers of operation, 237
    excision of, early history of, 238
    excision of, how to perform, 238
    excision of, history of operation, 231
    excision of, when justifiable, 237
    extension into neighboring organs, 229
    general character of, 218
    generalization of, 220
    generally painful, 229
    hæmorrhage from, 229
    indications for colotomy, 248, 250
    insidiousness of, 228
    involvement of lymphatics in, 230
    location of, 228
    microscopic anatomy of, 219
    mode of development, 220
    of sigmoid flexure, excision of, 243
    of sigmoid flexure, diagnosis of, 231
    operation of crushing, 250
    opium for, 248
    osteoid, 225
    palliative treatment of, 247
    partial removal of, 248
    peculiar feel of, 230
    prevention of obstruction by, 248
    proctotomy for, 249
    regulation of passages in, 247
    results of excision, 232

Cancer, secondary ulceration in, 229
    secondary deposits of, 231
    significance of pain, 229
    symptoms of, 229
    treatment of, **231**
Carcinoma, melanotic, **224**
Cartilaginous tumors (see **enchondroma**)
Cauliflower excrescences, 143
Cellulitis, **gangrenous, around anus, 75**
Chancre, 169
    frequency of, 169
Chancroid, **167**
    as cause of stricture, 168
    great extent of, 167
    spontaneous cure of, 167
Clamp and cautery in treatment of collapse, 118
    for removing hæmorrhoids, 108
Clark, A., villous polypus, 136
Coccyx, excision of, for imperforate anus, 39
Coccygodynia, 281
Colectomy, **242**
    analysis of cases, 246
    conclusions concerning operation, 247
    history of cases, 243
Colloid cancer, **222**
    case of, **223**
    distinction from encephaloid, 254
    malignancy of, 223
    mode of progress, 224
Colonoscope, 60
Colotomy, accidents attending, 42
    after-treatment of cases, 217
    choice between lumbar and inguinal operation, 43
    description of operation, 41
    for stricture, 214
    history of, 40
    in cases of imperforate anus, 40
    limits to the operation, 215
    re-establishment of natural passages after, **253**
    statistics of operation, **215**
Columnæ recti, 8
Concretions, intestinal, 252
Condylomata, 146
    as sign of syphilitic ulceration, 172
    definition of, 146

Condylomata, description of, 146
    diagnosis of, 147
    non-syphilitic nature of, 146
    undefined use of term, **138**
    treatment of, 148
Congenital cysts, 156
    malformations, **rules** for treatment, 37
    stricture, 31
Constipation, 272
    causes, 273
    infantile, 273
    treatment, 274
Cripps on cancer, 218
Curling, villous polypus, 137
Cusack, clamp for hæmorrhoids, 108
Cysts, 151
    congenital, 156
    containing fœtal remains, 155
    containing hair, 152
    dermoid, 151, 153
Czarny, case of colectomy 266

Defecation, physiology of, 19
Depraved sexual appetite accounting **for** foreign bodies in rectum, 262
Depressor for rectal examinations, 59
Dermoid cysts, 151
Diagnosis in rectal disease, **48**
    cases illustrating errors, **48**
    necessity of physical examination, 49
    position of patient, 52
    questions to be asked, 51
Digital examinations, 54
Dilatation of stricture, 199
Discharge caused by polypus, 141
Divulsion for stricture, 200
Douglas' pouch, 6
Dumarquay on treatment of fissure, **177**
Dupuytren, operation for closing artificial anus, 46
Dupuytren's operation, modified **by** Barker, 47
Dysentery, 166
    treatment by injections of nitrate of silver, 180

Eczema marginatum, 270
Edward II., murder of, 259

Ecrasement linéaire in the treatment of fistula, 83
Elastic ligature for fistula, 84
Emmet, rare case of spina bifida, 155
Encephaloid, 222
    diagnosis of, 222
    enucleation of, 250
    rough test for, 222
Enchondroma, 151
Enemata for constipation, 275
    nutritive, 286
    nutritive, how to administer, 287
Enterotome, 47
Epithelioma, 220
    comparative frequency of, 227
    cylindrical, 226
Ergot, injections in prolapse, 115
Erythema, 272
Esthiomène, 164
Examination table, 50
Excision of cancer (see cancer), 231
Extirpation of anus, control of passages after, 28

Fæcal vomiting, 284
Fatty tumors, study by Virchow, 151
    tumors (see lipomata), 150
Fenestrated speculum, 59
Ferrand, case of prolapse treated by injections, 115
Fibromata, 149
Fibrous polypus, 149
Fissure, 160
    associated with polypi, 162, 178
    associated with spasm of sphincter, 160
    division of sphincter, 177
    due to hard fæces, 166
    due to congenital narrowness, 160
    due to leucorrhœa, 160
    due to herpes, 160
    due to inflammation of a sinus of Morgagni, 161
    general location of, 161
    in children, 178
    nothing peculiar in ulceration, 160
    ointment for, 176
    pain accompanying, 174
    painless and painful, 160

Fissure, passage of bougies for, 176
    shape and appearance of, 162
    stretching sphincter for, 177
    subcutaneous division of sphincter, 177
    symptoms of, 173
    theories concerning causation, 160
    tolerant and intolerant, 160
    treatment, 176
    without spasm, 160
Fistula, blind external, 78
    blind internal, 78
    cases in which operation is contra-indicated, 81
    causes of blind internal, 80
    caused by stricture, 187
    classification of, 78
    complete, 78
    deep, 78
    diagnosis of deep from superficial, 76
    dressings after operation, 88
    due to ulceration, 80
    due to previous abscess, 78
    hæmorrhage in operation, 88
    how to detect internal orifice, 78
    how to treat in phthisical patient, 82
    knife of author, 86
    in connection with abscess, 79
    in connection with phthisis, 81
    multiple, 80
    scarifying old tracks, 87
    spring-scissors for, 87
    spontaneous cure, 81
    structure of track, 79
    subcutaneous, 78
    submucous, 79
    submuscular, 79
    superficial, symptoms of, 80
    symptoms of blind internal, 81
    treatment, 81
    treatment by cauterization, 83
    treatment by écrasement linéaire, 83
    treatment of blind internal, 89
    treatment of deep, 89
    treatment of horseshoe variety, 88
    treatment of multiple tracks, 88

Fistula, treatment by elastic ligature, 84
   treatment by incision, 85
   treatment by simple ligature, 83
   treatment when there is no internal orifice, 87
   value of history in diagnosis, 80
   with deep and extensive tracks, 80
   with double track, 79
Flattened fæces in diagnosis, 188
Fœtal inclusions, 154
Forceps for ligating hæmorrhoids, 107
Foreign bodies, 256
   bodies causing abscess, 73
   bodies causing ulceration, 159
   bodies causing perforation, 259
   bodies, complications of, 258, 264
   bodies, general treatment of, 258
   bodies introduced per anum, 260
   bodies, prognosis, 263
   bodies, pelvic abscess caused by, 259
   bodies remaining long time in rectum, 263
   bodies, removal by laparoenterotomy, 265
   bodies swallowed, 256
   bodies swallowed, prognosis, 258
   bodies, treatment of, 264
   bodies, ulceration caused by, 268
Fournier, ano-rectal syphiloma, 149
Fungus hæmatodes, 222

Galvano-cautery for fistula, 83
Gangrenous cellulitis, 75
Gangrene of rectum after confinement, 173
Gariel, pessary of, 54
Glandular polypus, 139
Gonorrhœa as a cause of stricture, 167
Gonorrhœal proctitis, 69
Gorget, 86
Gosselin on chancroidal stricture, 169
Gummata, 148
   authorities upon, 148
   rarity of, 148
Gussenbauer, cases of colectomy, 244

Hæmorrhage from rectum, how to control, 63
   secondary, after application of caustics to prolapse, 117
Hæmorrhoid, arterial, 97
   associated with uterine disease, 101
   associated with pregnancy, 101
   capillary, 96
   curative treatment, 101
   cases and results of carbolic acid injections, 104
   cases unsuitable for opertion, 101
   definition, 91
   description of external, 92
   division into external and internal, 91
   division of external into venous and cutaneous, 92
   due to phymosis, 101
   external, acute inflammation of, 94
   external, formed by dilatation of vein, 93
   external, in a child, 101
   external, treatment of cutaneous, 96
   external, treatment when inflamed, 95
   first passage after operation upon, 107
   forceps to be used in ligating, 107
   formed by extravasation of blood, 92
   internal, diagnosis of, 98
   internal, varieties of, 96
   internal, method of cure by carbolic acid injections, 105
   Internal, operation by clamp and cautery, 106
   palliative treatment of internal, 98
   suppuration of external, 73
   suppuration of internal, 72

Hæmorrhoid, symptomatic, 101
    symptoms of **internal**, 98
    treatment by **caustics**, 102
    treatment of **external**, by incision, 93
    treatment by ligature, 106
    treatment when strangulated, 100
    treatment by injections of carbolic acid, 103
    ulceration following operation, 108
    various operations for, 109
    venous, 97
Hairy cyst, 152
Helmuth's **ligature holder**, 84
Herpes, 272
Hilton, white line between external and internal sphincters, 5
Hodges, on pilo-nidal sinus, 153
Hydatids, 154

Impaction of fæces, 252
    of fæces, causes of, 252
    of fæces, cases, 253
    of fæces, dilatation of sphincter for removal, 256
    of fæces, diarrhœa caused by, 253
    of fæces, errors in **diagnosis**, 253
    of fæces, location **of**, 253
    symptoms of, 253
    treatment of, 255
Intestinal obstruction, treatment, 198
    obstruction, mechanism of, 191
    obstruction, coming on suddenly, 190
    obstruction, fatal when calibre of bowel is considerable, 190
    concretions, 252
Invagination, 124
    acute, 129
    change in evacuations, 128
    chronic, 128
    diagnosis of, 129
    degrees of, 125

Invagination, extravasation caused by, 127
    frequency with which different parts are affected, 125
    fæcal vomiting in, 128
    immediate **effects of**, 126
    laparatomy for, 133
    of large intestine, 128
    pain in, 128
    pathological changes in, 126
    peritonitis caused by, 127
    relative frequency of, 126
    sloughing of included portion, 126
    symptoms of, 127
    taxis for, 130
    treatment by injections, 130
    treatment by puncture, 132
    terminations of, 126
    tumor caused by, 128
    ulceration and perforation caused by, 127
Imperforate anus, 31
    anus, rules **for treatment**, 39
Incontinence of fæces, 77
    of fæces, cure, 77
    of fæces, treatment, 77
Incision of fistula, 85
Inflammation of rectum, 66
Inferior hæmorrhoidal arteries, 13
    hæmorrhoidal veins, 13
Inspection of anus, 53
Instrument case, 52
Iodine as cure for fistula, 83
Irritable ulcer (see fissure)
Ischio-coccygeus muscle, 11

Kleberg, operation **for prolapse** with elastic ligature, 122
Kohlrausch, plica transversalis recti, 23
    case of stricture due to hypertrophy of valves, 184
Kronlein, case of attempt to re-establish anus after colotomy, 46

Lamp for rectal examinations, 51
Laparatomy for obstruction, 133

Laparo-enterotomy for foreign bodies, 255
Laugier's instrument for examining for stricture, 57
Levator ani, 11
    ani, spasm of, 13
Licorice powder, compound, 90
Lieberkühn, follicles of, 9
Ligament, anterior sacro-coccygeal, 7
    pubo-prostatic, 12
Lipomata, 150
    cretaceous formations on, 151
    as cause of invagination, 151
    attached far up, 151
    divided into pedunculated and non-pedunculated, 150
    study of by Virchow, 151
Little, examination table, 50
Littre, operation of, 40
Lumbar colotomy, dangers in children, 43
    nerve-centre governing sphincters, 17
Lupus exedens, 164
    exedens, treatment, 165
Lymphatics of rectum, 19

Mackenzie, on the treatment of dysentery by injections of nitrate of silver, 180
Malformations of rectum and anus, 30
Manual examination of rectum, 61
Marshall, case of colectomy, 245
Martini, case of colectomy, 246
Mason, on chancroidal stricture, 168
Mathieu, rectal supporter, 115
Melanoma, 224
    analysis of cases, 225
    cases and literature, 224
    diagnosis of, 225
    duration of, 225
    general character of, 225
    location of, 225
    malignancy of, 224
Meso-rectum, 5
Middle hæmorrhoidal arteries, 13
    hæmorrhoidal veins, 13
Molk on fœtal inclusions, 154
Mollière, ablation of prolapse, 121
    experiments in producing prolapse, 113

Morgagni, columns of, 8
    sinuses of, 8, 22, 24
Mucous membrane, glandular layer, 9
    membrane, muscular layer, 9
    membrane, valves of, 21
Muscles of rectum, 10
Muscularis mucosæ, 8

Nerves of rectum, 16
Neuralgia of rectum, 280
Nitric acid for hæmorrhoids, 102
Non-malignant growths, 135
    stricture, excision of, 214
Nutritive suppositories, 288
    enemata, 286

Operations, hæmorrhage after, 63
    preparations for, 62
    retention of urine after, 65
Osteoma, 225
Osteo-carcinoma, 225
Osteo-sarcoma, 225
Owen, attempt to re-establish anus after colotomy, 45

Packing the rectum for hæmorrhage, 64
Paget, syphilitic ulceration, 171
Pain, anatomical explanation of, 17
Papilloma, granular, 134
Papillomata, 143
Paquelin's thermo-cautery, 63
Pederasty, 5, 159
Pelvis, measurements in children, 38
Pelvi-rectal abscess, 73
Perforation of bowel by foreign body, 259
Peristalsis, reverse, 284
Peritoneum, contained in prolapse, 118
    in polypus, 142
    relations to rectum, 5
Peritonitis due to stricture, 187
    from perforation, 129
Pig's tail in rectum, 259
Pilo-nidal sinus, 153
Plica transversalis recti, 23
Polyadenomata, 139
Polypus, 135
    adenomatous, 138
    composed of hypertrophied villi, 135
    diagnosis, 141

Polypus, diagnosis from hæmorrhoid, 142
    difficulty **of distinguishing from cancer,** 141
    danger of hæmorrhage in extirpation, 142
    discharge mistaken for dysentery or cancer, 141
    double pedicle, 138
    examination for, 141
    fatal extirpation, 142
    fibrous or hard, 139
    large vessels in pedicle, **142**
    of adults, **135**
    recurrence of, 140
    reduction by taxis, 113
    sarcomatous, 139
    secondary **diseases associated with,** 141
    soft and hard, 135
    spontaneous expulsion, 141
    symptoms, 140
    treatment, 142
    villous, 136
Polypi of childhood, 135
    associated with fissure, 162
Posterior umbilicus, 153
Proctitis, 66
    causes, varieties, and symptoms, 68
    treatment, 70
Proctoplasty, 38
Proctotomy, for imperforate anus, **38**
    author's knife for, 205
    after-treatment of, 213
    cases of, 206, 209
    cases suitable for, 213
    favorable results of, 213
    in cancer, 211
    indications for, **213**
    internal, 202
    external, 204
    literature of, 213
Prolapse, a cause of stricture, 160
    ablation of old and extensive, 122
    causes of, 112
    changes which **occur** in, 119
    containing peritoneum, 118
    of second degree, 118
    of mucous membrane alone, 112
    palliative treatment, 114

Prolapse, production of, by inflating submucous tissue, 113
    reduction when inflamed, 121
    reduction in case of circular **slough,** 121
    removal of first variety, 115
    removal with **elastic ligature,** 123
    strangulation of, 119
    third and fourth varieties, 124
    treatment by clamp and cautery, 118
    treatment by cauterization, 116
    treatment by injections, 115
    treatment of first variety, 113
    treatment when operation is contra-indicated, 119
    varieties of, 110
Pruritus ani, 269
    ani, **symptoms, 269**
    ani, **causes, 269**
    ani, treatment, **270**
Puncture of intestine for relief of obstruction, 132

Quain, bleeding tumor, 136
    stricture due to hypertrophy of valves, 184

Raised mucous patch, 147
    mucous patch, development of, 147
    mucous patch, diagnosis from simple wart, 148
Rectal absorption, 283
    alimentation, 282
    depressor, 59
    supporter, 115
    touch, how to practise, **54**
Rectitis, hyperplastic, 149
    fibro-sclerous, 149
Recto-coccygeus muscle, 10
Recto-vesical muscular fibres, **7**
Rectum, arteries of, 13
    congenital malformation, 30
    curves of, **2**
    definition of, 1
    development, 30
    divisions, 3
    dressings after operation, 65

Rectum, ending by abnormal anus in
perineal or anal regions, 34
ending in blind pouch, 33
ending in bladder, urethra, or
vagina, 36
ending of longitudinal fibres, 7
excision of (see cancer)
fixed position of, 1
hæmorrhage from, 67
imperforate, 38
layers of wall, 5
length, 1
lymphatics of, 19
mucous membrane, 8
muscles of, 10
muscular coats, 6
normally empty state, 27
nerves, 16
operations upon, 62
rules for examination, etc., 48
relations of different divisions, 3
submucous tissue, 7
total absence, 37
upper limit of, 2
variations in position, 3
veins, 13
Retention of urine after operation, 65
Retractor recti muscle, 10
Roché, case of fatal rupture of prolapse, 120
Rodent ulcer, 165
ulcer, diagnosis, 166
Rupture of bowel in prolapse, 120
of rectum causing abscess, 74

Sarcoma, alveolar, 222
Sarcomatous polypus, 139
Scirrhus, 221
extent, 222
rarity, 228
Scrofula, as cause of ulceration, 164
Secondary hæmorrhage after nitric acid
to hæmorrhoids, 102
Sigmoid flexure, variations in position, 43
Smith, case of divulsion, 231
clamp for hæmorrhoids, 108
removal of severe prolapse, 122
Sodomy, appearance of rectum, 159
cause of ulceration, 159
cause of vegetations, 144
medico-legal proofs, 159

Sodomy (see pederasty)
Spasm of levator ani after operations, 13
of sphincter, 277
of sphincters associated with fissure, 160
of sphincters associated with impaction, 279
Specula, varieties of, 58
Sphincter, dilatation, 60
external, 10
functions, 28
internal, 10
nerve control of, 17
third, 7, 19
Spina bifida, 155
bifida, diagnosis, 156
Stimson on cancer, 219
Storer, method of rectal exploration, 53
Strangulation of prolapse, 119
Stricture, attempts at spontaneous cure, 186
bougies for detection of, 192
cause of abscess, 74
cause of fistula, 187
cause of peritonitis, 187
cause of ulceration, 172
congenital, 31
dangers in examination, 55, 57
due to pressure from without, 181
due to hypertrophy of valves, 184
due to gonorrhœa, 167
due to chancroid, 168
due to traumatism, 185
diet for, 197
difficulty in diagnosis when high up, 58
following removal of hemorrhoids, 108
general treatment, 197
how to measure extent, 57
manual examination for diagnosis, 62, 193
mechanism of production of flattened fæces, 189
non-malignant, 181
non-malignant, alternate diarrhœa and constipation, 188
non-malignant, cause of, 181

Stricture, non-malignant, change in wall of bowel capable of producing, 183
non-malignant, changes in bowel above and below, 186
non-malignant, constitutional remedies for, 193
non-malignant, dangers of examination, 191
non-malignant, diagnosis of, 191
non-malignant, difficulty in diagnosis, 191
non-malignant, divided into venereal, non-venereal, cicatricial, and fibrous, 183
non-malignant, pathological anatomy, 186
non-malignant, probable causes of when extensive, 188
non-malignant, symptoms, 188
non-malignant, treatment by dilatation, 199
non-malignant, usual seat of, 187
non-malignant, value of flattened fæces in diagnosis, 188
non-venereal, 184
spasmodic at anus, 181
spasmodic at rectum, 182
spasmodic, explanation of supposed cases, 182
sudden death in, 216
sigmoid flexure, treatment of, 217
syphilitic, but not ulcerative, 183
syphilitic, specific treatment for, 194
use of long cylindrical speculum for examination, 193
treatment by division, 202
treatment by divulsion, 200
value of dilatation, 200
venereal, 183
venereal but not syphilitic, 183
Superior hæmorrhoidal artery, 13
hæmorrhoidal veins, 13

Suppositories, nutritive, 288
Sustentator tunicæ mucosæ, 8
Syme, treatment of fissure, 177
Syphilis and vegetations, 143
diagnostic marks of ulceration, 171
character of secondary ulceration, 170
tertiary ulcerations, 170
late manifestations causing stricture, 183
Syphilitic stricture, 183
stricture, specific treatment, 194
Sympathetic nerves of rectum, 18
hæmorrhoids, 101

Talma, autopsy on, 186
Tensor fasciæ pelvis, 10
Thermo-cautery, 63
Third sphincter, 19
Transversus perinæi, 13
Tubercular ulceration (see ulcer)
Tumor, peculiar bleeding (Quain), 136
Tumors, non-malignant, 135
Trousseau on fissure, 160

Ulcer, caused by applications to fissure, 159
caused by foreign bodies, 258
chancroidal, 167
dysenteric, 166
follicular, 164
of hæmorrhoid, 168
scrofulous, 164
simple, 159
tubercular, 162
tubercular, authorities on, 163
tubercular, cause of fistula, 163
tubercular, cause of hæmorrhage, 163
tubercular, cause of stricture, 163
tubercular, characters of, 163
tubercular, course of, 163
tubercular, distinction between true tubercular and ulceration in a tubercular person, 162
tubercular, location of, 162
tubercular, treatment of, 164
venereal, 167
Ulceration, causing abscess, 74
caused by sodomy, 159
diagnosis of, 175

Ulceration, diet in, 179
    division of sphincter **for**, 180
    following operation **for** hæmorrhoids, 108, 159
    from application of nitric acid to prolapse, 159
    from childbirth, 159
    from foreign bodies, **159**
    from hard fæces, 158
    from surgical operations, 159
    gravity of, 175
    harm done by exercise, **179**
    local remedies for, 179
    non-malignant, 158
    non-tubercular, 164
    occurring during the course of syphilis, 172
    suppositories and enemata for, 179
    syphilitic, **diagnosis** from tubercular, **172**
    syphilitic, cases **of, 172**
    symptoms, 174
    treatment, 176
    treatment by absolute rest, 178
    treatment by large injections of nitrate of silver, 179
    within rectum, treatment of, 178

Valves of rectum, 21
    of rectum, hypertrophy of, 29
Vance, rare case of fissure, 161
Van Buren, operation for prolapse, 117
    spasmodic stricture, **183**
    speculum, 58

Veins **of** rectum, 13
Vegetations, 143
    causation, 144
    due to sodomy, 144
    relation to syphilis, 144
    microscopic anatomy **of,** 143
Venereal stricture, 183
Venereal ulceration, 167
Vienna paste for hæmorrhoids, **102**
Vidal, cases of **prolapse treated by injections, 116**
Villous polypi, 136
    tumor, 136
Virchow, on fatty tumors, 151
Vomiting of fæces, 284

Warts, 143
    causing symptoms of fissure, **145**
    diagnosis, 145
    due to pregnancy, 144
    due to gonorrhœa, 144
    due to leucorrhœa, 144
    mistaken for syphilitic condylomata, 145
    non-contagiousness of, **144**
    non-inoculability of, 144
    powerlessness of specific treatment for, **144**
    symptoms, 144
    treatment, **145**
    within rectum, 145
Wounds of rectum, 281
    of rectum, complications, **282**
    of rectum, treatment, 282

Zappula, case of supposed syphilitic stricture cured by antisyphilitic **treatment, 194**

www.ingramcontent.com/pod-product-compliance
Lightning Source LLC
Chambersburg PA
CBHW031903220426
43663CB00006B/751